Lecture Notes in Physics

Founding Editors

Wolf Beiglböck

Jürgen Ehlers

Klaus Hepp

Hans-Arwed Weidenmüller

Volume 1045

Series Editors

Roberta Citro, Salerno, Italy

Peter Hänggi, Augsburg, Germany

Betti Hartmann, London, UK

Morten Hjorth-Jensen, Oslo, Norway

Maciej Lewenstein, Barcelona, Spain

Satya N. Majumdar, Orsay, France

Luciano Rezzolla, Frankfurt am Main, Germany

Angel Rubio, Hamburg, Germany

Wolfgang Schleich, Ulm, Germany

Stefan Theisen, Potsdam, Germany

James D. Wells, Ann Arbor, MI, USA

Gary P. Zank, Huntsville, AL, USA

The series Lecture Notes in Physics (LNP), founded in 1969, reports new developments in physics research and teaching - quickly and informally, but with a high quality and the explicit aim to summarize and communicate current knowledge in an accessible way. Books published in this series are conceived as bridging material between advanced graduate textbooks and the forefront of research and to serve three purposes:

- to be a compact and modern up-to-date source of reference on a well-defined topic;
- to serve as an accessible introduction to the field to postgraduate students and non-specialist researchers from related areas;
- to be a source of advanced teaching material for specialized seminars, courses and schools.

Both monographs and multi-author volumes will be considered for publication. Edited volumes should however consist of a very limited number of contributions only. Proceedings will not be considered for LNP.

Volumes published in LNP are disseminated both in print and in electronic formats, the electronic archive being available at springerlink.com. The series content is indexed, abstracted and referenced by many abstracting and information services, bibliographic networks, subscription agencies, library networks, and consortia.

Proposals should be sent to a member of the Editorial Board, or directly to the responsible editor at Springer:

Dr Lisa Scalone
lisa.scalone@springernature.com

Abdulla Rakhimov · Shukhrat Mardonov

Theory of Quantum Bose Liquids Beyond Bogoliubov Approximation

Springer

Abdulla Rakhimov
Department of Theoretical Physics
Institute of Nuclear Physics
Tashkent, Uzbekistan

Shukhrat Mardonov
Department of Research
New Uzbekistan University
Tashkent, Uzbekistan

ISSN 0075-8450　　　　　ISSN 1616-6361　(electronic)
Lecture Notes in Physics
ISBN 978-3-032-05095-3　　　ISBN 978-3-032-05096-0　(eBook)
https://doi.org/10.1007/978-3-032-05096-0

This Springer imprint is published by the registered company Springer Nature Switzerland AG
The registered company address is: Gewerbestrasse 11, 6330 Cham, Switzerland

*Dedicated to our FAMILIES and to our
teachers: B. Ahmedov, F. C. Khanna,
M. Musakhanov,
E. Sherman, and V. Yukalov*

Preface

The study of Bose–Einstein condensation (BEC) has provided profound insights into the behavior of quantum systems at ultracold temperatures, offering a window into the enigmatic world of quantum fluids and beyond. This book embarks on a comprehensive journey through the theoretical, experimental aspects of BEC, integrating fundamental concepts with advanced developments that underline its significance in contemporary physics. The chapters are meticulously structured to guide readers through an escalating complexity of topics, making this book a valuable resource for students, researchers, and professionals in the field.

Chapter 2 introduces the fundamental properties of quantum fluids, drawing a distinction between gases and liquids and laying the groundwork for understanding ultracold quantum liquid droplets. The chapter emphasizes the critical role of the chemical potential in determining the properties of these systems, highlighting its relevance in both theoretical modeling and experimental realization. Bose–Einstein condensation has been an important topic in quantum physics for a long time since the first predictions in 1924, and especially after the experimental observations in 1995. While the pioneer works of Bose and Einstein concern only the noninteracting gas, in reality, the particles do interact and the rigorous understanding of interacting systems remains a very challenging problem in theoretical physics. The theory of interacting Bose gases essentially started in 1947 when Bogoliubov proposed an approximation and used it to predict the excitation spectrum of Bose gases. In particular, Bogoliubov theory gives a satisfactory explanation of Landau's criterion for superfluidity.

However, Bogoliubov theory has, mainly, two limitations:

(i) The theory assumes that the dimensionless gas parameter $\gamma = \rho a_s^3$ (where ρ and a_s are the number density and the s-wave scattering length, respectively) is very small, $\gamma \sim 10^{(-5)}$;

(ii) It does not take into account finite temperature fluctuations.

For this reason, Bogoliubov theory is able to explain the phenomenon of superfluidity, e.g., in helium four only qualitatively ($\gamma(^4He) \approx 0.2$), letting alone the dependence of the superfluid fraction on temperature. As to dilute gases, due to recent development of the Feshbach resonance technique, experimenters are able to create BEC with much larger values of the gas parameter, for which next order corrections to Bogoliubov theory should be developed and applied.

The correction beyond the theory may be important even in the weakly interacting regime of atomic gases. Particularly, as it has been predicted by Petrov, due to the correction, referred in the literature as the LHY term, a new state of matter-dilute droplets may arise in a two component Bose mixture with an attractive interspecies interaction.

One of the widely used methods, based on the field-theoretical description is Optimized Perturbation Theory (OPT). This theory sometimes appears in the literature under different names such as modified perturbation theory, variational Gaussian approximation, linear delta expansion, and so on. It has been successfully employed in various branches of physics, chemistry and applied mathematics. The present monograph introduces and demonstrates the application of OPT for the description of BEC of atomic gases and magnons (triplons in quantum magnets).

The monograph can be roughly divided into three main parts. In the first part, we make an introduction to field theoretical methods of many-particle systems, focusing on Bose–Einstein condensation. Here, mainly addressing students, we explain the meaning of such concepts as Spontaneous Symmetry Breaking (SSB), Green functions, Matsubara formalism, path integrals, etc. Then a reader can pass to the second part, where the first (Chaps. 7, 8) and the second order (Chap. 17) of OPT are outlined in detail. Practical application of the theory to BEC of atomic gases, including two-component Bose mixtures, as well as to quantum magnets, can be found in the next part (Chaps. 9–16). For better assimilation of the material, each chapter is provided with tasks.

Although the monograph is intended to theorists in the physics of condensed matter, in our opinion, a theorist should at least superficially imagine how a certain physical quantity is measured in experiments. To this end, we take a short mental tour of the experimenter's laboratories in Chap. 5. in order to outline their work in a nutshell.

The target audience for the present monograph is researchers and postgraduate students studying theoretical physics.

Tashkent, Uzbekistan
July 2025 Abdulla Rakhimov
Shukhrat Mardonov

Acknowledgements We would like to express our deepest gratitude to F. Abdullaev, I. Askerzade, B. Bayzakov, E. Ibragimova, A. Nasirov, E. Sherman, and V. Yukalov for their invaluable discussions, insightful suggestions, and unwavering support during the preparation of this book. Their expertise and constructive feedback have been instrumental in shaping the ideas and concepts presented herein. This work would not have been possible without their generous contributions of time and knowledge, and we are truly honored to have had the opportunity to collaborate with such esteemed colleagues.

Financial support: The work is financed by the state budget of the Republic of Uzbekistan.

Professor Shukhrat Mardonov is partly supported by Grant FZ-2020092957 of the Ministry of Higher Education, Science and Innovation of the Republic of Uzbekistan.

Contents

1 Introduction 1
    References 5

2 Quantum Fluids 7
    2.1   Gases Versus Liquids 7
    2.2   Ultracold Quantum Liquid Droplets 8
    2.3   The Importance of the Chemical Potential 9
    References 10

3 A System of Noninteracting Particles-Ideal Gas 13
    3.1   Degenerate (Ideal) Bose Gas 14
          3.1.1   BEC Phase 16
          3.1.2   Thermodynamics of BEC 18
          3.1.3   Normal Phase $T > T_c$ and the Jump
                    in the Gradient of the Heat Capacity 19
          3.1.4   Near T_c 21
    3.2   Landau Criterion 23
    References 24

4 BEC of Interacting Particles 27
    4.1   Introduction 27
    4.2   Second Quantization 29
          4.2.1   Second Quantization in Many Particle Physics 30
          4.2.2   Ideal Gas of Bosonic Quasiparticles 33
    4.3   Spontaneous Symmetry Breaking and Goldstone Modes 34
          4.3.1   SSB in Particle Physics 35
          4.3.2   Continuous Symmetry and Goldstone Theorem 37
    4.4   Green Functions 38
          4.4.1   Matsubara Formalism 40
    4.5   Hugenholtz-Pines Theorem 41
          4.5.1   Real Field Representation 43
    4.6   Green Function of an Ideal Bose Gas 46

4.7   Green Function of Interacting Bose Particles 47
       4.7.1   Bilinear Approximation with the Contact Potential 49
       4.7.2   Application of Hugenholtz-Pines Relation 50
4.8   Physical Observables in the Bilinear Approximation 50
       4.8.1   Condensed and Uncondensed Fractions 50
       4.8.2   Bogolyubov Approximation 52
4.9   Zero Temperature 52
       4.9.1   Methods of Regularization 53
       4.9.2   Power Series in γ for n_0 55
       4.9.3   Elementary Excitations and Sound Velocity 57
4.10   Conclusion 59
References 59

5   Experiments in a Nutshell 63
5.1   Introduction 63
5.2   Magnetic Trap 64
5.3   Optical Trapping 65
5.4   Structure Factor 66
       5.4.1   Neutron Scattering 66
       5.4.2   Bragg Scattering 68
       5.4.3   Static Structure Factor 70
5.5   The Sound Velocity 71
5.6   Pair Correlation Function 72
       5.6.1 T_c via $g(r)$ 73
5.7   Measurement of Condensed Fraction 74
5.8   Tan's Contact 75
5.9   Quantum Liquid Droplets 77
References 78

6   Path Integrals for Bosons 81
6.1   The Principle of Least Action and Gross-Pitaevskii
       Equation 81
       6.1.1   Classical Mechanics 82
       6.1.2   Classical Statistical Mechanics 83
       6.1.3   Quantum Statistical Mechanics 83
       6.1.4   Gross-Pitaevskii Equation 84
6.2   Path Integrals in Quantum Mechanics: Physical
       Interpretation 85
6.3   Path Integrals in Quantum Mechanics: Mathematics 86
6.4   Imaginary Time and Field Formulation 89
6.5   Evaluation of Path Integrals 89
       6.5.1   Determinant G^{-1} and Lee-Huang Yang Term 93
       6.5.2   Hubbard-Stratanovich Transformation for Bosons 94

6.6 Expected Values 98
6.6.1 Expected Values in One Loop Approximation
for Uniform Systems 99
6.6.2 Grand Thermodynamic Potential Ω 101
6.7 Conclusion 103
References 103

7 Optimized Perturbation Theory 105
7.1 Introduction 105
7.2 OPT: How Does It Work? 106
7.3 OPT: General Ideology 108
7.4 Eigenvalue Problem: Anharmonic Oscillator 110
7.4.1 Standart Perturbation Theory 111
7.4.2 OPT Calculations 112
7.5 Relativistic Finite Temperature Field Theory 114
7.6 Conclusion 116
References 117

8 BEC in Optimal Perturbation Theory 119
8.1 Introduction 119
8.2 Thermodynamic Potential 119
8.3 Fixing Trial Parameters 123
8.4 Hugenholtz-Pines Theorem and Hohenberg-Martin
Dilemma 125
8.4.1 Hohenberg-Martin Dilemma 126
8.5 Solution of Hohenberg–Martin Dilemma: Yukalov Receipt 127
8.5.1 The Role of Anomalous Density 127
8.6 The Effective Action and Thermodynamic Potential
in Yukalov Prescription 129
8.6.1 Critical Temperature and Normal Phase 131
8.7 Condensed Phase, $T \leq T_c$ 132
8.7.1 The Main Equation in Dimensionless Form 133
8.7.2 Thermodynamic Quantities 135
8.7.3 Superfluid Density 136
8.8 Structure Factor and Correlation Function 138
8.9 Conclusion 140
References 140

9 Tan's Contact in Mean-Field Theories 141
9.1 Introduction 141
9.2 Bilinear Approximation 142
9.3 OPT 143
9.4 Bogolyubov Approximation 144
9.5 Results and Discussions 145
9.6 Conclusion 147
References 147

10 The Condensate Phase and Its Relation to the Anomalies Density 149
  10.1 Introduction 149
  10.2 Variational Perturbation Theory with Arbitrary Phase 151
  10.3 How to Measure $|\sigma|$? 153
  10.4 Numerical Results and Discussions 155
  10.5 Conclusion 156
  References 156

11 Two-Component Bose Mixture 157
  11.1 Introduction 157
  11.2 OPT for Two-Component Bose Mixtures 159
    11.2.1 Normal and Anomalous Densities 164
    11.2.2 Particular Cases of Present OPT 164
  11.3 Miscible-Immiscible Transition 165
    11.3.1 Condensed and Normal Phases 166
    11.3.2 Intermediate Summary 171
  11.4 Balanced Symmetric Bose Mixtures 172
    11.4.1 A. Zero Temperature 172
  11.5 Summary 175
  References 175

12 Quantum Liquid Droplets in Two-Component Bose Mixtures 177
  12.1 Introduction 177
  12.2 $\mathcal{E}_{LHY}$ and $\mathcal{E}_{MFA}$ at Zero Temperature 177
    12.2.1 $\mathcal{E}_{MFA}$ and Phase Locking 178
  12.3 Droplet Formation 179
  12.4 Extended Gross-Pitaevskii Equation in the Effective Single-Component Approximation 180
    12.4.1 One Component GP Equation 182
  12.5 Properties of the Isotropic Droplet 183
  12.6 The Problem of Imaginary Sound Velocity 185
    12.6.1 Ota and Astrakharchik 185
    12.6.2 Bosonic Pairing Theory 186
    12.6.3 Application of Beliaev Theory 187
    12.6.4 Optimized Perturbation Theory 187
  12.7 Summary and Concluding Remarks 189
  References 190

13 Bose-Einstein Condensation of Triplons in Quantum Magnets 193
  13.1 Introduction 193
  13.2 Dimerized Quantum Magnets (DQM) 194
  13.3 Triplon Condensation and the Chemical Potential 196
    13.3.1 Magnons Versus Triplons 198
  13.4 Effective Hamiltonian in Bond Operator Formalism 199
  13.5 Summary and Concluding Remarks 199
  References 200

14 Thermodynamics of Dimerized Quantum Magnets ... 201
   14.1 Introduction ... 201
   14.2 OPT for the Triplon Gas ... 201
      14.2.1 Normal Phase $T > T_c$ ... 204
      14.2.2 BEC Phase $T \leq T_c$ ... 205
      14.2.3 Critical Temperature and the Density ... 207
   14.3 Critical Behavior of C_H and C ... 208
      14.3.1 A. $T \to T_c + 0$ Region ... 208
      14.3.2 B. $T \to T_c - 0$ Region ... 209
   14.4 Experimental Observation of Tan's Contact for Triplons ... 210
   14.5 Summary ... 211
   References ... 211

15 Magnetocaloric Effect and Grüneisen Parameter of Quantum Magnets with Spin Gap ... 213
   15.1 Introduction ... 213
   15.2 Magnetocaloric Effect ... 213
      15.2.1 Magnetic Grüneisen Parameter ... 215
   15.3 Critical Properties of Γ_H ... 217
      15.3.1 Near Critical Temperature T_c ... 217
      15.3.2 Quantum Critical Point ... 217
   15.4 Low Temperature Expansion ... 219
   15.5 Divergence of Γ_H Near Quantum Critical Point ... 221
      15.5.1 Zero Temperature Sound Velocity Near QCP ... 221
   15.6 Numerical Results for Realistic Systems ... 222
   15.7 Summary ... 223
   References ... 223

16 Weak Anisotropy in Spin-Gapped Quantum Magnets ... 225
   16.1 Introduction ... 225
   16.2 The Hamiltonian and the Thermodynamic Potential Including EA and DM Interactions ... 227
   16.3 Effects of EA Interaction, $\gamma \neq 0$, $\gamma' = 0$ ... 229
      16.3.1 Violation of the Phase Invariance Due to Anisotropies ... 230
      16.3.2 HP Relation for a Quasicondensate ... 231
      16.3.3 Condensed Phase $T < T_c$ ... 232
      16.3.4 Normal Phase $T > T_c$ ... 233
      16.3.5 The Shift of the Critical Temperature Due to Exchange Anisotropy ... 234
   16.4 Effects of DM Anisotropy, $\gamma = 0$, $\gamma' \neq 0$ ... 236
   16.5 Critical Temperature of a Crossover ... 237
   16.6 Magnetization Curves for $TlCuCl_3$ ... 238
   16.7 Summary ... 239
   References ... 240

17 Optimized Perturbation Theory in δ^2 Order 243
    17.1 Introduction 243
    17.2 Zero Temperature Effective Action and Partition Function 244
    17.3 $\Omega(T = 0)$ in the First Order of δ 248
    17.4 The Second Order Calculations 249
    17.5 Momentum Space 252
    17.6 Summary 254
    References 254

Appendix A: Chemical Potential in Boltzmann Distribution 255

Appendix B: Laser Cooling 257

Appendix C: Functionals and Functional Derivatives 259

**Appendix D: The Green Function and Densities in Two-Component
Bose Mixtures** 263

**Appendix E: Effective Hamiltonian in Bond-Operator
Representation** 267

Acronyms

BCS	Bardeen–Cooper–Schriffer
BEC	Bose–Einstein Condensate
DM	Dzyaloshinsky–Moriya
DQM	Dimerized Quantum Magnets
EA	Exchange Anyzotropy
FAC	Fastest Apparent Convergence
GPE	Gross–Pitaevskii Equation
HM	Hohenberg–Martin dilemma
HP	Hugenholtz–Pines
HST	Hubbard–Stratanovich
LDA	Linear Delta Approach
MFA	Mean Field Approximation
PMS	Principle of Minimal Sensitivity
QCP	Quantum Critical point
RF	Radio Frequency
SPT	Standart Perturbation Theory
SSB	Spontaneous Symmetry Breaking
TOF	Top Of Flight
VPT	Variational Perturbation Theory
ZM	Zero Mode

Properties and Parameters

Some experimental characteristics of Bose-Einstein condensate

Kinds of atom	^{87}Rb, ^{7}Li, ^{23}Na, ^{52}Cr, ^{4}He, ...
Critical temperatures	50 nK–2 μK
Density	10^{11}–10^{15}cm^{-3}
# of atoms	$100(^{7}\text{Li})$–$10^{6}(^{23}\text{Na})$–$10^{9}(^{4}\text{He})$
Lifetime	A few seconds to several minutes
Spatial size: The shape at diameters	10–15 μm
Cigar-shaped at length	30 μm and diameter 15 μm

Physical constants and other definitions

Bohr radius	$a_B = \frac{4\pi\varepsilon_0\hbar^2}{m_e e^2} = \frac{\hbar}{m_e c\alpha} \approx 5.292 \times 10^{-9}$ cm
Reduced Plank constant	$\hbar = h/2\pi \approx 1.054.. \times 10^{-27}$ g cm^2 s^{-1}
Vacuum permittivity	$\varepsilon_0 = 8.854 \times 10^{-12}$ F $\cdot$ m^{-1}
Electron rest mass	$m_e \approx 0.9109 \times 10^{-34}$ g
Elementary charge	$e = 1.602^{-19}$ C
Speed of Light	$c = 2.998 \times 10^{10}$ cm $\cdot$ s^{-1}
Fine structure constant	$\alpha \approx 7.297 \times 10^{-3}$
Avogadro's Constant	$N_A = 6.02214076 \times 10^{23}$ mol^{-1}
Condensate initial width	$a_0 = (\hbar/M\omega)^{1/2}$
Mass of atom	M (^{87}Rb $= 87/N_A \approx 1.444 \times 10^{-22}$ g)
Harmonic oscillator frequency	$\omega(\sim 10\text{–}100$ Hz)
Self-interaction parameter	$g = 4\pi\hbar^2|a_s|/Ma_z$ (for 2D)
s-wave scattering length	$|a_s| \approx 100a_B \approx 5 \times 10^{-7}$ cm
Condensate extension along z-axis	a_z ($\tilde{1}0^{-4}$ cm for 2D)
Boltzmann constant	$k_B = 8.617 10^{-5}$ ev/K

Introduction

1

Theoretical Foundations: The theoretical foundations of Bose-Einstein condensation emerged from the intersection of quantum mechanics, statistical mechanics, and the study of ideal gases. In the early 20th century, physicists were grappling with the implications of quantum theory, which described the behavior of particles at the microscopic level. At the same time, statistical mechanics provided a powerful framework for understanding the collective behavior of large ensembles of particles.

In 1924, Indian physicist Satyendra Nath Bose made a significant breakthrough when he developed a new statistical method for counting particles with integer spin. Bose's approach, which later became known as Bose-Einstein statistics, treated particles with identical quantum states as indistinguishable entities. He sent his findings to Albert Einstein, who recognized the importance of Bose's work and extended it to predict the behavior of ideal gases at low temperatures.

Einstein's theoretical analysis revealed that at sufficiently low temperatures, the thermal de Broglie wavelength of particles would become comparable to the inter-particle spacing. Under these conditions, a large fraction of particles would occupy the lowest quantum state, leading to a macroscopic occupation of the ground state. This phenomenon, known as Bose-Einstein condensation, represented a new phase of matter with unique quantum properties.

Einstein's prediction of Bose-Einstein condensation was a profound departure from classical physics and challenged existing notions of matter and phase transitions. It demonstrated that at ultra-low temperatures, quantum effects could dominate over classical behavior, leading to novel collective phenomena.

Despite the theoretical elegance of Bose and Einstein's work, the experimental realization of Bose-Einstein condensation remained elusive for several decades. It was not until the advent of advanced cooling and trapping techniques in the 1980s and 1990s that physicists were able to cool atoms to temperatures close to absolute zero and observe BEC in the laboratory.

© The Author(s), under exclusive license to Springer Nature Switzerland AG 2026   1
A. Rakhimov and S. Mardonov, *Theory of Quantum Bose Liquids Beyond Bogoliubov Approximation*, Lecture Notes in Physics 1045,
https://doi.org/10.1007/978-3-032-05096-0_1

The theoretical foundations laid by Bose and Einstein provided the conceptual framework for understanding the behavior of particles at low temperatures and paved the way for the experimental discovery of BEC. Their pioneering work marked a pivotal moment in the history of physics and continues to inspire research in quantum mechanics and condensed matter physics to this day.

Overall, the theoretical foundations of Bose-Einstein condensation represent a triumph of theoretical insight and mathematical rigor, demonstrating the power of quantum mechanics to uncover new and unexpected phenomena in the natural world.

Experimental Realization: The experimental realization of Bose-Einstein condensation represented a significant milestone in the field of ultracold atomic physics. Despite the theoretical predictions made by Bose and Einstein in the 1920s, it took several decades for experimental techniques to advance to the point where BEC could be observed in the laboratory.

One of the key challenges in achieving BEC was the need to cool atoms to temperatures close to absolute zero, where quantum effects become dominant. Traditional methods of cooling, such as evaporative cooling and laser cooling, were not sufficient to reach the required temperatures for BEC. However, in the 1980s and 1990s, groundbreaking advancements in laser cooling and trapping techniques paved the way for the experimental realization of BEC.

One of the pioneering experiments in achieving BEC was conducted by physicists Eric Cornell and Carl Wieman at the Joint Institute for Laboratory Astrophysics (JILA) in Boulder, Colorado. Using a combination of laser cooling and magnetic trapping techniques, Cornell and Wieman were able to cool a dilute gas of rubidium-87 atoms to temperatures just a few billionths of a degree above absolute zero. By carefully tuning the trapping parameters, they observed [1] the formation of a BEC in 1995, confirming the predictions made by Bose and Einstein several decades earlier.

Around the same time, another research group led by Wolfgang Ketterle at the Massachusetts Institute of Technology (MIT) also achieved BEC using a different experimental approach [2]. Ketterle's team used a combination of laser cooling and evaporative cooling techniques to cool a gas of sodium atoms to temperatures low enough to observe BEC. Their experiments, conducted in 1995, provided further confirmation of the existence of BEC and opened up new avenues for studying the properties of this exotic state of matter.

The experimental realization of BEC sparked a flurry of research activity in the field of ultracold atomic physics. Researchers around the world began exploring the properties of BEC, including its coherence, superfluidity, and quantum dynamics. BEC also provided a platform for investigating fundamental phenomena such as quantum phase transitions and quantum coherence effects.

The groundbreaking experiments conducted by Cornell, Wieman, Ketterle, and others paved the way for numerous technological advancements and practical applications of BEC. These include precision measurement devices, atomic clocks, and quantum sensors. Moreover, BEC has emerged as a promising platform for quantum information processing, with potential applications in quantum computing and quantum communication.

Therefore, the experimental realization of Bose-Einstein condensation represented a triumph of experimental ingenuity and technological innovation. The achievement of ultracold temperatures and the observation of BEC opened up new frontiers in the study of quantum matter and paved the way for a wide range of applications in physics and technology. The discovery of BEC stands as a testament to the power of collaboration between theory and experiment in advancing our understanding of the natural world.

Further Developments and Applications: Since its discovery in the mid-1990s, Bose-Einstein condensation has continued to captivate the interest of physicists worldwide, leading to numerous further developments and applications. Researchers have explored various aspects of BEC, including its properties, dynamics, and potential practical applications, with exciting implications for both fundamental physics and technological innovation. One of the most significant areas of research following the discovery of BEC has been the study of its fundamental properties. Physicists have investigated the coherence and superfluidity of BEC, revealing fascinating quantum phenomena such as quantized vortices and collective excitations known as phonons. These studies have provided valuable insights into the behavior of quantum fluids and the nature of phase transitions at ultra-low temperatures.

Moreover, researchers have explored the use of BEC as a platform for quantum simulation and computation. By manipulating the quantum states of BEC using external fields and laser techniques, scientists can emulate complex quantum systems that are difficult to study using classical computers. This has led to advances in areas such as quantum optimization, quantum simulation of condensed matter systems, and the development of quantum algorithms for solving specific computational problems.

BEC has also found practical applications in a variety of fields, including precision measurement and sensing. For example, BEC-based atomic clocks have achieved unprecedented levels of accuracy and stability, making them invaluable tools for applications such as global navigation systems, telecommunications, and fundamental tests of the laws of physics. Similarly, BEC-based sensors have demonstrated remarkable sensitivity to magnetic fields, gravitational waves, and inertial forces, opening up new possibilities for high-precision measurements in science and engineering.

In addition to its applications in metrology and sensing, BEC holds promise for technological innovations in other areas. For instance, researchers are exploring the use of BEC for quantum information processing, where the unique properties of quantum systems could enable exponential speedup in certain computational tasks. BEC-based quantum computers and simulators could revolutionize fields such as cryptography, materials science, and drug discovery, offering new insights and capabilities that were previously unthinkable.

Furthermore, BEC has potential applications in other areas of physics, such as the study of exotic quantum phenomena and the exploration of novel quantum materials. By manipulating the properties of BEC and studying its interactions with external fields and environments, researchers can gain deeper insights into fundamental aspects of quantum mechanics and the behavior of matter at the nanoscale.

In conclusion, the discovery of Bose-Einstein condensation has spurred a wealth of further developments and applications in physics and technology. From the exploration of fundamental quantum phenomena to the development of cutting-edge technologies, BEC continues to push the boundaries of our understanding and capabilities. As researchers continue to unlock the secrets of BEC and harness its potential, we can expect even more exciting discoveries and innovations in the years to come.

For experiments, the most common atoms employed are alkalis. The reason behind this choice is their simple internal level structure (one valence electron) that a hydrogen-like equation can describe. This picture simplifies their manipulation as compared to dipole atoms or two-electron atoms.

The aim of the book: In the dynamic and ever-expanding field of quantum physics, the exploration of Bose-Einstein condensates (BEC) stands as a cornerstone that bridges the intricate worlds of theoretical predictions and experimental observations. This book delves into the profound complexities of quantum fluids, particularly focusing on ultra-cold quantum liquid droplets and the behavior of gases and liquids at quantum scales. Through a meticulous examination of the chemical potential's role, the text navigates through the phenomena of noninteracting particles, ideal Bose gases, and the thermodynamics associated with BEC phases.

Our journey through the text is not merely theoretical. It is grounded in rigorous mathematical frameworks such as secondary quantization, spontaneous symmetry breaking, and various approximation methods that play crucial roles in understanding the real-world implications of quantum mechanics. Moreover, the book explores the experimental techniques that allow us to observe and quantify the predictions made by quantum theories, such as magnetic and optical trapping, neutron and Bragg scattering, and the measurement of sound velocity and condensed fractions.

Each chapter is designed to build upon the previous, starting with fundamental concepts before advancing to complex theories and their applications. The book addresses significant theoretical insights and experimental methods, shedding light on phenomena like superfluidity, phase transitions, and the critical behaviors near phase boundaries. It not only elucidates the fundamental aspects of quantum mechanics but also introduces advanced topics such as the interaction of bosonic particles, the Gross-Pitaevskii equation, and path integrals, offering a comprehensive understanding of quantum fields.

The later sections of the book are dedicated to applying these concepts to specific systems such as triplons in quantum magnets, exploring both theoretical and practical aspects. This includes detailed discussions on the thermodynamics of dimerized quantum magnets, the magnetocaloric effect, and the critical properties of quantum systems under various external influences.

As we progress, the text challenges the reader to consider the implications of quantum mechanics beyond idealized scenarios, engaging with the material complexities that influence experimental outcomes and theoretical frameworks alike. This book aims to equip students, researchers, and academicians with the knowledge to push the boundaries of what is currently known and to inspire further research into the quantum realm.

Present lectures are not just a recounting of established knowledge but an invitation to actively engage with the ongoing debates and challenges within the field of quantum physics. It provides a detailed, nuanced view of the current state of research and the potential future directions in the study of quantum fluids and Bose-Einstein condensates.

References

1. M.H. Anderson, J.R. Ensher, M.R. Matthews, C.E. Wieman, E.A. Cornell, Observation of Bose-Einstein condensation in a dilute atomic vapor. Science **269**(5221), 198–201 (1995). https://doi.org/10.1126/science.269.5221.198
2. K.B. Davis, M.O. Mewes, M.R. Andrews, N.J. van Druten, D.S. Durfee, D.M. Kurn, W. Ketterle, Bose-Einstein condensation in a gas of sodium atoms. Phys. Rev. Lett. **75**, 3969–3973 (1995). https://doi.org/10.1103/PhysRevLett.75.3969

Quantum Fluids

2

2.1 Gases Versus Liquids

The term fluid can mean a gas or a liquid. And what does it mean by a quantum fluid?

In the present lectures, in general, we consider quantum objects such as quantum liquids, quantum magnets, quantum gases etc. On the freshman level, one may define a quantum fluid as a system of many particles in whose behavior effects of quantum mechanics and quantum statistics are very important [1]. In quantum objects, even in the absence of mutual interactions, the particles have a profound influence on each other due to spin-statistics theorems. When do the quantum effects become important? The simple answer is given as $\lambda_T^3 \sim V/N$. That is when the thermal wavelength of a particle with mass m and the energy $k_B T$

$$\lambda = \frac{h}{\sqrt{2\pi m k_B T}} \tag{2.1}$$

becomes compatible with the volume per particle V/N. This criterion is well fulfilled, say, for electrons due to their small mass, in any realistic solid or liquid at any temperature T. A reader can find in literature [2] a more accurate criterion $N/N_E >> 1$ in terms of the number of states N_E below an energy E.

Spin statistics dramatically divides particles (even atoms and molecules) into two classes: bosons, who has an integer spin ($S = 0, 1, 2...$) and fermions with half-integer spin ($S = 1/2, 3/2...$).[1] For a compound particle (atom molecule), its total angular momentum $\mathbf{J} = \mathbf{L} + \mathbf{S}$ defines its statistics. This classification is based on a natural assumption that a total wave function of N particles may be symmetric or

---

[1] We shall not discuss anyons.

© The Author(s), under exclusive license to Springer Nature Switzerland AG 2026
A. Rakhimov and S. Mardonov, *Theory of Quantum Bose Liquids Beyond Bogoliubov Approximation*, Lecture Notes in Physics 1045,
https://doi.org/10.1007/978-3-032-05096-0_2

antisymmetric upon interchange of any two of these particles. Physically this means that particles of a system with an antisymmetric wave function cannot occupy the same single-particle quantum state. Experimental observations show that this rule is good not only for electrons but also for all fermions [3]. As to a system of bosons, here number of particles in a single-particle quantum state is not restricted. This makes possible an interesting phenomena-Bose-Einstein condensation which displays itself as an occupation of the lowest-energy single-particle state by macroscopic number of particles.[2]

What is the main difference between gases and liquids? From a bird's eye view gas is a state of matter characterized by molecules that are widely spaced, move freely, and have no fixed shape or volume, while the liquid is a state of matter characterized by molecules that are close together, move more slowly, and have a definite volume but no fixed shape. Besides, liquids cannot be practically compressed, while gases can be compressed easily.

On the microscopical level, their difference is originated in the nature of the interatomic potential properties of, especially, dilute gases can be described by a simple contact interaction:

$$V(\mathbf{r}_1 - \mathbf{r}_2) = g\delta(\mathbf{r}_1 - \mathbf{r}_2), \tag{2.2}$$

while for liquids, a more complicated potential, say van der Waals potential, depicted in Fig. 2.1a is needed. From the figure, it is seen that the liquid phase appears only when a sensitive balance between attractive and repulsive energy terms take place.

The question arises, "Can the contact interaction either lead to the formation of a liquid state?" A cursory answer is, naturally, "No, cannot". In fact, if the coupling constant g in (2.2) is positive, then the repulsive interaction prevents the atoms from getting closer, otherwise, ($g < 0$), the system collapses. However, in 2015, D. Petrov found a positive answer to this question, predicting a new exotic phase of matter-quantum liquid droplets [4].

2.2   Ultracold Quantum Liquid Droplets

D. Petrov has shown that the quantum liquid droplet[3] appears as a liquid and survives due to quantum fluctuations, which are responsible for its stabilization. This is illustrated in Fig. 2.1b. Here, in contrast to the Van der Waals liquids, the system is so dilute that the atoms do not observe the details of the interatomic potential. However, they remain self-bound due to the compensation of attractive mean field forces and the repulsive effect of quantum fluctuations. Further, droplets were experimentally observed in a two-component Bose mixture [6] and dipolar Bose gases [7]. The experiments show that the droplet really exhibits the properties of a liquid, such

---

[2] See [1] for a more rigorous definition.
[3] Further, for simplicity, we use the term *droplet*.

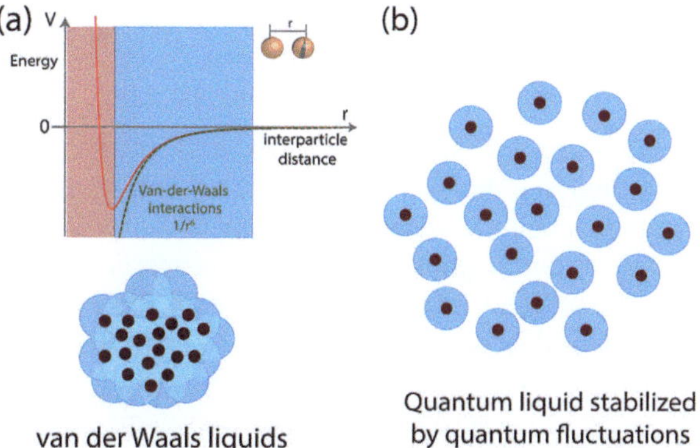

Fig. 2.1 Schematic representation of a "conventional" liquid. **a** Particles are bound together at an inter-particle distance where the repulsive core of the potential (red) is compensated by the attraction between particles given by the van der Waals interaction (blue). **b** Ultra-dilute liquid stabilized by quantum fluctuations. The particles are bound at a length scale where the details of the two-body potentials are negligible. This liquid is a direct consequence of pure quantum mechanical effects. Reprinted with permission from [5]. © 2018, C. Cabrera. All rights reserved

as surface tension, surface excitation and gas-to-liquid phase transition. A typical droplet contains several thousand atoms (even ~ 800 in dipolar gases), has a tiny size, (radial size $\equiv \sigma_r \leq 500nm$), lifetime of several hundreds of ms, which confirms a strong stabilization mechanism inside.

We shall outline such experiments in Chap. 5 and discuss the two-component Bose mixture of Bose-Einstein condensates in Chap. 11.

Here we note that, in statistical physics, mechanical stability of the system may also be verified by the condition

$$\partial\mu/\partial\rho \geq 0. \tag{2.3}$$

where μ and ρ are the chemical potentials and the density, respectively. In the next section, we briefly discuss the importance of chemical potential to describe many particle systems along with temperature, entropy, free energy, etc.

2.3 The Importance of the Chemical Potential

By definition, the chemical potential μ of a system in equilibrium is the energy which is required to remove (or add) one particle from the system: $\mu = (\partial E/\partial N)_{S,N,V}$. Although in most textbooks (see e.g. [8]) it is introduced for a system with a variational number of particles, in realistic calculations μ is of utmost important also when N is fixed.

It is well known that physical systems in equilibrium [9, 10] are studied in statistical mechanics through statistical ensembles. For example, the canonical ensemble that exchanges only energy with the surroundings is called NVT. On the other hand, a grand canonical ensemble also makes it possible to exchange with a parameter μ, which may be interpreted as a chemical potential and referred to as a $\mu V T$ system. In the former case, N, or equivalently, the number density $\rho = N/V$ is fixed, so the chemical potential $\mu \equiv \mu(\rho, T)$ has to be calculated from the equation of state. In the latter case, that is for $\mu V T$ systems, μ is fixed as an external input parameter which defines N through the thermal equation [11]. Actually, in dealing with gases or fluids, one has to calculate μ since the number of atoms is fixed. On the other hand, dealing with quantum magnets in terms of quasiparticles (magnons, triplons) you can easily determine (and fix) μ as a function of the external magnetic field H_{ext} but have to make rather complicated calculations to find the total number of quasiparticles N [12,13].

In field theories, used to describe many-particle systems, the chemical potential may be considered as a Lagrange multiplier. For example, in representative ensembles, one may introduce two chemical potentials, μ_0 and μ_1 in accordance with two existing constraints [14]. One of them is used to fix the condensate fraction, and the second one to satisfy the Hohenberg-Pines theorem. We shall discuss these points in Chap. 8 in detail.

References

1. A. Leggett, *Quantum Liquids: Bose Condensation and Cooper Pairing in Condensed-Matter Systems*. Series of Oxford Graduate Texts (OUP Oxford, 2006). https://books.google.co.uz/books?id=HnlPAwAAQBAJ
2. W. Dickhoff, D. Van Neck, *Many-body Theory Exposed! Propagator Description of Quantum Mechanics in Many-Body Systems* (World Scientific Publishing Company, 2005). https://books.google.co.uz/books?id=gzk8DQAAQBAJ
3. W. Pauli, "Towards the unification, part 2: simplified equations, covariant derivative, photons. Pauli. Z. Physik **31**, 765 (1925)
4. D.S. Petrov, Quantum mechanical stabilization of a collapsing Bose-Bose mixture. Phys. Rev. Lett. **115**, 155302 (2015). https://link.aps.org/doi/10.1103/PhysRevLett.115.155302
5. C.R.C. Cabrera, *PhD Thesis: Quantum Liquid Droplets in a Mixture of Bose-Einstein Condensates* (Polytechnic University of Catalonia, Institute of Photonic Sciences, Catalonia, 2018)
6. C.R. Cabrera, L. Tanzi, J. Sanz, B. Naylor, P. Thomas, P. Cheiney, L. Tarruell, Quantum liquid droplets in a mixture of Bose-Einstein condensates. Science **359**(6373), 301–304 (2018). https://www.science.org/doi/abs/10.1126/science.aao5686
7. I. Ferrier-Barbut, H. Kadau, M. Schmitt, M. Wenzel, T. Pfau, Observation of quantum droplets in a strongly dipolar Bose gas. Phys. Rev. Lett. **116**, 215301 (2016). https://link.aps.org/doi/10.1103/PhysRevLett.116.215301
8. L. Landau, E. Lifshitz, *Statistical Physics: Volume 5* (Elsevier Science, 2013), no. v. 5. https://books.google.co.uz/books?id=VzgJN-XPTRsC
9. A. Rakhimov, Tan's contact as an indicator of completeness and self-consistency of a theory. Phys. Rev. A **102**, 063306 (2020). https://link.aps.org/doi/10.1103/PhysRevA.102.063306
10. A. Rakhimov, T. Abdurakhmonov, B. Tanatar, Critical behavior of Tan's contact for bosonic systems with a fixed chemical potential. J. Phys.: Condensed Matter **33**(46), 465401 (2021). https://dx.doi.org/10.1088/1361-648X/ac1ec6
11. U. Marzolino, μPT statistical ensemble: systems with fluctuating energy, particle number, and volume. Sci. Rep. **11**(1), 15096 (2021). https://doi.org/10.1038/s41598-021-94013-x
12. V. Zapf, M. Jaime, C.D. Batista, Bose-Einstein condensation in quantum magnets. Rev. Mod. Phys. **86**, 563–614 (2014). https://link.aps.org/doi/10.1103/RevModPhys.86.563

13. A. Rakhimov, S. Mardonov, E. Sherman, Macroscopic properties of triplon Bose–Einstein condensates. Ann. Phys. **326**(9), 2499–2516 (2011). https://www.sciencedirect.com/science/article/pii/S0003491611001047
14. V.I. Yukalov, Representative ensembles in statistical mechanics. Int. J. Mod. Phys. B **21**(01), 69–86 (2007). https://doi.org/10.1142/S0217979207035893

A System of Noninteracting Particles-Ideal Gas

In general, to describe a statistical system in equilibrium from "first principles", one has to construct an appropriate Hamiltonian (or Lagrangian) operator which could be schematically presented as

$$H = H_{kin} + H_{\text{int}} + H_{ext}, \tag{3.1}$$

where H_{kin} is the kinetic term, H_{int} the interaction potential between, say, atoms, and H_{ext} is the external field like a trapping potential or a magnetic field for $\mu V T$ systems. By definition, an ideal gas is a system of an infinite number of noninteracting particles ($H_{ext} = H_{\text{int}} = 0$) confined in the volume V, so that $\rho = N/V \neq \infty$. Although such a system can not be realized in nature, studying idealized cases may serve as a checkpoint for realistic calculations and give at least qualitative information on the thermodynamic properties of real systems.

From textbooks [1,2] we know that a particle of a mass m in an ideal gas moves freely as a plane wave: $\psi = \exp(i\mathbf{kr})$[1] with the energy $\varepsilon_k = \mathbf{k}^2/2m$ so that the possible values of momentum $\mathbf{k}$ form a continuum. Therefore, the sum over the momentum can be replaced by an integration

$$\sum_{\mathbf{k}} \rightarrow \frac{V}{(2\pi)^3} \int d^3\mathbf{k}. \tag{3.2}$$

---

[1] In the most part of this book we adopt the units $k_B = 1$ for the Boltzmann constant, $\hbar = 1$ for the Planck constant.

A. Rakhimov and S. Mardonov, *Theory of Quantum Bose Liquids Beyond Bogoliubov Approximation*, Lecture Notes in Physics 1045, https://doi.org/10.1007/978-3-032-05096-0_3

The number of particles with momentum in the interval $\mathbf{k} + d\mathbf{k}$ may be defined by the famous Boltzmann distribution

$$f_{BL} = \frac{1}{e^{(\varepsilon_k - \mu)/T}}. \tag{3.3}$$

Clearly, here T and m are supposed to be given parameters. As to μ-chemical potential, it should be determined from a thermodynamic equation, which is in this case equal to the normalization condition:

$$N = \sum_k \frac{1}{e^{(\varepsilon_k - \mu)/T}}. \tag{3.4}$$

Task 3.1 Evaluate the chemical potential of an ideal gas of helium-four atoms in normal conditions ($T = 273.15\,K$, $p = 760\,\text{mmHg}$) by using Eq. 3.4.[2]

3.1　Degenerate (Ideal) Bose Gas

The famous Boltzmann distribution 3.3 proposed long before quantum mechanics, does not take into account quantum statistics, which is sensitive to the spin of particles. We know from textbooks [1] that the mean distribution of a free boson with the energy $\varepsilon_k = \mathbf{k}^2/2\,m$ in an ideal gas is

$$f_B(\varepsilon_k) = \frac{1}{e^{\beta(\varepsilon_k - \mu)} - 1}, \tag{3.5}$$

where for a NVT system the chemical potential $\mu(N, T, V)$ should be evaluated as a solution to following equation

$$N = \sum_k \frac{1}{e^{\beta(\varepsilon_k - \mu)} - 1}, \tag{3.6}$$

or equivalently

$$\rho = \frac{1}{(2\pi)^3} \int \frac{d^3k}{e^{\beta(\varepsilon_k - \mu)} - 1}. \tag{3.7}$$

This formula can also be used to evaluate the critical temperature T_c of BEC. To make this term clear, let us refresh in our mind the concept of BEC by considering the qualitative evaluation of a Bose system with increasing temperature.

- At exactly zero temperature $T = 0$, all particles of an ideal gas occupy the energy level with $\varepsilon_k = 0$. We call such particles condensed particles (red balls in Fig. 3.1).

---

[2] Solution of some tasks can be found in the Appendices.

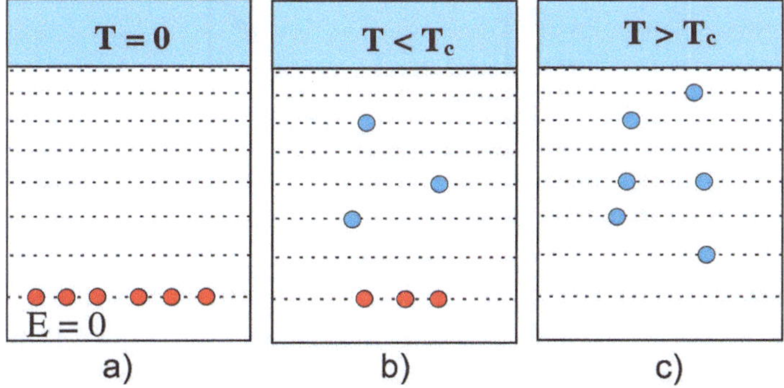

Fig. 3.1 Illustration of energy level occupations in the boxed ideal Bose gas [3]. **a** At $T = 0$, all particles lie in the ground state, $E = 0$. **b** For $T < T_c$, some particles are in excited states, but there is still macroscopic occupation of the ground state. **c** For $T > T_c$ all particles are out of the level with $E = 0$. Reprinted with permission from [3]. © 2016, C. Barenghi and N.G. Parker. All rights reserved

- What will happen when we increase the temperature? Then, due to the temperature fluctuations, some particles (atoms) start to leave the lowest level and move to higher levels (blue balls in Fig. 3.1). We call such particles uncondensed (depleted) particles. So, at finite temperatures, we have in general both kinds of particles: condensed and uncondensed ones. Let's denote the number of former and latter as N_0 and N_1 respectively. Clearly, $N_0 + N_1 = N$ at any temperature, since the total number of particles is constrained-otherwise BEC cannot occur [1].
- Further increase of temperature up to a critical $T = T_c$ leads to a phase transition so that there is no longer a condensed particle above the critical temperature $N_0(T \geq T_c) = 0$ by definition of T_c.

What about the chemical potential $\mu(T)$? Well, it cannot be positive at any temperature, $\mu(T) \leq 0$, since f_B in Eq. 3.5, and hence the denominator of this equation must be positive even at $\varepsilon_k = 0$, i.e. $\exp(-\mu/T) \geq 1$. If the density in 3.7 is kept fixed, while the temperature is reduced, the absolute value of μ is expected to decrease, to allow the density to remain constant. The limit $\mu = 0$ is then reached at $T = T_c$, so $|\mu|$ can not decrease any more. Thus, we come to the conclusion that for an ideal Bose gas

$$\begin{cases} \mu = 0, \, if \quad T \leq T_c, \\ \mu < 0, \, if \quad T > T_c. \end{cases} \tag{3.8}$$

Now it is clear that T_c can be evaluated from 3.7 as a function of the density by setting there $\mu = 0$:

$$\begin{aligned} \rho &= \frac{1}{(2\pi)^3} \int \frac{d^3k}{e^{\varepsilon_k/T_c} - 1} = \frac{1}{2\pi^2} \int_0^\infty \frac{k^2 dk}{e^{\varepsilon_k/T_c} - 1} \\ &= \frac{\sqrt{2}(mT_c)^{3/2}}{2\pi^2} \int_0^\infty \frac{\sqrt{x}dx}{e^x - 1} = \frac{\sqrt{2}(mT_c)^{3/2}\sqrt{\pi}\zeta(3/2)}{4\pi^2}, \end{aligned} \tag{3.9}$$

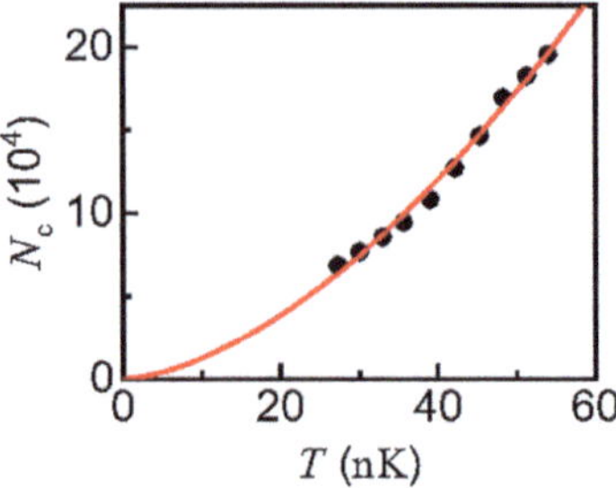

Fig. 3.2 The number of condensed atoms vs the temperature from Eqs. 3.11–3.13. It is seen that experimental points [4], denoted by the filled circles, lie very close to the curve $N_0 \sim T^{3/2}$. Reprinted under CC-BY-4.0 [4]. ©2021, N. Navon et al.

where $\zeta(3/2) = 2.612$ - Riemann ζ function. Inverting this equation and restoring natural units, we come to the following formula for T_c

$$T_c = \frac{2\rho^{2/3}\pi(\hbar c)^2}{mc^2 k_B(\zeta(3/2))^{2/3}} \approx \frac{3.31\rho^{2/3}(\hbar c)^2}{mc^2 k_B}, \tag{3.10}$$

where $\hbar c = 197.33 ev \cdot nm, k_B = 8.617 \cdot 10^{-5} ev/K, m \equiv m_{atom} \approx Am_p, m_p c^2 \approx 940 \cdot 10^6 ev$ and ρ is the number density in units $1/nm^3$.

Does this formula have any practical significance? Yes, of course. In fact, in a very dilute and weakly interacting regime, a system of ultracold Bose atoms can be considered as an ideal gas. In this regime, almost all particles may be assumed to be condensed, $\rho \approx \rho_0 = N_0/V$, and hence the Eq. 3.10 is presented as $N_0 \propto T^{3/2}$. Surprisingly, the experiments [4], made in an optical trap, where the gas is homogeneous, confirm the validity of this relation, as it is shown in Fig. 3.2.

However, for a strongly interacting system, such as the liquid helium the relation (3.10) does not work. In fact, in this case, setting $A = 4$, $\rho = 22nm^{-3}$ one obtains $T_c(^4He)|_{id.gas} = 3.12K$. Comparing this value with the experimental one $T_c(^4He)|_{exper.} = 2.17K$ one may see this. On the other hand, this situation may be improved by introducing an effective mass. Using here another value for the mass of an atom of ^4He as $m_{effective} = 1.44m$ one could obtain the desired $T_c = 2.17$K.

Task 3.2 Estimate T_c for ideal gas of ^{39}K with $\rho = N/V = 2 \cdot 10^{14}/cm^3$.

In actual experiments on dilute Bose gases, the critical temperature is about nano-Kelvins, as it is seen, e.g. from Fig. 3.2. How can experimenters achieve such low temperatures? You may find a qualitative answer in Appendix B, but for the details you have to read, for example, some review articles [5–8].

Below, after having fixed T_c we start to discuss two possible phases, condensed $T < T_c$ and normal $T > T_c$ ones separately.

3.1.1   BEC Phase

Clearly, at exactly zero temperature, all particles are condensed: $n_0 = N_0/N = 1$ and $n_1 = N_1/N = 0$, which means that there is no quantum depletion in an ideal gas. However, no one has cancelled yet temperature fluctuations for $T \neq 0$. They make some particles leave the ground state and form a set of uncondensed particles, in spite of $T < T_c$. Their number is given by Eq. 3.4 with $\mu = 0$

$$N_1(T < T_c) = \sum_k \frac{1}{e^{\beta \varepsilon_k} - 1}, \tag{3.11}$$

such that the condensed fraction N_0 is defined from the normalization condition as $N_0(T < T_c) = N - N_1(T < T_c)$. On the other hand, by definition $N_0(T = T_c) = 0$ and hence

$$N_1(T = T_c) = \sum_k \frac{1}{e^{\varepsilon_k/T_c} - 1}. \tag{3.12}$$

Now, using (3.9), (3.11) and (3.12) one may directly obtain a well known textbook formulas [1]

$$\frac{N_1(T < T_c)}{N_1(T = T_c)} = \frac{N_1(T < T_c)}{N} = \frac{T^{3/2}}{T_c^{3/2}} = n_1(T < T_c),$$
$$n_0(T < T_c) = 1 - n_1(T < T_c) = 1 - \frac{T^{3/2}}{T_c^{3/2}}. \tag{3.13}$$

These simple relations surprisingly well describe the experimental data (see Fig. 3.3).

Task 3.3 Using the condition $N_1(T < T_c) < N$ in the BEC phase show that $\lambda_T \geq d$ where d is the average inter-particle distance, $d \sim \rho^{-1/3}$ and $\lambda_T = \sqrt{2\pi/mT}$ is the thermal wavelength.

Note that this condition shows that at $T > T_c$, atoms behave like point particles, but at $T < T_c$ they behave like matter waves, whose overlap $\lambda_T \sim d$ signals the onset of BEC near T_c (see Fig. 3.4).

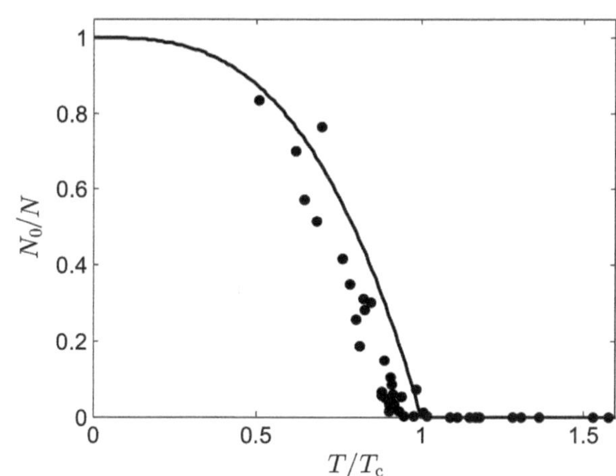

Fig. 3.3 Variation of condensate fraction N_0/N (see Eq. 3.13) with temperature for a harmonically-trapped BEC, with the ideal-gas predictions (solid line) compared to experimental measurements (points) from Ref. [9] (circles), with $T_c = 280$ nK. Reprinted with permission from [9]. ©1996, American Physical Society. All rights reserved

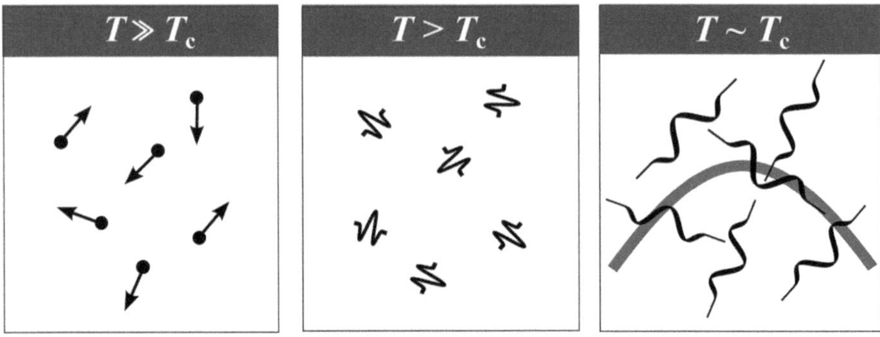

Fig. 3.4 Schematic of the transition between a classical gas and a Bose-Einstein condensate. At high temperatures ($T >> T_c$), the gas is a thermal gas of point-like particles. At low temperatures (but still exceeding T_c), the de Broglie wavelength $\lambda_{dB} = h/p$ (p-momentum) becomes significant, yet smaller than the average spacing d. At T_c, the matter waves overlap ($\lambda_{dB} \sim d$), marking the onset of Bose-Einstein condensation. Reprinted with permission from [3]. © 2016, C. Barenghi and N.G. Parker. All rights reserved

3.1.2 Thermodynamics of BEC

For further references, we calculate other thermodynamic characteristics of an ideal Bose gas starting from the total energy of the whole system. Using Eq. 3.5 one obtains

$$E(T \le T_c) = \sum_k \frac{\varepsilon_k}{e^{\beta \varepsilon_k} - 1} = \frac{3\sqrt{2} V T^{5/2} m^{3/2} g_{5/2}(1)}{8\pi^{3/2}} = \frac{3V T \zeta(5/2)}{2\lambda_T^3}, \quad (3.14)$$

where $\zeta(5/2) \approx 1.342$. Here we have introduced the Bose function

$$g_\nu(z) = \frac{1}{\Gamma(\nu)} \int_0^\infty \frac{x^{\nu-1} dx}{z^{-1} e^x - 1}, \quad (3.15)$$

which is often used by the BEC community in practical calculations. The heat capacity $C_v = (\partial E/\partial T)_V$ and the entropy defined through $C_v = T(\partial S/\partial T)$ are given as

$$C_v(T \le T_c) = \left(\frac{\partial E}{\partial T}\right)_V = \frac{5E}{2T},$$
$$S(T \le T_c) = \int \frac{C_v}{T} dT = \frac{5E}{3T}. \quad (3.16)$$

For further reference, using Eqs. 3.10 and 3.14 we can rewrite C_v in BEC phase as

$$\frac{C_v(T \le T_c)}{N} = 1.926\, t^{3/2}, \quad (3.17)$$

where $t = T/T_c$. Due to $\mu = 0$ the free energy $F = E - TS$ and the grand thermodynamic potential $\Omega = F - \mu N$ coincide:

$$\Omega(T < T_c) = F(T < T_c) = -\frac{VT\zeta(5/2)}{\lambda_T^3} = -\frac{2}{3}E < 0, \qquad (3.18)$$

and hence, the pressure does not depend on the volume:

$$P(T < T_c) = -\frac{\Omega}{V} = \frac{T\zeta(5/2)}{\lambda_T^3}. \qquad (3.19)$$

Note that, at exactly $T = 0$, all these quantities are equal to zero, which means that there is no zero mode for an ideal Bose gas. At finite temperature, the total energy E is positive, but the free energy F, which reaches its minimum in the equilibrium, is negative, $F(T < T_c) = \Omega(T < T_c) < 0$. It is worth keeping in mind that for an ideal Bose gas at low temperatures $E \sim T^{5/2}$, $C_v \sim T^{3/2}$, $S \sim T^{3/2}$ and $P \sim T^{5/2}$.

3.1.3 Normal Phase $T > T_c$ and the Jump in the Gradient of the Heat Capacity

This phase is usually not discussed in the majority of textbooks. However, it has become very important when one deals with quantum magnets [10] or BCS-BEC crossover.[3]

In the previous subsection, we have stated that N_1, i.e. the number of uncondensed particles, increases with $T \to T_c$, and becomes equal to N at $T \geq T_c$. The region $T > T_c$ is often referred to as a normal phase, where the particle distribution is given by

$$N_1(T > T_c) = N = \sum_k \frac{1}{e^{\beta(\varepsilon_k - \mu)} - 1} = \sum_k \frac{1}{e^{\beta\varepsilon_k}z^{-1} - 1}, \qquad (3.20)$$

where $z = \exp(\beta\mu) < 1$ is the fugacity. This equation may be used in two different ways: For NVT systems (mainly atomic gases), it is a nonlinear equation with respect to the chemical potential $\mu(T, N)$, while for μVT systems, it is just an equality for evaluation of the number of particles for a given μ defined through the external field (quantum magnets, photon gas in a dye-filled optical microcavity [11]). Let's discuss only ideal atomic Bose gases.

The Eq. 3.20 can be easily presented in the following compact form:

$$g_{3/2}(z) = \rho\lambda_T^3, \qquad (3.21)$$

---

[3] The BCS-BEC crossover is a phenomenon in condensed matter physics where a system transitions from the Bardeen-Cooper-Schrieffer (BCS) state, which describes a superconductor, to the BEC state, which describes a superfluid.

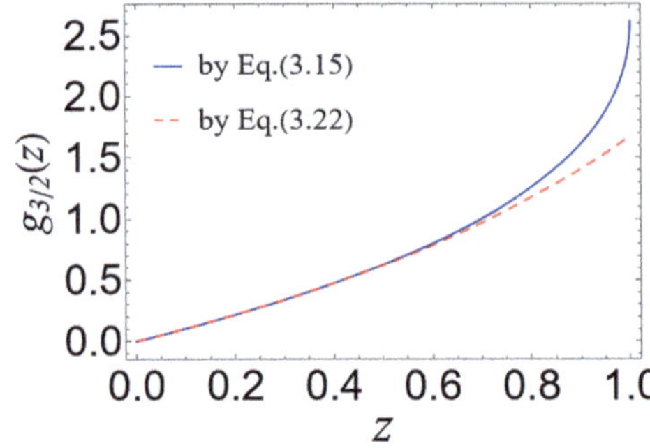

Fig. 3.5 Bose function $g_{3/2}(z)$ versus z: exact numerical (solid line) and approximated by Eq. 3.22 (dashed line)

where we introduce Bose function in accordance with Eq. 3.15, which has following power series in $z < 1$

$$g_{3/2}(z) = z + \frac{z^2}{2^{3/2}} + \frac{z^3}{3^{3/2}} + \frac{z^4}{4^{3/2}} + O(z^5). \tag{3.22}$$

In Fig. 3.5 we plot the function $g_{3/2}(z)$ (exact numerical and its power series). It is seen that it is an increasing function, and the power series up to z^4 is good for $z \leq 0.5$.

As an example let's calculate μ for, say, ^{39}K with the very dilute density $\rho = N/V = 0.2 \cdot 10^{-6}/nm^3$ at rather high temperature $T = 0.1K$. Clearly, for these parameters we have $\lambda_T = \sqrt{2\pi/mT} \approx 0.14nm$, $d \sim 1/\rho^{1/3} \approx 171nm$. So, $\lambda_T <<$ d and we are really out of the BEC phase. Evidently, RHS of Eq. 3.21 is rather small $\rho\lambda_T^3 \approx 0.56 \cdot 10^{-9}$, and hence we can use $g_{3/2}(z) \approx z$ to obtain $z = 0.56 \cdot 10^{-9}$, $\mu/k_B T = ln(z) = -21.3 < 0$.

In the previous subsection, we have calculated some thermodynamic characteristics in the BEC phase. Now, for completeness, we bring analytic formulas for them in the normal phase and, particularly, show that the heat capacity has no jump at $T = T_c$, i.e. $\lim_{\epsilon \to 0} C_v(T_c + \epsilon) = \lim_{\epsilon \to 0} C_v(T_c - \epsilon)$, but its derivative $(\partial C_v/\partial T)$ has a jump at the critical temperature.

Starting from the energy per volume E/V we can find immediately

$$\frac{E(T > T_c)}{V} = \frac{1}{V} \sum_k \frac{\varepsilon_k}{e^{\beta\varepsilon_k}z^{-1} - 1} = \frac{3T\, g_{5/2}(z)}{2\lambda_T^3}, \tag{3.23}$$

in terms of the Bose function, which has the power series

$$g_{5/2}(z) = z + \frac{z^2}{2^{5/2}} + \frac{z^3}{3^{5/2}} + \frac{z^4}{4^{5/2}} + O(z^5). \tag{3.24}$$

Note that, putting here $z = 1$, $g_{5/2}(1) = \zeta(5/2) = 1.342$ we come back to $E(T < T_c)$ in (3.14). To calculate the entropy, it is convenient to use $S = -(\partial\Omega/\partial T)_\mu$ where the grand thermodynamic potential for an ideal gas is given by [12]:

$$\Omega(T) = T \sum_k \ln(1 - ze^{-\beta\varepsilon_k}). \tag{3.25}$$

So,

$$S(T) = -\beta \left[\Omega + \sum_k \frac{\mu - \varepsilon_k}{e^{\beta \varepsilon_k} z^{-1} - 1} \right]. \tag{3.26}$$

Note that the relations (3.25) and (3.26) are true for any temperature for an ideal Bose gas. Further, integrating by parts in Eq. 3.25 ($\sqrt{x} \equiv 2/3 d(x^{3/2})$) one can express Ω, S as well as F in terms of Bose functions:

$$\Omega(T > T_c) = -\frac{V T g_{5/2}(z)}{\lambda_T^3} = -\frac{2}{3} E(T > T_c) = -PV,$$

$$S(T > T_c) = \frac{V}{\lambda_T^3} \left[\frac{5 g_{5/2}(z)}{2} - g_{3/2}(z) \ln z \right], \tag{3.27}$$

$$F(T > T_c) = E - TS = \Omega + N\mu = \frac{V T}{\lambda_T^3} \left[-g_{5/2}(z) + g_{3/2}(z) \ln z \right].$$

As to the heat capacity in the normal phase, evaluation of $C_v(T > T_c)$ directly from the above equations by using, say, $C_v = T(\partial S/\partial T)$ is a little cumbersome business. Instead, it is easier and more preferable to use the following approximation:

$$\frac{C_v(T > T_c)}{N} = 1.496 + 0.341 t^{-3/2} + 0.089 t^{-3}, \tag{3.28}$$

obtained by Wang [13], which nicely works even up to $t \equiv T/T_c \sim 100$.

Task 3.4 Derive the textbook formula [2]

$$\frac{C_v(T > T_c)}{N} = \frac{15 V g_{5/2}(z)}{4 \lambda_T^3} - \frac{9 g_{3/2}(z)}{4 g_{1/2}(z)}. \tag{3.29}$$

Hint: To evaluate dz/dT differentiate both sides of (3.20) and use $dN/dT = 0$, $z(\partial g_\nu(z)/\partial z) = g_{\nu-1}(z)$. Note that the Bose function $g_{1/2}(z)$ is not regular: $g_{1/2}(z \to 1) = \infty$

3.1.4 Near T_c

What happens with thermodynamic quantities close to the critical temperature? To answer this question, it is convenient to discuss the difference

$$\Delta A = \lim_{\epsilon \to 0} A(T_c - \epsilon) - \lim_{\epsilon \to 0} A(T_c + \epsilon) \equiv A(T_c - 0) - A(T_c + 0), \tag{3.30}$$

for a physical quantity $A(T)$. Now using equations given in the last two subsections, and bearing in mind that $\lim_{\epsilon \to 0} z(T_c - \epsilon) = \lim_{\epsilon \to 0} z(T_c + \epsilon) = 1$, one may easily come to the following conclusions: $\Delta E = \Delta F = \Delta \Omega = \Delta S = \Delta C_v = 0$. On the

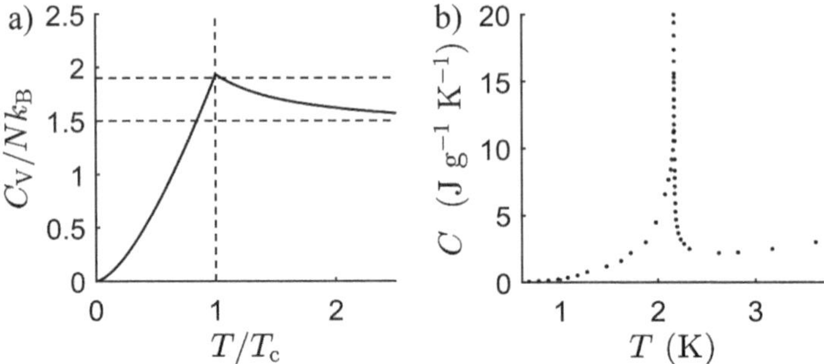

Fig. 3.6 **a** Heat capacity C_v of the ideal Bose gas as a function of temperature T. **b** Experimental heat capacity of liquid Helium near the λ-point of $T \approx 2.2$ K. Both curves show a similar cusped structure. Reprinted with permission from [3]. © 2016, C. Barenghi and N.G. Parker. All rights reserved

other hand, differentiating the Eqs. 3.17 and 3.28 with respect to T one can easily see that

$$\Delta(\partial C_v/\partial T) \approx \frac{3.66}{T_c} \neq 0. \tag{3.31}$$

Therefore the heat capacity is continuous at $T = T_c$ but its derivative has a jump i.e.

$$(\partial C_v/\partial T)|_{T=T_c-\epsilon} \neq (\partial C_v/\partial T)|_{T=T_c+\epsilon}.$$

This discontinuity implies that an ideal Bose gas really exhibits a phase transition at a temperature $T = T_c$. In Fig. 3.6 we present C_v of an ideal Bose gas (a) and the experimental C_v data for liquid Helium (b). It is seen that both curves show a similar cusped structure. The similarity which is seen in Fig. 3.6, as well as the behavior of the heat capacity at small T as $C_v(t << 1) \sim t^{3/2}$ is key evidence in linking superfluid helium to Bose-Einstein condensation.

Therefore, the ideal gas model could serve as a simple theory of BEC. However, this model fails to predict the phenomenon of superfluidity, which is the ability of the Bose-Einstein condensed gas to flow without experiencing friction. The main reason for this failure is hidden in the energy dispersion E_k, which is, in fact, just the dependence of the energy of a particle E on its momentum **k**. People working on high-energy physics used to think that this dependence may be only quadratic ($E_k = k^2/2m$) or relativistic ($E_k^2 = k^2c^2 + m^2c^4$). This is true, but when a particle is moving in a medium, such dependence may be changed, becoming a very important physical observable quantity, which carries information on the excitation spectrum of the system as a whole. An excitation spectrum of a quantum liquid, similarly to the spectrum of a crystal, can be described in terms of quasiparticles-phonons: when we excite the system, some phonons with energy dispersion E_k are created and vice versa. In real experiments, the function E_k can be accurately measured by inelastic neutron scattering.

So, what kind of energy dispersion is good for the occurrence of superfluidity? The answer was obtained by Landau nearly eighty years ago [14].

3.2   Landau Criterion

To study the physical meaning of superfluidity, we consider a small body of mass M inside, say, superfluid helium, with no internal degrees of freedom, that is moving with a total momentum $\mathbf{K} = M\mathbf{v}$ and kinetic energy $E = \mathbf{K}^2/2M$. This body (slab) can experience friction by creating excitations in the gas, which must occur in such a manner that both momentum and energy are conserved in the process. Friction occurs when part of the kinetic energy of the slab can be converted into heat by the creation of a quasiparticle excitation. Now let's consider a new configuration in which a quasiparticle with momentum $\mathbf{k}$ (in the original rest frame of the liquid) and energy E_k is present. In the new configuration, the slab then has velocity $\mathbf{v}' = \mathbf{v} + \mathbf{k}/M$, momentum $\mathbf{K}' = M\mathbf{v} + \mathbf{k} = \mathbf{K} + \mathbf{k}$ and energy $E' = \mathbf{K}'^2/2M + E_k = E + (\mathbf{kv}) + E_k$ where the recoil energy $\mathbf{k}^2/2M$ can be neglected. The most favorable case is for $\mathbf{k}$ in the opposite direction of $\mathbf{v}$, and creation of a quasiparticle with momentum $\mathbf{k}$ is therefore allowed when $E' < E$ or $E_k < kv$. We see now that dissipation can occur only for velocities

$$v > v_c = \left(\frac{E_k}{k}\right)_{min}, \tag{3.32}$$

where v_c is referred to as the Landau critical velocity. If $v_c \neq 0$ there will be superflow for $v < v_c$.

This depends on the form of the excitation spectrum E_k. For example, for an ideal gas $E_k = \mathbf{k}^2/2m$ and hence there would not be superfluidity, since v_c(ideal gas)$= (k^2/2mk)_{min} = 0$. On the other hand, if $E_k \sim sk$, where s is the macroscopic speed of sound, then $v_c \sim s \neq 0$ and such a type of gas can experience superfluidity. In Fig. 3.7 we present experimental data [15] for spectrum E_k of quasiparticle excitations in superfluid helium at $T \sim 1.1$ K. It is seen that, for very small momentum, the dispersion is linear, $E_k \sim sk$, in accordance with the Landau criterion. The next example is the BEC of Cooper pairs, which occurs in atomic Fermi gases with an attractive interaction. Their excitation spectrum is gapped: $E_k^2 = (\varepsilon_k - \mu)^2 + \Delta^2 = (\Delta^2 + (k - k_0)^2/2m)^2$ with the energy gap $E_{k=k_0} = \Delta$. So, the minimum of E_k/k is also nonzero, yielding again a critical velocity below which the gas is superfluid.

Task 3.5. Evaluate critical velocity for Cooper pairs.

The superfluidity of atomic Fermi gases is analogous to conventional superconductivity in metals, where the electrons can form Cooper pairs due to a mutual attractive interaction, generated by lattice vibrations, i.e. by phonons.

Here it should be underlined that, strictly speaking, the relation between Bose-Einstein condensation and superfluidity is a long-standing and not completely understood problem. The current understanding is that Bose-Einstein condensation is neither necessary nor sufficient for superfluidity. A reader can find more detailed discussion on this subject, e.g. in Ref. [16]. However, there is no doubt that superfluidity can not take place in the system without mutual interaction between its particles. There-

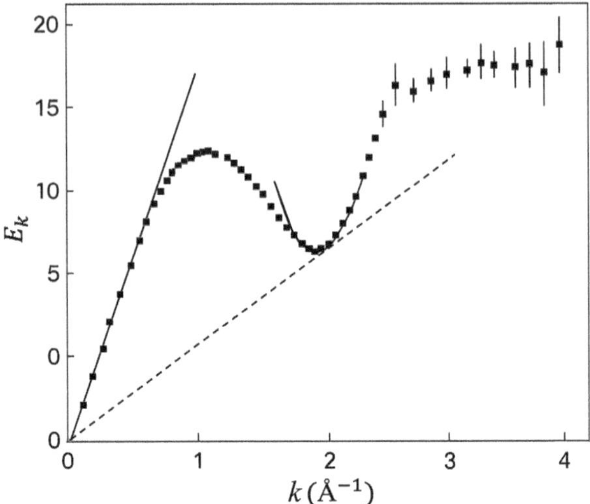

Fig. 3.7 Spectrum of quasiparticle excitations in He-II at $T = 1.1$ K. Experimental data from [15]. The solid and dashed curves indicate the Landau critical velocities for the phonon and roton excitations, respectively. A D B Woods and R A Cowley, reprinted with permission from [15]. ©1973. IOP Publishing Ltd. All rights reserved

fore, one has to take into account mutual interactions to describe the experimentally observed phenomena of superfluidity. The remaining part of the present lectures concerns the problem of considering interactions in the realm of actual theories.

References

1. L. Landau, E. Lifshitz, *Statistical Physics: Volume 5* (Elsevier Science, 2013), no. v. 5. https://books.google.co.uz/books?id=VzgJN-XPTRsC
2. K. Huang, *Introduction to Statistical Physics*. Series Civil and Environmental Engineering, 2nd edn. (CRC Press, 2009). https://books.google.co.uz/books?id=rKmC3bMEVxIC
3. C. Barenghi, N.G. Parker, *A Primer on Quantum Fluids*. Series Springer Briefs in Physics (Springer, Cham, 2016). https://doi.org/10.1007/978-3-319-42476-7
4. N. Navon, R.P. Smith, Z. Hadzibabic, Quantum gases in optical boxes. Nat. Phys. **17**(12), 1334–1341 (2021). https://doi.org/10.1038/s41567-021-01403-z
5. I. Bloch, J. Dalibard, W. Zwerger, Many-body physics with ultracold gases. Rev. Mod. Phys. **80**, 885–964 (2008). https://link.aps.org/doi/10.1103/RevModPhys.80.885
6. J. Weiner, V.S. Bagnato, S. Zilio, P.S. Julienne, Experiments and theory in cold and ultracold collisions. Rev. Mod. Phys. **71**, 1–85 (1999). https://link.aps.org/doi/10.1103/RevModPhys.71.1
7. F. Dalfovo, S. Giorgini, L.P. Pitaevskii, S. Stringari, Theory of Bose-Einstein condensation in trapped gases. Rev. Mod. Phys. **71**, 463–512 (1999). https://link.aps.org/doi/10.1103/RevModPhys.71.463
8. N.P. Proukakis, D.W. Snoke, P.B. Littlewood, *Universal Themes of Bose-Einstein Condensation*. (Cambridge University Press, 2017). https://doi.org/10.1017/9781316084366
9. J.R. Ensher, D.S. Jin, M.R. Matthews, C.E. Wieman, E.A. Cornell, Bose-Einstein condensation in a dilute gas: Measurement of energy and ground-state occupation. Phys. Rev. Lett. **77**, 4984–4987 (1996). https://link.aps.org/doi/10.1103/PhysRevLett.77.4984
10. A. Rakhimov, S. Mardonov, E. Sherman, Macroscopic properties of triplon Bose–Einstein condensates. Ann. Phys. **326**(9), 2499–2516 (2011). https://www.sciencedirect.com/science/article/pii/S0003491611001047

11. J. Schmitt, Dynamics and correlations of a Bose–Einstein condensate of photons. J. Phys. B: Atomic. Mol. Opt. Phys. **51**(17), 173001 (2018). https://dx.doi.org/10.1088/1361-6455/aad409

12. A. Rakhimov, I.N. Askerzade, Thermodynamics of noninteracting bosonic gases in cubic optical lattices versus ideal homogeneous Bose gases. Int. J. Mod. Phys. B **29**(18), 1550123 (2015). https://doi.org/10.1142/S0217979215501234

13. F.Y.-H. Wang, Specific heat of an ideal Bose gas above the Bose condensation temperature. Am. J. Phys. **72**(9), 1193–1194 (2004). https://doi.org/10.1119/1.1764562

14. L. Landau, E. Lifshitz, L. Pitaevskii, *Statistical Physics: Theory of the Condensed State*. Series Course of theoretical physics. (Elsevier Science, 1980). https://books.google.co.uz/books?id=igpBAQAAIAAJ

15. A.D.B. Woods, R.A. Cowley, Structure and excitations of liquid helium. Rep. Prog. Phys. **36**(9), 1135 (1973). https://dx.doi.org/10.1088/0034-4885/36/9/002

16. V.I. Yukalov, Principal problems in Bose-Einstein condensation of dilute gases. Laser Phys. Lett. **1**(9), 435–461 (2004). https://onlinelibrary.wiley.com/doi/abs/10.1002/lapl.200410097

BEC of Interacting Particles

4

4.1 Introduction

To go far from the ideal gas model one has to develop a theory that deals with two kinds of interactions: a trapping external potential $V_{trap}(\mathbf{r}_i)$ and an interatomic potential $V_{int}(\mathbf{r}_i - \mathbf{r}_j)$, between the atoms localized in space coordinates $\mathbf{r}_i$ and $\mathbf{r}_j$. The trapping potential is needed to confine a large amount of atoms in a small volume, such that the atoms can not reach its physical walls. This may be achieved by imposing a magnetic field (magnetic trap) or (and) laser beams with a special configuration (optical trap). Nowadays it is possible to create a variety of configurations: from parabolic to torus and even square box [1] (see next chapter).

As to interatomic interactions, in general, they may involve two and even three particles. The origin of three body interaction $V_3(\mathbf{r}_i, \mathbf{r}_j, \mathbf{r}_k)$ in many particle system is following. In actual experiments, gaseous BECs can be realized for alkali atoms, whose ground states are known as metals. The first step in solidification of atomic gases is to form a liquid containing molecules. The average lifetime of atomic BEC is around a minute! During this time the system transfers into its metallic phase by itself in accordance with laws of nature. Note that, you can hardly find a free atom in the air, since atoms of any gas prefer to form a molecule. However, for a molecule to be formed, at least three atoms must collide for the laws of energy and momentum conversation. In such process two atoms form a molecule while a third one carries away extra energy and momentum, that are released upon molecular formation. In a rarefied gas, the probability of three atoms being found within the range of interaction is very small, and therefore BECs can actually be formed in an excited many-body state similar to the resonance state. Therefore it is thanks to three body interactions

A. Rakhimov and S. Mardonov, *Theory of Quantum Bose Liquids Beyond Bogoliubov Approximation*, Lecture Notes in Physics 1045, https://doi.org/10.1007/978-3-032-05096-0_4

a macroscopic system[1] of atoms in BEC turns into small ice, as it should be at very low temperatures.

As long as three body inelastic collisions are negligible, BECs with repulsive interactions can exist in a thermodynamic stable state. However, it should be noted that BECs with even two body attractive interaction can only exist in a metastable state even if the inelastic collisions are negligible, as it is prone to collapse due to dynamical instability. For completeness, we note also that, in two-component Bose gases attractive interspecies interaction may lead to the occurrence of stable droplets [2]. In the present lectures, we neglect three-body interactions and deal with mainly two body ones.

It is well known that interatomic interaction is given by rather complicated potential such as van der Waals force or especially for helium by Aziz potential [3]. However, luckily due to the diluteness of the gaseous system of atoms, assumed to be condensed experimentally, a theorist may use so-called contact interaction

$$V(\mathbf{r}_1 - \mathbf{r}_2) = g\delta(\mathbf{r}_1 - \mathbf{r}_2), \quad V_k = \int d\mathbf{r} e^{-i\mathbf{k}\mathbf{r}} V(\mathbf{r}) = g, \tag{4.1}$$

where $\delta(\mathbf{r})$ is three dimensional Dirac delta function and $g = const$ is the interaction strength. All other corrections to this simple potential are referred in the literature as finite range effects (see e.g. Ref. [4]). It is easy to show that the two body scattering problem with the potential (4.1) may be solved analytically [5,6] leading to the result

$$a_s = \frac{gm}{4\pi\hbar^2}, \tag{4.2}$$

where a_s is the s-wave scattering length. At very low temperatures, and hence low energies E, the s- wave scattering length determines the amplitude and the total cross-sections of elastic scattering as $a_s = -f(\theta, E)_{E\to 0}$ and $\sigma(E \to 0) = 8\pi a_s^2$, respectively. From Eq. (4.1) one can immediately understand that, the sign of a_s determines the nature of interaction: when $a_s < 0$ the interaction potential is attractive which may cause a collapse of BEC, and otherwise, it is repulsive and desirable.

The question arises: How does an experimenter measure a_s, or σ which describe the scattering of two atoms inside a tiny condensate? Well, in particle physics one finds a_s as a fitting parameter to cover experimentally measured cross-section. In many-body physics an experimenter uses another strategy, which includes measurements based on bound state molecular and photo association spectroscopy [7].For the most common atomic species, ^{87}Rb and ^{23}Na, $a_s \approx 5.8nm$ and $a_s \approx 2.8nm$, respectively, provided that the gas parameter $\gamma \equiv \rho a_s^3 \ll 1$. But, strictly speaking, the value of scattering length can range from very small values, as happens in spin-polarized hydrogen, where a_s is of the order of the Bohr radius, to relatively large values, say, in ^{87}Rb. In some cases, it is negative, as happens in the case of ^7Li or

---

[1] Actually this "macroscopic" system has a typical size $200\mu m * 300\mu m$ and can be seen only by a microscope.

^{85}Rb. Moreover, the interaction strength g between the same atoms, and hence a_s, can be tuned by the experimenter! How do they do that?

From particle physics, we used to think that the interaction potential between two particles is fixed by nature. So, nobody can change either nucleon-nucleon or electron-electron interactions! However, the difference between an elementary particle (say, proton) and an atom is that a particle is a structureless object (at least at very low energies), while atom is not. The interaction parameter, particularly a_s depends on the condition, in which internal state is it in. The latter can be changed by using a special technique referred to as a method of Feshbach resonances [8]. The Feshbach resonance occurs when the energy of a bound state of an interatomic potential is equal to the kinetic energy of a colliding pair of atoms. In ultracold atomic experiments, the resonance is controlled via the magnetic field B and the kinetic energy may be assumed approximately zero. Since the channels differ in internal degrees of freedom such as spin and angular momentum, their difference in energy is dependent on B by the Zeeman effect. This effect modifies the scattering length as

$$a_s = a_0 \left(1 + \frac{\Delta}{B - B_0} \right), \tag{4.3}$$

where a_0 is the background scattering length, B_0 is the magnetic field strength where resonance occurs, and Δ is the resonance width. This allows for measurement and manipulation of the scattering length to zero or arbitrarily high values by tuning the external magnetic field B. Therefore, concluding the present section we state that the interaction of two ultracold atoms may be modelled by the contact potential (4.1) whose unique parameter g may be varied by hand. The next question is "How does this interaction modify the energy dispersion and thermodynamics of a real gas in comparison with the ideal gas?". In the next section, we shall seek for the answer in the framework of the Hamiltonian formalism. But, before doing this we have to refresh in our mind a method of second quantization.

4.2   Second Quantization

It is interesting to note that, since the pioneering work by Bogolyubov [9] theorists in low-temperature statistical physics exploit successfully methods of field theories [10, 11], originally developed for high energy particle physics. In this field, where the number of particles is not fixed, they introduce creation (annihilation) operators to describe asymptotic $|in\rangle$ and $|out\rangle$ states, presenting the interaction Hamiltonian also in terms of such operators. The principle of such identity is given via relations for commutators as $[a(\mathbf{k}), a^\dagger(\mathbf{k})]_- = \delta(\mathbf{k} - \mathbf{k}')$ and $[c_s(\mathbf{k}), c_{s'}^\dagger(\mathbf{k})]_+ = \delta_{s,s'}\delta(\mathbf{k} - \mathbf{k}')$ for bosons (with operators $a(a^\dagger)$) and fermions (with operators $c(c^\dagger)$), respectively. The ground state corresponds to the vacuum $|0\rangle$, so that, by definition, $a_t(\mathbf{k})|0\rangle = c_s(\mathbf{k})|0\rangle = 0$. For example for the elastic pion-nucleon scattering

$$\pi(\mathbf{k}) + N(\mathbf{p}) \rightarrow \pi(\mathbf{k}') + N(\mathbf{p}'), \tag{4.4}$$

the initial state is $|in\rangle = a_t^\dagger(\mathbf{k})c_s^\dagger(\mathbf{p})|0\rangle$ and the final state is $|out\rangle = a_{t'}^\dagger(\mathbf{k}')c_{s'}^\dagger(\mathbf{p}')|0\rangle$, where $a_t^\dagger(\mathbf{k})|0\rangle$ corresponds to one pion state with momentum $\mathbf{k}$ and isospin projection t, while $c_s^\dagger(\mathbf{p})|0\rangle$ describes one nucleon state with momentum $\mathbf{p}$ and the spin projection s. Now one may rewrite the pion-nucleon interaction Hamiltonian in terms of creation-annihilation operators and evaluate the amplitude of this scattering applying relevant Feynman rules [12]. Here, a reader may ask "And what does statistical physics have to do with it?".

Clearly, there is no scattering problem in equilibrium statistical physics. As to in medium two-body atomic scattering, it can be described by the usual scattering theory in the framework of first quantization. Thus, people are mostly interested in thermodynamic quantities such as total and free energies of the whole system, the entropy, the energy spectrum of excitations etc. In general, these quantities can be found from the grand thermodynamic potential, Ω given by

$$\Omega = -T \ln Z, \qquad Z = Tre^{-\beta H}, \tag{4.5}$$

where Z is referred to as the partition function, and the trace should be taken within all over phase space of the system, which is a hard task. On the other hand, it is well known that a trace of any operator could be easily evaluated, when the operator has a diagonal form. It turns out that, it is more convenient to diagonalize a physical operator when it is presented in the formalism of second quantization. Moreover, diagonalization of the Hamiltonian of a system leads directly to its total energy. In fact, by common rules of operator algebra, the eigenvalues of any operator coincide with its diagonal elements after a certain transformation of its base vectors.[2]

4.2.1 Second Quantization in Many Particle Physics

The second quantization in statistical physics is similar to the one in the field theory of high energy physics [13]. Below we outline this procedure for bosons,[3] which makes it possible to exploit field theoretical methods in statistical physics [14].

Let us consider a system of N non-interacting bosons. Each of them may occupy a state $|p_i\rangle$ with a wave function $\psi_{p_i}(\mathbf{r})$. Let in a state ψ_{p_i} there are N_i pieces of particles. Clearly for a bosonic system N_i may be arbitrary, providing that $\sum_i N_i = N$. The wave function of the whole system, with say, N_1 particles are in the first state, N_2 particles are in the second state etc., is given by

$$\psi_{N_1,N_2\ldots} \equiv |N_1, N_2 \ldots\rangle = (N_1!, N_2! \ldots N!)^{1/2} \sum \psi_{p_1}(\mathbf{r}_1)\psi_{p_2}(\mathbf{r}_2)\ldots\psi_{p_N}(\mathbf{r}_N), \tag{4.6}$$

where the sum is taken over all permutations of these suffixes $p_1, p_2 \ldots p_N$. The total number of terms in the sum equals to $(N_1!, N_2! \ldots /N!)^{-1}$. Now we introduce

---

[2] Actually the Schrodinger equation $H\psi = E\psi$ is just a problem for finding eigenvalues of the operator H.

[3] For fermions it has its own complicated specifications.

an annihilation operator $\hat{a}_i$, such that when acting on the whole function $\psi_{N_1,N_2...}$ it decreases the suffix N_i by unity and at the same time multiplies by $\sqrt{N_i}$:

$$\hat{a}_i|N_1, N_2 \ldots, N_i \ldots\rangle = \sqrt{N_i}|N_1, N_2 \ldots, N_i - 1 \ldots\rangle. \tag{4.7}$$

For example, $\hat{a}_3$ diminishes by one the number of particles in the third state, creating a new whole state without one particle. Mathematically this is presented as:

$$\langle N_1, N_2 \ldots, N_i - 1 \ldots|\hat{a}_i|N_1, N_2 \ldots, N_i \ldots\rangle \equiv \langle N_i - 1|\hat{a}_i|N_i\rangle = \sqrt{N_i}. \tag{4.8}$$

By definition, all other matrix elements of $\hat{a}_i$ are zero, since for example $\hat{a}_4|N_3\rangle = 0$. Now taking hermitian conjugate of (4.8) we may introduce a creation operator as

$$\langle N_i|\hat{a}_i^\dagger|N_i - 1\rangle = \sqrt{N_i}, \tag{4.9}$$

which means that when acting on the whole function it adds one particle in i-th state:

$$\hat{a}_i^\dagger|N_1, N_2 \ldots, N_i \ldots\rangle = \sqrt{N_i + 1}|N_1, N_2 \ldots, N_i + 1 \ldots\rangle. \tag{4.10}$$

These are illustrated in Fig. 4.1 where e.g. in the first box on the left $N_1 = 3$, $N_2 = 2$, $N_3 = 3$ etc.

It is understood that the product of operators $\hat{a}_i^\dagger \hat{a}_i$ ($i = 3$ in Fig. 4.1) acting on the whole, state leaves unchanged all filling numbers $N_1, N_2 \ldots$ and hence

$$\hat{a}_i^\dagger \hat{a}_i = N_i, \quad \hat{a}_i \hat{a}_i^\dagger = N_i + 1. \tag{4.11}$$

From this equation, we immediately find following commutation rule

$$\hat{a}_i \hat{a}_i^\dagger - \hat{a}_i^\dagger \hat{a}_i = 1, \tag{4.12}$$

which is expected for bosonic operators. Moreover, since any $\hat{a}_i$ or $\hat{a}_i^\dagger$ acts only on the i-th state, operators with different suffixes commute

$$\hat{a}_i \hat{a}_k - \hat{a}_k \hat{a}_i = 0, \quad \hat{a}_i \hat{a}_k^\dagger - \hat{a}_k^\dagger \hat{a}_i = 0, \quad (i \neq k). \tag{4.13}$$

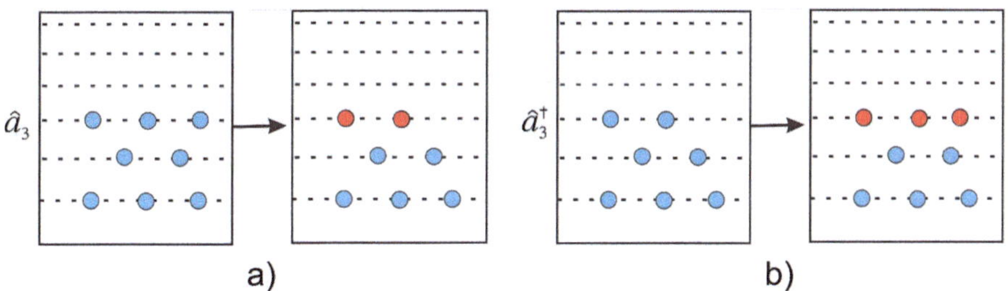

Fig. 4.1 Illustration of action of $\hat{a}_i$ ($i = 3$) and $\hat{a}_i^\dagger$ ($i = 3$) on the many body states. The annihilation operator $\hat{a}_3$ just removes one particle from the third state, while the creation operator $\hat{a}_3^\dagger$ adds one more particle into the third state. All other states remain unchanged

Now let $\hat{f}_\alpha^{(1)}$ is a one body operator, pertaining to the α-th particle acting only on the one body function $\psi(\mathbf{r}_\alpha)$:

$$\hat{F}^{(1)} = \sum_\alpha \hat{f}_\alpha^{(1)} = \sum_{i,k} \langle k|\hat{f}|i\rangle \hat{a}_i^\dagger \hat{a}_k$$

$$\langle k|\hat{f}|i\rangle = \int d\mathbf{r}\psi_k^*(\mathbf{r})\hat{f}\psi_i(\mathbf{r}). \tag{4.14}$$

Similarly, for a two-body operator $f^{(2)}$ acting on variables r_α, r_β, we introduce following representation in terms of creation/annihilation operators:

$$\hat{F}^{(2)} = \frac{1}{2}\sum_{i,j,k,m} \langle km|\hat{f}^{(2)}|ij\rangle \hat{a}_k^\dagger \hat{a}_m^\dagger \hat{a}_i \hat{a}_j$$

$$\langle km|\hat{f}^{(2)}|ij\rangle = \int d\mathbf{r}_1 d\mathbf{r}_2 \psi_k^*(\mathbf{r}_1)\psi_m^*(\mathbf{r}_2)\hat{f}^{(2)}\psi_i(\mathbf{r}_1)\psi_j(\mathbf{r}_2). \tag{4.15}$$

This formalism, which seems rather complicated, can be put in a somewhat more compact form by introducing filed operators-ψ operators

$$\hat{\psi}(\mathbf{r}) = \sum_i \psi_i(\mathbf{r})\hat{a}_i, \qquad \hat{\psi}^\dagger(\mathbf{r}) = \sum_i \psi_i^*(\mathbf{r})\hat{a}_i^\dagger, \tag{4.16}$$

where one may use also plane waves as $\psi_k(\mathbf{r}) \sim \exp(i\mathbf{kr})$.

Task 4.1: Using commutation rules (4.12) and (4.13) derive following commutation rules for the fields $\hat{\psi}$:

$$\hat{\psi}(\mathbf{r})\hat{\psi}(\mathbf{r}') - \hat{\psi}(\mathbf{r}')\hat{\psi}(\mathbf{r}) = 0,$$

$$[\hat{\psi}(\mathbf{r}), \hat{\psi}^\dagger(\mathbf{r}')]_- = \hat{\psi}(\mathbf{r})\hat{\psi}^\dagger(\mathbf{r}') - \hat{\psi}^\dagger(\mathbf{r}')\hat{\psi}(\mathbf{r}) = \delta(\mathbf{r} - \mathbf{r}'). \tag{4.17}$$

In this way, any physical operator can be represented in terms of $\hat{\psi}$. Particularly,

- Hamiltonian of bosons in the external field $U_{ext}(\mathbf{r})$ with the mutual two body interaction $U(\mathbf{r}, \mathbf{r}')$ is

$$\hat{H} = \int d\mathbf{r}\left[\frac{\nabla\hat{\psi}^\dagger(\mathbf{r})\nabla\hat{\psi}(\mathbf{r})}{2m} + U_{ext}(\mathbf{r})\hat{\psi}^\dagger(\mathbf{r})\hat{\psi}(\mathbf{r})\right]$$

$$+ \frac{1}{2}\int\int d\mathbf{r}d\mathbf{r}'\hat{\psi}^\dagger(\mathbf{r})\hat{\psi}^\dagger(\mathbf{r}')U(\mathbf{r}, \mathbf{r}')\hat{\psi}(\mathbf{r})\hat{\psi}(\mathbf{r}'), \tag{4.18}$$

where we used the identity $\int d\mathbf{r}\hat{\psi}^\dagger(\mathbf{r})\nabla^2\hat{\psi}(\mathbf{r}) = -\int d\mathbf{r}\nabla\hat{\psi}^\dagger(\mathbf{r})\nabla\hat{\psi}(\mathbf{r})$. In momentum space, the first term in (4.18) corresponds to the energy dispersion $\varepsilon_k = \mathbf{k}^2/2m$.

- Momentum :

$$\hat{\mathbf{P}} = -i \int d\mathbf{r} \hat{\psi}^{\dagger}(\mathbf{r}) \nabla \hat{\psi}(\mathbf{r}). \tag{4.19}$$

- Angular momentum (α-th component)

$$\hat{L}_{\alpha} = -i\epsilon_{\alpha\beta\gamma} \int d\mathbf{r} r_{\beta} \hat{\psi}^{\dagger}(\mathbf{r}) \frac{\partial \hat{\psi}(\mathbf{r})}{\partial r_{\gamma}}, \tag{4.20}$$

where $\epsilon_{\alpha\beta\gamma}$ is the antisymmetric tensor.

These operators are used to find the average value of a physical quantity, e.g. $E_{tot} = \langle \hat{H} \rangle$

Task 4.2 : Prove that in the laboratory system, where the system of particles is in the rest, the total momentum is zero: $P_{tot} = \langle \hat{P} \rangle = 0$

And at last, we should bring the normalization condition for the field operators:

$$N = \int d\mathbf{r} \langle \hat{\psi}^{\dagger}(\mathbf{r}) \hat{\psi}(\mathbf{r}) \rangle, \tag{4.21}$$

which can be proven by using Eqs. (4.11) and (4.16).

4.2.2 Ideal Gas of Bosonic Quasiparticles

It is well known from textbooks that excitations and thermodynamic properties of crystals are described in terms of unreal particles-phonons more conveniently. Similar quasiparticles-magnons or triplons serve as a handy tool in quantum magnetism. As to our sheep-Bose-Einstein condensation of atomic gases, their collective excitations in the BEC phase can be described with their quasiparticles, bogolons, named after Nikolay Bogolyubov [9], who introduced them in 1947 to establish a connection between superfluidity and BEC. In the first approximation, which is good for weakly interacting dilute gases, one may neglect their mutual interaction, considering the system as an ideal gas. Consequently, in momentum space, the Hamiltonian of an ideal gas of bosonic quasiparticles has the following form in the second quantization formalism:

$$\hat{H} = \sum_{k} E_k \hat{b}_{\mathbf{k}}^{\dagger} \hat{b}_{\mathbf{k}}, \tag{4.22}$$

where $\hat{b}_{\mathbf{k}}^{\dagger}$ and $\hat{b}_{\mathbf{k}}$ are creation and annihilation operators of the quasiparticle with momentum $\mathbf{k}$, respectively with following properties

$$[\hat{b}_{\mathbf{k}}, \hat{b}_{\mathbf{k}}^{\dagger}] = \delta(\mathbf{k} - \mathbf{k}'), \qquad \langle \hat{b}_{\mathbf{k}} \hat{b}_{-\mathbf{k}} \rangle = \langle \hat{b}_{\mathbf{k}}^{\dagger} \hat{b}_{-\mathbf{k}}^{\dagger} \rangle = 0,$$

$$\langle \hat{b}_{\mathbf{k}}^{\dagger} \hat{b}_{\mathbf{k}} \rangle \equiv f_B(E_k) = \frac{1}{e^{\beta E_k} - 1}. \tag{4.23}$$

Here, one should be attentive to the summation over momentum $\sum_k$. In practical calculations in $D = 3$ one may use the formula

$$\sum_k F(\mathbf{k}^2) = \frac{4\pi V}{(2\pi)^3} \int_0^\infty k^2\, F(\mathbf{k}^2) dk, \tag{4.24}$$

for isotropic systems of atomic gases. However for a crystal with the lattice parameters a_x, a_y, a_z, the integration is usually taken in the finite volume of the first Brillouin zone [15]

$$\sum_k F(k_x, k_y, k_z) = \frac{V}{(2\pi)^3} \int_{-\pi/a_x}^{\pi/a_x} dk_x \int_{-\pi/a_y}^{\pi/a_y} dk_y \int_{-\pi/a_z}^{\pi/a_z} dk_z\, F(k_x, k_y, k_z). \tag{4.25}$$

An effective mass of a quasiparticle can be defined as $m = [\partial^2 E_k/\partial k^2]^{-1}$ where, in general, the function E_k has rather complicated form than the simple $\mathbf{k}^2/2m$ one.

4.3 Spontaneous Symmetry Breaking and Goldstone Modes

Symmetries play an important role in nature. The presence of symmetry in a given system makes it easier construction of the proper theoretical model. Sometimes a glance at the form of the symmetry reveals the possible behavior of the particle (or the whole system), at least qualitatively.

Let's consider a particle in $1D$ effective potential, illustrated in Fig. 4.2, from the point of view of the symmetry with respect to $x \to -x$ (parity symmetry). The effective potentials $U_2(x) = x^2 + x^4$ (Fig. 4.2b) and $U_3(x) = -x^2 + x^4$ (Fig. 4.2c) are symmetric, while $U_1(x) = \exp(-x^2)(x^2 - x^3)$ (Fig. 4.2a) is not under this transformation. Where the particle can be found most probably? Clearly, according to general rules of physics, the particle prefers to stay in its ground state, which corresponds to the minimum of the effective potential. Thus the particle will be localized in the well with the center $x \sim 1.55$ in Fig. 4.2a, or near $x \sim 0$ in Fig. 4.2b. But what about $U_3(x)$ in Fig. 4.2c which has two minima? Actually, it may occupy the left well or the right one with equal probability. If by some chance it ended up, say, in the left well, then it cannot move to the right well because of the potential barrier near $x = 0$.[4]

In this situation we have got dilemma: On the one hand, the system in Fig. 4.2c is symmetric under the transformation $x \to -x$, but on the other hand, there is no symmetry: The left well is occupied by the particle, but the right one is empty. So, how do we qualify this strange symmetry? In 1960 Yoichiro Nambu introduced a special term-"Spontaneous symmetry breaking (SSB)" and showed that such phenomenon can be found in all fields of physics: particle physics, high energy, ferromagnetism, superconductivity, BEC etc. Therefore, we have seen that a symmetry can be broken explicitly as in Fig. 4.2a or spontaneously (Fig. 4.2c). Moreover, an interesting point is that one can cause SSB by hand, by changing the physical parameters of the system

---

[4] Strictly speaking it can move to another well only by tunneling.

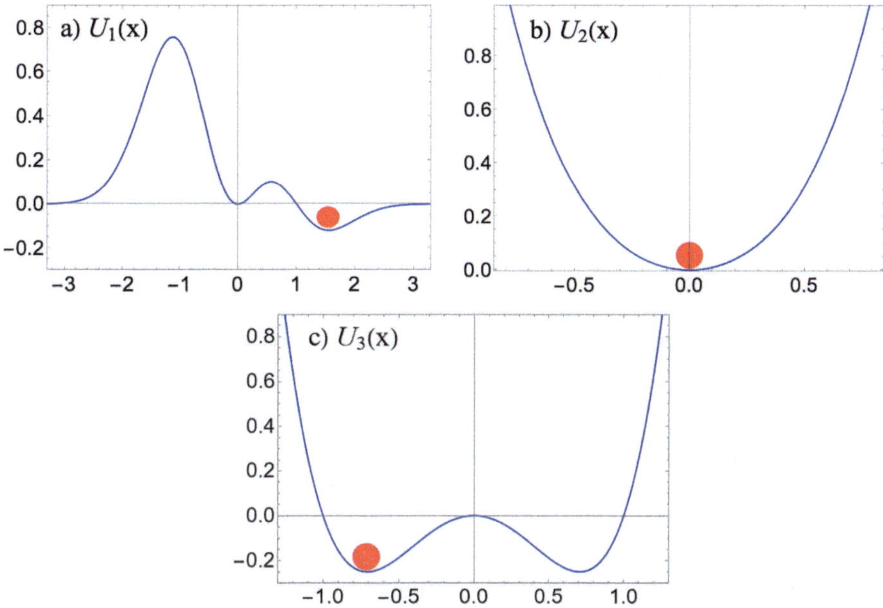

Fig. 4.2   1D effective potentials. **a** $U_1(x) = \exp(-x^2)(x^2 - x^3)$, **b** $U_2(x) = x^2 + x^4$ and **c** $U_3(x) = -x^2 + x^4$

e.g. temperature, pressure etc. In our toy model, this can be done by changing the parameter α of the effective potential $U(x) = \alpha x^2 + x^4$. If somehow we change the sign of α then we say that we have broken $x \to -x$ symmetry spontaneously. Now we refresh in our mind SSB in particle physics, where it has fundamental consequences.

4.3.1   SSB in Particle Physics

We know that the field of free scalar particles is described by the following Lagrangian:

$$\mathcal{L}_0 = \frac{1}{2}(\partial_\mu \phi)(\partial^\mu \phi) - \frac{1}{2}m^2\phi^2. \tag{4.26}$$

where $(\partial_\mu \phi)(\partial^\mu \phi) = (\partial\phi/\partial t)^2 - (\partial\phi/\partial \mathbf{r})^2$ $(\mu = 0, 1, 2, 3)$ and m is the particle mass. The Lagrangian leads to the Klein-Gordon equation $\partial_\mu \partial^\mu \phi + m^2 \phi = 0$. Now let's switch on the self-interaction between particles considering $\tilde{m}$ as a free parameter. Then the Lagrangian takes the form of standard $\lambda\phi^4$ theory:

$$\mathcal{L} = \frac{1}{2}(\partial_\mu \phi)(\partial^\mu \phi) - \frac{1}{2}\tilde{m}^2\phi^2 - \frac{\lambda\phi^4}{4} \equiv \frac{1}{2}(\partial_\mu \phi)(\partial^\mu \phi) - V(\phi), \tag{4.27}$$

with $\lambda > 0$ and $V(\phi) = \tilde{m}^2\phi^2/2 + \lambda\phi^4/4$. Clearly, this Lagrangian is symmetric (invariant) under the transformation $\phi \to -\phi$. However, this symmetry can be vio-

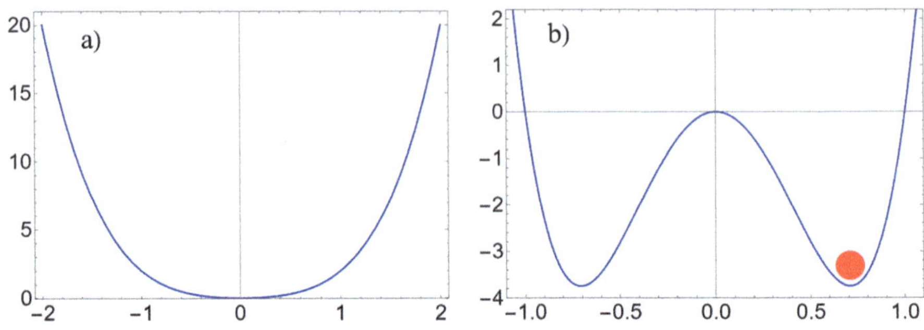

Fig. 4.3 Potentials in particle physics for **a** $\tilde{m}^2 > 0$ and **b** $\tilde{m}^2 < 0$ cases

lated depending on the sign of $\tilde{m}^2$. To make this statement clearly we discuss the two cases separately:

(1) Let $\tilde{m}^2 > 0$. Then $m = \tilde{m}$, so this case corresponds to the particle with the real mass m. Further, plotting $V(\phi)$ vs ϕ one can immediately see that the ground state corresponds to $\phi = 0$ (see Fig. 4.3a).

(2) Let $\tilde{m}^2 < 0$. Since now we have the wrong sign in the mass term we cannot say anything definite about the mass of such particles for a while. Moreover, the potential $V(\phi)$ has two minima, as illustrated on Fig. 4.3b. To localize them we differentiate $V(\phi) = -|\tilde{m}^2|\phi^2/2 + \lambda\phi^4/4$ with respect to ϕ :

$$\frac{\partial V}{\partial \phi} = -|\tilde{m}^2|\phi + \lambda\phi^3 = 0, \qquad \phi_{min} = \pm\frac{|\tilde{m}|}{\sqrt{\lambda}} \equiv \pm v. \qquad (4.28)$$

On the one hand, this is natural for a mathematician, but, on the other hand, there is something unexpected for a physician: The ground state is not unique-there are two absolutely equivalent ground states in such a system.

Why are we so interested in a ground state? Actually, in accordance with the fundamental principles of nature, it is the ground state, where a physical system prefers to "live forever", so that any further excitation or perturbation takes place near it. Now let's assume that by some reason the particle has occupied say, the right minimum, which corresponds to $\phi = +v = +|\tilde{m}|/\sqrt{\lambda}$. In other words, the particle does not exist on the left well, but it does on the right one. This clearly means that our original symmetry with respect to $\phi \to -\phi$ has been broken. Such a situation has been referred to as spontaneous symmetry breaking since its discovery in 1961 by Nambu and Jona-Lasinio.[5]

In accordance with the uncertainty principle any physical quantity always deviates near its mean value even when the system is in its ground state. These deviations are referred to as quantum or temperature fluctuations at zero $T = 0$ or finite temperature $T \neq 0$, respectively. Mathematically this can be presented as[6]

---

[5] The Nobel Prize in Physics 2008 was awarded to Yoichiro Nambu "for the discovery of the mechanism of spontaneously broken symmetry in subatomic physics".

[6] Assume that the particle is on the right well.

$$\phi(x) = v + \eta(x) \tag{4.29}$$

where $v = |\tilde{m}|/\sqrt{\lambda}$ corresponding to the mean value $\phi(x) = v = const$ and $\eta(x)$ describes fluctuations. In practical calculations one may use a perturbation scheme by using (4.29) so that the fluctuations will be integrated out. For this purpose one preliminary inserts (4.29) into the Lagrangian (4.27) to rewrite it as

$$\mathcal{L}' = \frac{1}{2}(\partial_\mu \eta)(\partial^\mu \eta) - \lambda v^2 \eta^2 - \lambda v \eta^3 - \frac{\lambda \eta^4}{4} + \dots . \tag{4.30}$$

Now comparing this with (4.26) we see that there comes a mass term $\sim v^2 \lambda \eta^2$ with a right sign, displaying the real mass $m_\eta^2 = 2v^2 \lambda$, i.e. $m = v\sqrt{2\lambda}$. Therefore, we come to the following conclusion: When the original parameter $\tilde{m}^2 > 0$, the particles mass equals to $m = \tilde{m}$, otherwise $(\tilde{m}^2 < 0)$ $m = v\sqrt{2\lambda} = \sqrt{2}|\tilde{m}|$.

4.3.2 Continuous Symmetry and Goldstone Theorem

There are various kinds of symmetries, the spontaneous breaking of which leads to different phenomena. In the previous subsection, we discussed the Lagrangian with the discrete symmetry: $\phi \to -\phi$. Below we consider a self-interacting complex scalar field with the Lagrangian

$$\mathcal{L} = (\partial_\mu \phi)^*(\partial^\mu \phi) - \tilde{m}^2 \phi^* \phi - \lambda(\phi^* \phi)^2, \tag{4.31}$$

which is invariant under the transformation $\phi \to \exp(i\alpha)\phi$, with $\alpha = constant$. In the literature, this type of symmetry is classified as $U(1)$ gauge symmetry. To study the ground state of such a system we introduce two real fields via

$$\phi = \frac{1}{\sqrt{2}}(\phi_1 + i\phi_2) \qquad \phi^* = \frac{1}{\sqrt{2}}(\phi_1 - i\phi_2). \tag{4.32}$$

Inserting (4.32) into (4.31) one can see that this time there is a circle of minima of the potential in the (ϕ_1, ϕ_2) plane of radius v such that

$$\phi_1^2 + \phi_2^2 = v^2, \qquad v^2 = -\frac{\tilde{m}^2}{\lambda}. \tag{4.33}$$

Again, we translate the field ϕ to minimum energy position which, without loss of generality, we may take as the point $\phi_1 = v$ and $\phi_2 = 0$. Now introducing fluctuating fields η and ξ by

$$\phi = \frac{1}{\sqrt{2}}(v + \eta(x) + i\xi(x)), \tag{4.34}$$

we rewrite the Lagrangian as follows

$$\mathcal{L}' = \frac{1}{2}(\partial_\mu \xi)^2 + \frac{1}{2}(\partial_\mu \eta)^2 + \tilde{m}^2 \eta^2 + \text{const.} + \text{cubic and quartic terms in } \eta, \xi. \tag{4.35}$$

Now, the third term has the form of mass term $-m_\eta^2 \eta^2$ for the η-field. Thus the $\eta-$ mass is $m_\eta = \sqrt{-2\tilde{m}^2}$. The first term in (4.35) represents the kinetic energy of the ξ-field, but there is no corresponding mass term for ξ. That is the theory also contains a massless scalar, which is known as a Goldstone boson. Our Lagrangian serves as a simple example for the Goldstone theorem, which states that massless scalars occur whenever a continuous symmetry of a physical system is spontaneously broken. The branch of the energy spectrum which corresponds to the massless Boson is referred as a Goldstone mode. The question arises: "Why there was no massless particle in our previous example with the one-component scalar field?". The point is that the theorem works for a continuous symmetry, not for a discrete one.

SSB is so general phenomenon in the nature and physical theories, that one can find it almost everywhere: Condensed matter (BEC, superconductivity, ferromagnetism), high energy particle physics (Quantum Chromodynamics, Standard model of electroweak interactions etc.) [16]. Note that the Large Hadron Collider in CERN was built for several billions of euros mainly for hunting a Higgs boson, which is directly related to SSB. And at last, we must underline that, spontaneous symmetry breaking is the necessary and sufficient condition for Bose-Einstein condensation [17]. In this case one deals with the breaking of global gauge symmetry $U(1)$, and the transformation similar to (4.29) is referred in the literature as a Bogolyubov shift, where the mean value of the field v can be interpreted as an order parameter, commonly introduced in Ginzburg-Landau theory [18].

4.4   Green Functions

Green functions play an utmost important role in theoretical physics. Method of Green functions has been successfully used to study scattering processes and boundary states [10, 19]. In perturbation theory, they are the "core" of Feynman diagrams referred to as propagators. (see the book by Fetter and Walechka [20] for the details).

In quantum mechanics, one may define one propagator in the following simple way. Let the Hamiltonian, schematically, consists of free H_0 and interaction potential V:

$$H = H_0 + V \tag{4.36}$$

In general H_0, being defined as an unperturbated Hamiltonian, may include some interactions also, e.g. a trapping external potential. Then the operators of a free $\widehat{G}_0$ and an exact $\widehat{G}$ propagators are defined as:

$$\widehat{G}_0 = \frac{1}{E - H_0 + i\epsilon}, \qquad \widehat{G} = \frac{1}{E - H + i\epsilon}, \tag{4.37}$$

where $\epsilon \to 0$ and E is the total energy of the system being an eigenvalue of H.[7]

From Eqs. (4.36)–(4.37) it is easy to derive the Lippmann-Schwinger equation for the Green functions [19]:

$$\widehat{G} = \widehat{G}_0 + \widehat{G}_0 V \widehat{G}. \tag{4.38}$$

So, when the exact Green function is known, the spectrum of the system, i.e. the dispersion can be found from its poles, or more exactly, as the solutions of equation $Det(\widehat{G}^{-1}) = 0$, as it is directly seen from Eq. (4.37). In scattering theory a similar Lippmann-Schwinger integral equation for the T-matrix holds:

$$T = V + V\widehat{G}_0 T, \tag{4.39}$$

which is very convenient to evaluate e.g. amplitude of the elastic scattering $f_E(\mathbf{k}', \mathbf{k}) = -2m\pi^2 \langle \mathbf{k}'|T(E)|\mathbf{k}\rangle$ and hence the S wave scattering length as $a_s = -f_E(\mathbf{k}, \mathbf{k})|_{E \to 0}$.

Now it is time to pass to the field theory. In quantum mechanics $\widehat{G}$ and $\widehat{G}_0$ "work" on the basis of vectors of wave functions ψ. Here, as it outlined in previous sections one should use second quantization, replacing the function ψ by the $\hat{\psi}$ operator. So, for the real scalar field the Green's function is defined as $G(x, x') =< \mathcal{T}\psi(x)\psi(x') >$, where $x \equiv (t, \mathbf{r})$, and $\mathcal{T}$ means time ordering [12]. For the complex fields, discussed in Sect. 4.4, we can introduce a vector $(\psi, \psi^\dagger)$ such that $G = i\mathcal{T} < (\psi, \psi^\dagger) \cdot (\psi, \psi^\dagger)^\dagger >$, that is [20]

$$\begin{aligned}
G_{11}(x, x') &= -i < \mathcal{T}\psi(x)\psi^\dagger(x') >, \\
G_{12}(x, x') &= -i < \mathcal{T}\psi(x)\psi(x') >, \\
G_{21}(x, x') &= -i < \mathcal{T}\psi^\dagger(x)\psi^\dagger(x') >, \\
G_{22}(x, x') &= -i < \mathcal{T}\psi^\dagger(x)\psi(x') > .
\end{aligned} \tag{4.40}$$

Therefore, in general, the operator $\widehat{G}$ has the form of a 2×2 matrix:

$$\widehat{G} = \begin{pmatrix} \widehat{G}_{11} & \widehat{G}_{12} \\ \widehat{G}_{21} & \widehat{G}_{22} \end{pmatrix}. \tag{4.41}$$

Transferring the Eqs. (4.40) directly to the condensed matter physics raises the question "How do we take into account the temperature?". The answer is not trivial. Actually, in the literature, there are mainly two ways: Thermo Field Dynamics (TFD) and Matsubara formalism. The former is a real-time formalism in Minkowski space [21,22], while the latter is an imaginary time formalism in Euclidean space produced by the Wick rotation $t \to -i\tau$. The main difference between them is the

---

[7] Actually, the small ϵ will be needed in order to use in the integrals via a well-known formula $1/(x - x_0 \pm i\epsilon) \to P(1/(x - x_0)) \mp i\pi\delta(x - x_0)$.

following: In TFD the temperature is "inserted" into the wave functions (temperature-dependent vacuum) while the operators are temperature-independent. In Matsubara formalism we have the opposite picture: the temperature dependence comes through temperature-dependent operators, while the ψ operators do not depend on temperature. Below we outline Matsubara formalism since it is more commonly used. Note that, the inclusion of temperature is important not only in condensed matter physics but also in very high energy physics [23], where hadron properties may be modified due to medium effects [24].

4.4.1   Matsubara Formalism

Referring the reader to the Refs. [23,25] for details, below we give the main points of the Matsubara formalism for bosons. So, when we apply field theoretical methods to condensed matter physics or high energy physics we have to deal with loop corrections, which include space or momentum integrals. To exploit the Matsubara method one uses following transformations:

- For $4D$ space integrals

$$t \rightarrow -i\tau \qquad \int dt d^3 r \rightarrow (-i) \int_0^\beta d\tau d^3 r. \qquad (4.42)$$

- For the momentum space integrals:

$$k_0 \rightarrow i\omega_n, \qquad \int \frac{d^4 k}{(2\pi)^4} \rightarrow \frac{1}{\beta} \sum_{\omega_n} \int \frac{d^3 k}{(2\pi)^3}. \qquad (4.43)$$

- Keep in mind that for bosons the field operator has periodicity: $\phi(\beta, \mathbf{r}) = \phi(0, \mathbf{r})$, and $\omega_n = 2\pi T n$ with n-integer, $n = [-\infty \ldots \infty]$.

In particular, Fourier transformations have the form

$$\phi(\tau, \mathbf{r}) = \frac{1}{\sqrt{V\beta}} \sum_k \sum_n \phi(\omega_n, \mathbf{k}) \exp(i\omega_n \tau + i\mathbf{k}\mathbf{r}),$$

$$G_{ij}(\tau, \mathbf{r}; \tau', \mathbf{r}') = \frac{1}{V\beta} \sum_{k,k'} \sum_{n,m} \exp(i\omega_n \tau) \exp(-i\omega_m \tau') \qquad (4.44)$$

$$\times \exp(i\mathbf{k}\mathbf{r}) \exp(-i\mathbf{k}'\mathbf{r}') G_{ij}(\omega_n, \omega_m, \mathbf{k}, \mathbf{k}').$$

It should be underlined that, in contrast to TFD, Matsubara formalism is good only for equilibrium systems, since due the reduction of the time variable ($t \rightarrow -i\tau$) one has to loose the control over the time evolution. TFD is valid both for stationary and nonstationary processes, however, its practical usage is rather complicated because of doubling Hilbert space by introducing fictitious particles [26].

4.5 **Hugenholtz-Pines Theorem**

Now coming back to our task on BEC one may think that it would be enough just to rewrite the Eq. (4.40) by setting $t \to -i\tau$. However, this is not enough and not a trivial business. In fact here one has to keep in mind that, BEC is based on the SSB mechanism.

So, assume that our Hamiltonian (or Lagrangian) is defined in terms of some ψ field operators as a sum of "free" and "interacting " parts:

$$H \equiv H_0(\psi^\dagger, \psi) + U(\psi^\dagger, \psi). \tag{4.45}$$

Because of SSB, we have to separate fluctuating fields via the Bogolubov shift

$$\psi = v + \tilde{\psi}, \qquad \psi^\dagger = v + \tilde{\psi}^\dagger, \tag{4.46}$$

and define the thermal Green's function as[8]

$$G(x, x') = -\langle \begin{bmatrix} \tilde{\psi}(x) \\ \tilde{\psi}^\dagger(x) \end{bmatrix} \cdot [\tilde{\psi}^\dagger(x'), \tilde{\psi}(x')] \rangle \tag{4.47}$$

or in details

$$\begin{aligned}
G_{11}(x, x') &= -\langle \mathcal{T} \tilde{\psi}(x) \tilde{\psi}^\dagger(x') \rangle, \\
G_{12}(x, x') &= -\langle \mathcal{T} \tilde{\psi}(x) \tilde{\psi}(x') \rangle, \\
G_{21}(x, x') &= -\langle \mathcal{T} \tilde{\psi}^\dagger(x) \tilde{\psi}^\dagger(x') \rangle, \\
G_{22}(x, x') &= -\langle \mathcal{T} \tilde{\psi}^\dagger(x) \tilde{\psi}(x') \rangle,
\end{aligned} \tag{4.48}$$

where $x = (\tau, \mathbf{r})$ and the order parameter v is assumed to be a real constant. By definition, the free Green function $\widehat{G}_0$ corresponds to the case when the interaction potential is neglected, i.e. $\widehat{G}_0 = \widehat{G}_0|_{U=0}$. The analogue of the Lippmann-Schwinger equation in the field theory is the following Beliaev-Dyson [27] equation:

$$\widehat{G} = \widehat{G}_0 + \widehat{G}_0 \widehat{\Sigma} \widehat{G}, \qquad (\widehat{G})_{ij} = (\widehat{G}_0)_{ij} + \sum_{k,n=1,2} (\widehat{G}_0)_{ik}(\widehat{\Sigma})_{kn}(\widehat{G})_{nj}, \tag{4.49}$$

which is illustrated in Fig. 4.4.

The operator $\widehat{\Sigma}$ is referred to as the self-energy operator. Its diagonal and off-diagonal elements are normal $\widehat{\Sigma}_n$ and anomalous $\widehat{\Sigma}_{an}$ self-energies, respectively.

In practice, $\widehat{\Sigma}$ should be guessed or evaluated somehow approximately and then the whole Green function can be found say, iteratively:

$$\widehat{G} = \widehat{G}_0 + \widehat{G}_0 \widehat{\Sigma} \widehat{G}_0 + \widehat{G}_0 \widehat{\Sigma} \widehat{G}_0 \widehat{\Sigma} \widehat{G}_0, \tag{4.50}$$

---

[8] It seems that there is no unique convention on the sign of G in the literature. We adopt definition used in books [6,14].

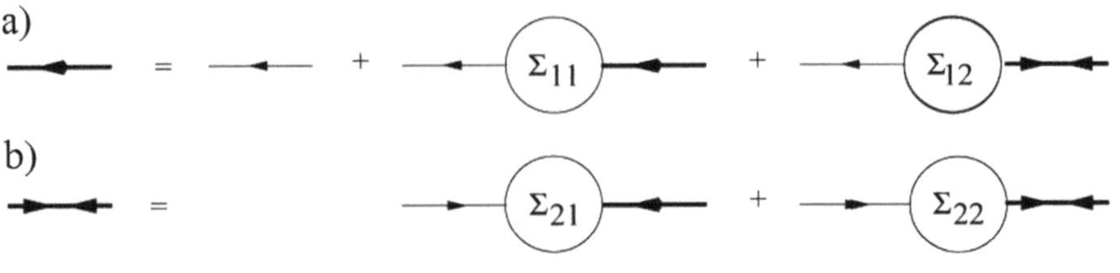

Fig. 4.4 Exact Beliaev-Dyson equations for the interacting **a** normal and **b** anomalous Green's functions. The thin lines represent G_0, whereas the thick lines represent G_{11} (or G_{22}) (one arrowhead) and G_{12} (two arrowheads)

where, certainly, $\widehat{G}_0$ and $\widehat{\Sigma}$ are supposed to be known. This equation may be served as a basis of the perturbation scheme. On the other hand, for describing the spectrum (or effective masses in particle physics) one usually uses its following compact equivalent form:[9]

$$\widehat{G}^{-1} = \widehat{G}_0^{-1} - \widehat{\Sigma}. \tag{4.51}$$

So, for example, if $\widehat{G}_0(z) = 1/(z - H_0 + i\epsilon)$ then $\widehat{G}(z) = 1/(z - H_0 - \widehat{\Sigma} + i\epsilon)$, where all information on the interaction is hidden in the self-energy Σ. The last two equations are good both in relativistic and nonrelativistic physics. Particularly, in relativistic particle physics Σ displays the mass renormalization being diagonal in most cases while in non-relativistic condensed matter physics, in general, it is not diagonal and related to the chemical potential. The last statement is realized by the theorem, proven by Hugenholtz and Pines more than 60 years ago [28], long before the experimental discovery of Bose-Einstein condensation.

To bring the formulation of this theorem we should pass to the momentum space

$$\tilde{\psi}(\tau, \mathbf{r}) = \frac{1}{\sqrt{V\beta}} \sum_{k} \sum_{n=-\infty}^{\infty} \tilde{\psi}(\omega_n, \mathbf{k}) \exp(i\omega_n \tau + i\mathbf{k}\mathbf{r}), \quad \omega_n = 2\pi n T,$$

$$G_{ij}(\tau, \mathbf{r}; \tau', \mathbf{r}') = \frac{1}{V\beta} \sum_{k} \sum_{n=-\infty}^{\infty} \exp(i\omega_n(\tau - \tau')) \exp(i\mathbf{k}(\mathbf{r} - \mathbf{r}'))$$

$$\times G_{ij}(\omega_n, \mathbf{k}), \tag{4.52}$$

where we assume the translation invariance, $G(x, x') = G(x - x')$, which holds for a homogeneous system. In momentum space, the Eq. (4.51) can be presented as

$$\Sigma_n(\omega_n, \mathbf{k}) = (G_0^{-1})_{11}(\omega_n, \mathbf{k}) - (G^{-1})_{11}(\omega_n, \mathbf{k}),$$

$$\Sigma_{an}(\omega_n, \mathbf{k}) = (G_0^{-1})_{12}(\omega_n, \mathbf{k}) - (G^{-1})_{12}(\omega_n, \mathbf{k}), \tag{4.53}$$

---

[9] This can be understood by means of series expansion: $1/(x - a) = 1/x + (1/x)a(1/x) + (1/x)a(1/x)a(1/x) + \cdots$.

Now we are on the state to bring forward the Hugenholtz-Pines (HP) theorem[10]

$$\mu = \Sigma_n(0,0) - \Sigma_{an}(0,0).$$
(4.54)

The theorem just states that, for homogeneous Bose gas the chemical potential is exactly equal to the difference of the diagonal and off-diagonal self-energies evaluated at zero momentum and zero frequency. Due to this theorem, one may observe the Goldstone modes in condensed matter physics also, similarly in particle physics.

4.5.1 Real Field Representation

Along with the basis of complex fields $\tilde{\psi}$, $\tilde{\psi}^\dagger$ the basis of two real fields ψ_1, ψ_2, introduced by

$$\tilde{\psi} = \frac{1}{\sqrt{2}}(\psi_1 + i\psi_2),$$
$$\tilde{\psi}^\dagger = \frac{1}{\sqrt{2}}(\psi_1 - i\psi_2),$$
(4.55)

is also widely used [31]. Let's denote the Green function in this basis as $\hat{D}$

$$\hat{D} = \begin{pmatrix} D_{11} & D_{12} \\ D_{21} & D_{22} \end{pmatrix},$$
(4.56)

where $D_{ij}(x, x') = < \mathcal{T}\psi_i(x)\psi_j(x') >$, $(i, j = 1, 2)$. After some algebraic manipulations one can find following relations between G_{ij}, defined in (4.48) and D_{ij}:

$$\hat{G} = -A\hat{D}A^{-1}, \qquad \hat{G}^{-1} = -A^{-1}\hat{D}^{-1}A,$$
(4.57)

where the transformation matrix A is

$$A = \frac{1}{\sqrt{2}}\begin{pmatrix} 1 & -i \\ 1 & i \end{pmatrix}, \qquad A^{-1} = \frac{1}{\sqrt{2}}\begin{pmatrix} 1 & 1 \\ i & -i \end{pmatrix}.$$
(4.58)

---

[10] A reader can find its proof elsewhere [6, 19, 28–30].

Particularly we have following explicit relations[11] :

$$
\begin{aligned}
- G_{11} &= \frac{1}{2}[D_{11} + D_{22} + i(D_{12} - D_{21}],\\
- G_{12} &= \frac{1}{2}[D_{11} - D_{22} - i(D_{12} + D_{21}],\\
- G_{21} &= \frac{1}{2}[D_{11} - D_{22} + i(D_{12} + D_{21}],\\
- G_{22} &= \frac{1}{2}[D_{11} + D_{22} - i(D_{12} - D_{21}],\\
- D_{11} &= \frac{1}{2}[G_{11} + G_{22} + G_{12} + G_{21}],\\
- D_{12} &= \frac{-i}{2}[G_{11} - G_{22} - G_{12} + G_{21}],\\
- D_{21} &= \frac{i}{2}[G_{11} - G_{22} + G_{12} - G_{21}],\\
- D_{22} &= \frac{1}{2}[G_{11} + G_{22} - G_{12} - G_{21}],
\end{aligned}
\tag{4.59}
$$

and vice versa

$$
\begin{aligned}
- G_{11}^{-1} &= \frac{1}{2}[D_{11}^{-1} + D_{22}^{-1} + i(D_{12}^{-1} - D_{21}^{-1})],\\
- G_{12}^{-1} &= \frac{1}{2}[D_{11}^{-1} - D_{22}^{-1} - i(D_{12}^{-1} + D_{21}^{-1})],\\
- G_{21}^{-1} &= \frac{1}{2}[D_{11}^{-1} - D_{22}^{-1} + i(D_{12}^{-1} + D_{21}^{-1})],\\
- G_{22}^{-1} &= \frac{1}{2}[D_{11}^{-1} + D_{22}^{-1} - i(D_{12}^{-1} - D_{21}^{-1})],
\end{aligned}
\tag{4.60}
$$

where e.g. $(\widehat{G}^{-1})_{11}$ is denoted just by G_{11}^{-1}.

Task 4.3: Derive and check the Eqs. (4.59) **and** (4.60)

Now it is time to see how the Hugenholtz-Pines theorem looks like in real field representation. To do so we naturally introduce the self energies via Green functions $\hat{D}$ as

$$
\widehat{\Pi}_{ij} = (\widehat{D}^{-1})_{ij} - (\widehat{D}_0^{-1})_{ij}
\tag{4.61}
$$

with $\hat{D}_0$ is a "free" Green function, such that Beliaev-Dyson equation has the form

$$
\widehat{D}^{-1} = \widehat{D}_0^{-1} + \widehat{\Pi} .
\tag{4.62}
$$

---

[11] Here we omit (x, x') for simplicity.

Using the Eqs. (4.53), (4.59), (4.60) and (4.61) one obtains the following relations between these two kinds of self energies:

$$
\begin{aligned}
\Sigma_n &= \frac{1}{2}(\Pi_{11} + \Pi_{22} + i(\Pi_{12} - \Pi_{21})), \\
\Sigma_{an} &= \frac{1}{2}(\Pi_{11} - \Pi_{22} - i(\Pi_{12} + \Pi_{21})),
\end{aligned}
\tag{4.63}
$$

and hence, from Eq. (4.54) establish the theorem in terms of $\widehat{\Pi}$ as

$$
\Pi_{22}(\omega_n = 0, \mathbf{p} = 0) + i\,\Pi_{12}(\omega_n = 0, \mathbf{p} = 0) = \mu,
\tag{4.64}
$$

Further, we will show that in most of cases the operator $\widehat{\Pi}$ is diagonal, and so, the Hugenholtz-Pines theorem takes rather a compact form in the real field representation:

$$
\Pi_{22}(0, 0) - \mu = 0,
\tag{4.65}
$$

Summarizing the last sections, we conclude that, SSB leads to Bose-Einstein condensate, whose collective excitations correspond to the poles of the propagator ($\widehat{G}$ or $\widehat{D}$) and the Hugenholtz-Pines theorem imposes certain conditions to small momentum behavior of these excitations. Below we outline an intuitive scheme[12] of extraction of the two-particle Green function in the BEC phase directly from the action $S(\psi, \psi^\dagger)$. We underline that this method is referred in the literature as a bilinear approximation, since in the evaluation of mean values in (4.48) only quadratic terms in fluctuating fields are included. So, suppose that for a homogeneous system we are given an Euclidean Lagrangian density $\mathcal{L}(\psi, \psi^\dagger)$, which defines the action as

$$
S(\psi, \psi^\dagger) = \int_0^\beta d\tau \int d\mathbf{r}\mathcal{L}(\psi(\tau, \mathbf{r}), \psi^\dagger(\tau, \mathbf{r})),
\tag{4.66}
$$

Then the thermal propagator can be determined by following the steps

- Make Bogolubov shift[13]

$$
\psi = \psi_0 + \lambda\tilde{\psi}, \qquad \psi^\dagger = \psi_0 + \lambda\tilde{\psi}^\dagger,
\tag{4.67}
$$

Then the action will be presented in powers of λ as $S = S_0 + \lambda S_1 + \lambda^2 S_2 + \lambda^3 S_3 + \cdots$.

- Pass to the real field formalism through equation (4.55) and to the momentum space.

---

[12] Its relevance will be understood further when we outline the path integral formalism.

[13] You will set $\lambda = 1$ at the end.

- Concentrate only on the quadratic part S_2. If it can be represented as

$$S_2 = \frac{1}{2} \sum_{\omega_n, \mathbf{p}} \sum_{i,j=1,2} \psi_i(\omega_n, \mathbf{p}) \mathcal{M}_{ij}(\omega_n, \mathbf{p}) \psi_j(\omega_n, \mathbf{p}), \tag{4.68}$$

Then the 2×2 matrix $\widehat{\mathcal{M}}$ can be identified as the inverse propagator: $\widehat{\mathcal{M}} = (\widehat{D})^{-1}$. If it is necessary, $\widehat{G}$ and $\widehat{G}^{(-1)}$ can be found from Eqs. (4.59)–(4.60) etc.

In the next sections, we demonstrate this procedure for evaluation of free $\widehat{D}_0$ and interacting Green functions.

4.6　Green Function of an Ideal Bose Gas

Here the Euclidean Lagrangian density is given by

$$\mathcal{L}_0 = \psi^\dagger(\tau, \mathbf{r}) \left[\partial_\tau - \frac{\nabla^2}{2m} - \mu \right] \psi(\tau, \mathbf{r}), \tag{4.69}$$

where μ is the chemical potential and m is the particle mass. After the shift (4.67) the action is given by

$$\begin{aligned}
S &= \int_0^\beta d\tau \int d\mathbf{r} \mathcal{L}_0 \\
&= \int_0^\beta d\tau \int d\mathbf{r} \left[-\mu\psi_0^2 - \mu\lambda\psi_0(\tilde{\psi} + \tilde{\psi}^\dagger) + \lambda^2 \tilde{\psi}^\dagger (\partial_\tau - \mu - \frac{\nabla^2}{2m})\tilde{\psi} \right] \\
&\equiv S_0 + \lambda S_1 + \lambda^2 S_2.
\end{aligned} \tag{4.70}$$

Now introducing real fields ψ_1, ψ_2 by the Eq. (4.55) we rewrite S_2 as follows:

$$\begin{aligned}
S_2 &= \frac{1}{2} \int_0^\beta d\tau \int d\mathbf{r}\tilde{\psi}^\dagger \left[\partial_\tau - \frac{\nabla^2}{2m} - \mu \right] \tilde{\psi} \\
&= \frac{1}{2} \int_0^\beta d\tau \int d\mathbf{r} \left[i(\psi_1 \partial_\tau \psi_2 - \psi_2 \partial_\tau \psi_1) \right. \\
&\quad - \frac{1}{2m}(\psi_1 \nabla^2 \psi_1 + \psi_2 \nabla^2 \psi_2) \\
&\quad \left. - \mu(\psi_1^2 + \psi_2^2) \right].
\end{aligned} \tag{4.71}$$

Here we used the following identities:

$$\int_0^\beta d\tau \psi_i(\tau) \partial_\tau \psi_i(\tau) = \int_0^\beta d\tau \psi_i d\psi_i = \frac{1}{2}(\psi_i^2(\beta) - \psi_i^2(0)) = 0, \tag{4.72}$$

and

$$\int d\mathbf{r}(\psi_1 \nabla^2 \psi_2 - \psi_2 \nabla^2 \psi_1) = \int d\mathbf{r}[\nabla(\psi_1 \Delta \psi_2 - \psi_2 \Delta \psi_1)] = 0, \tag{4.73}$$

where the first one holds due to the periodicity, $\psi(\tau = 0) = \psi(\tau = \beta)$, while the second one due to the Gauss theorem [14]. The Eq. (4.71) can be presented in a compact form as

$$S_2 = \frac{1}{2} \int_0^\beta d\tau \int d\mathbf{r} \sum_{i,j=1,2} \psi_i(\tau, \mathbf{r}) \left[-\delta_{ij}(\mu + \frac{\nabla^2}{2m}) + i\varepsilon_{ij}\partial_\tau \right] \psi_j(\tau, \mathbf{r}), \quad (4.74)$$

where ε_{ij} is the antisymmetric tensor: $\varepsilon_{11} = \varepsilon_{22} = 0$, $\varepsilon_{12} = 1$, $\varepsilon_{21} = -1$. In momentum space S_2 has a desired form:

$$S_2 = \frac{1}{2} \sum_{\omega_n, \mathbf{k}} \sum_{i,j=1,2} \psi_i(\omega_n, \mathbf{k})(D_0^{-1})_{ij}(\omega_n, \mathbf{k})\psi_j(\omega_n, \mathbf{k}), \quad (4.75)$$

with the inverse propagator:

$$D_0^{-1}(\omega_n, \mathbf{k}) = \begin{pmatrix} \varepsilon_k - \mu & \omega_n \\ -\omega_n & \varepsilon_k - \mu \end{pmatrix}, \quad (4.76)$$

where $\varepsilon_k = \mathbf{k}^2/2m$. Inverting (4.76) one can find $\hat{D}$ as

$$D_0(\omega_n, \mathbf{k}) = \frac{1}{(E_k^2 + \omega_n^2)} \begin{pmatrix} \varepsilon_k - \mu & -\omega_n \\ \omega_n & \varepsilon_k - \mu \end{pmatrix}, \quad (4.77)$$

where $E_k = |\varepsilon_k - \mu|$. As to the free propagator in the $(\tilde{\psi}^\dagger, \tilde{\psi})$ representation, found from Eqs. (4.59) and (4.77), it turns out to be diagonal:

$$G_0(\omega_n, \mathbf{k}) = \frac{1}{(E_k^2 + \omega_n^2)} \begin{pmatrix} -\varepsilon_k + \mu - i\omega_n & 0 \\ 0 & -\varepsilon_k + \mu + i\omega_n \end{pmatrix}, \quad (4.78)$$

and the inverse one is

$$G_0^{-1}(\omega_n, \mathbf{k}) = \begin{pmatrix} -\varepsilon_k + \mu + i\omega_n & 0 \\ 0 & -\varepsilon_k + \mu - i\omega_n \end{pmatrix} \quad (4.79)$$

4.7   Green Function of Interacting Bose Particles

The Lagrangian density of interacting identical bosons is given by

$$\mathcal{L} = \mathcal{L}_0 + \mathcal{L}_{int},$$

$$\mathcal{L}_0 = \psi^\dagger(\tau, \mathbf{r}) \left[\partial_\tau - \frac{\nabla^2}{2m} - \mu \right] \psi(\tau, \mathbf{r}), \quad (4.80)$$

$$\mathcal{L}_{int} = \frac{1}{2} \int d\mathbf{r}' |\psi(\tau, \mathbf{r}')|^2 U(|\mathbf{r} - \mathbf{r}'|)|\psi(\tau, \mathbf{r})|^2,$$

where $U(|\mathbf{r} - \mathbf{r}'|)$ is the spherical-symmetric two body interaction potential between bosons. Clearly, the Green function $\widehat{D}$ which takes into account the interaction will be a 2×2 matrix, similar to $\widehat{D}_0$ in the Eq. (4.77). So, before constructing $\widehat{D}^{-1}$ explicitly, lets observe the origin of each matrix element of $D_0^{-1}(\omega_n, \mathbf{k}$, which corresponds to the free Lagrangian $\mathcal{L}_0$.

- In the Eq. (4.76) off-diagonal element e.g. $(D_0^{-1})_{12} = \omega_n$ comes from the term $\psi^\dagger \partial_\tau \psi$ of the Lagrangian which produces the cross term $\sim \psi_1 \partial_\tau \psi_2$. Will $\mathcal{L}_{int}$ produce a cross term? Well, after the shift (4.67) the quadratic part of $\mathcal{L}_{int}$ (or the action) will include following terms: $\psi_0^2 \tilde{\psi}^\dagger \tilde{\psi}$, $\psi_0^2 (\tilde{\psi}^\dagger \tilde{\psi}^\dagger + \tilde{\psi} \tilde{\psi})$. It can be clearly understood that, due to the commutation of real fields ψ_1 and ψ_2, neither of these terms include the cross term such as $\psi_0^2 \psi_1 \psi_2$. Therefore, the presence of $\mathcal{L}_{int}$ can not modify off-diagonal matrix elements, and hence $(D^{-1})_{12} = (D_0^{-1})_{12} = \omega_n$, $(D^{-1})_{21} = (D_0^{-1})_{21} = -\omega_n$, unless two loop or higher order corrections are included [31].
- Therefore, $\mathcal{L}_{int}$ may modify only diagonal terms: $(D^{-1})_{ii} \neq (D_0^{-1})_{ii}$. Let's denote quadratic terms originated from $\mathcal{L}_{int}$ as $(1/2)Y_1\psi_1\psi_1$ and $(1/2)Y_2\psi_2\psi_2$. In general Y_i includes ψ_0^2 and interaction constants. Then, together with the contribution from $\mathcal{L}_0$ such as $\varepsilon_k - \mu$, the total contribution from $\mathcal{L}$ can be denoted as $\varepsilon_k + X_i$, so that

$$D^{-1}(\omega_n, \mathbf{k}) = \begin{pmatrix} \varepsilon_k + X_1 & \omega_n \\ -\omega_n & \varepsilon_k + X_2 \end{pmatrix}, \tag{4.81}$$

$$D(\omega_n, \mathbf{k}) = \frac{1}{(E_k^2 + \omega_n^2)} \begin{pmatrix} \varepsilon_k + X_2 & -\omega_n \\ -\omega_n & \varepsilon_k + X_1 \end{pmatrix}, \tag{4.82}$$

where $E_k^2 = (\varepsilon_k + X_1)(\varepsilon_k + X_2)$ and $X_i \sim -\mu + Y_i$ should be evaluated explicitly. Clearly, they depend on the explicit form $U(|\mathbf{r} - \mathbf{r}'|)$ as well as from the method of approximation.

For reference purposes, we provide also explicit expressions for G and G^{-1}:

$$G_{11}(\omega_n, \mathbf{k}) = \frac{-\varepsilon_k - i\omega_n - (X_1 + X_2)/2}{(E_k^2 + \omega_n^2)},$$

$$G^{-1}_{11}(\omega_n, \mathbf{k}) = -\varepsilon_k + i\omega_n - (X_1 + X_2)/2,$$

$$G_{12}(\omega_n, \mathbf{k}) = G_{21}(\omega_n, \mathbf{k}) = \frac{(X_1 - X_2)}{2(E_k^2 + \omega_n^2)}, \tag{4.83}$$

$$G^{-1}_{12}(\omega_n, \mathbf{k}) = G^{-1}_{21}(\omega_n, \mathbf{k}) = \frac{X_2 - X_1}{2},$$

$$G_{22}(\omega_n, \mathbf{k}) = G_{11}(-\omega_n, \mathbf{k}), \quad G^{-1}_{22}(\omega_n, \mathbf{k}) = G^{-1}_{11}(-\omega_n, \mathbf{k})$$

and also for the self energies:

$$\Sigma_n = \mu + \frac{X_1 + X_2}{2}, \qquad \Sigma_{an} = \frac{X_1 - X_2}{2}, \tag{4.84}$$

which follow from the Eqs. (4.79) and (4.83).

In the next section, we consider the simplest but widely used interaction $U(|\mathbf{r} - \mathbf{r}'|) = g\delta(\mathbf{r} - \mathbf{r}')$, which is referred as the zero range Fermi pseudo-potential, or shortly an s-wave contact potential.

4.7.1 Bilinear Approximation with the Contact Potential

Let the Lagrangian is given by (4.80) with

$$\mathcal{L}_{int} = \frac{g}{2}\left(\psi^{\dagger}(\tau, \mathbf{r})\psi(\tau, \mathbf{r})\right)^2, \tag{4.85}$$

where coupling constant is given from scattering theory by $g = 4\pi a_s/m$ with a_s is the s-wave scattering length. Having performed the steps outlined above, one finds

$$
\begin{aligned}
S_2 &= \frac{1}{2}\int_0^{\beta} d\tau \int d\mathbf{r}\left[i(\psi_1\partial_\tau\psi_2 - \psi_2\partial_\tau\psi_1) + \psi_1(-\frac{\nabla^2}{2m} - \mu + 3g\psi_0^2)\psi_1\right.\\
&\quad \left. + \psi_2(-\frac{\nabla^2}{2m} - \mu + g\psi_0^2)\psi_2\right]\\
&= \frac{1}{2}\int_0^{\beta} d\tau \int d\mathbf{r}\sum_{i,j}\left[i\varepsilon_{ij}\psi_i\partial_\tau\psi_j + \psi_i(-\frac{\nabla^2}{2m} + X_i)\psi_j\delta_{ij}\right]\\
&= \frac{1}{2}\sum_{\omega_n,\mathbf{k}}\sum_{i,j=1,2}\psi_i(\omega_n, \mathbf{k})(D^{-1})_{ij}(\omega_n, \mathbf{k})\psi_j(\omega_n, \mathbf{k}).
\end{aligned}
\tag{4.86}
$$

So, in this case D^{-1} is given by (4.81) with

$$X_1 = -\mu + 3g\rho_0, \qquad X_2 = -\mu + g\rho_0, \tag{4.87}$$

where we took into account that, the order parameter ψ_0 in (4.67) is related to the condensed fraction of BEC as $\rho_0 = \psi_0^2$. Note that for the noninteracting gas with $g = 0$, we have $X_1(id.gas) = X_2(id.gas) = -\mu$ as it has been expected from (4.76).

Task 4.4: Let $\mathcal{L}_{int}$ includes two and three body contact interactions as [31,32]

$$\mathcal{L}_{int} = g_2\left(\psi^{\dagger}(\tau, \mathbf{r})\psi(\tau, \mathbf{r})\right)^2 + g_3\left(\psi^{\dagger}(\tau, \mathbf{r})\psi(\tau, \mathbf{r})\right)^3. \tag{4.88}$$

Evaluate $\widehat{D}^{-1}(\omega_n, \mathbf{k})$ within the bilinear approximation.

4.7.2 Application of Hugenholtz-Pines Relation

Now the time has come to exploit the Hugenholtz-Pines theorem. From the Eqs. (4.61), (4.76) and (4.81) we obtain

$$\Pi(\omega_n, \mathbf{k}) = \begin{pmatrix} \mu + X_1 & 0 \\ 0 & \mu + X_2 \end{pmatrix}.$$ (4.89)

Finally, Eq. (4.65) leads to $\Pi_{22} = \mu$ that is, in general, $\underline{X_2 = 0}$. Note that, the Eq. (4.84) also leads to $X_2 = 0$ under the condition $\Sigma_n - \Sigma_{an} = \mu$. Therefore, having established that, X_2 should be exactly equal to zero, we may find the chemical potential in terms of the condensed fraction as $\mu = g\rho_0$, and using this in (4.87) conclude $X_1 = 2g\rho_0$. It should be underlined that, the last two equations hold only for the contact interaction in the bilinear approximation. The spectrum is simplified as $E_k = \sqrt{\varepsilon_k(\varepsilon_k + X_1)}$. We shall discuss the consequences of such a spectrum in the next section.

4.8 Physical Observables in the Bilinear Approximation

Now we are in the stage of evaluation of some physical parameters of BEC such as condensed (uncondensed) fractions, the sound velocity, superfluid density etc.

4.8.1 Condensed and Uncondensed Fractions

First of all, we have to fix the normalization condition for the fields in the equilibrium for the homogeneous system. Let N be the total number of particles with the density $\rho = N/V$. Then we require that

$$N = \int d\mathbf{r} \langle \psi^\dagger(\mathbf{r})\psi(\mathbf{r})\rangle.$$ (4.90)

Furthermore, in accordance with the physical meaning of the Bogolyubov shift (4.67) the order parameter ψ_0 is related to the number of condensed particles N_0 as

$$N_0 = \int d\mathbf{r}\psi_0^2,$$ (4.91)

with the density $\rho_0 = N_0/V$. Since, on the average, each particle is either in the condensate or out of the condensate, while the total number of particles is fixed, then following normalization condition holds:

$$N_0 + N_1 = N, \qquad \rho_0 + \rho_1 = \rho,$$ (4.92)

where N_1 is, naturally, the average number of uncondensed particles with the density $\rho_1 = N_1/V$. From the above normalization conditions, one can see that ρ_1 is directly related to the the mean value of fluctuating fields

$$\rho_1 = \frac{1}{V} \int d\mathbf{r} \langle \tilde{\psi}^\dagger(\mathbf{r}) \tilde{\psi}(\mathbf{r}) \rangle. \tag{4.93}$$

Although this expression looks rather abstract, we are able to represent it at least in terms of X_1 and X_2. In fact, using the definition of the propagator D_{ij}, the Eq. (4.93) can be presented as

$$\begin{aligned}
\rho_1 &= \frac{1}{2V} \int d\mathbf{r} \langle (\psi_1(\mathbf{r}) - i\psi_2(\mathbf{r}))(\psi_1(\mathbf{r}) + i\psi_2(\mathbf{r})) \rangle \\
&= \frac{1}{2V} \int d\mathbf{r} \sum_i D_{ii}(\tau, \mathbf{r}; \tau', \mathbf{r}')|_{\tau' \to \tau, \mathbf{r}' \to \mathbf{r}}.
\end{aligned} \tag{4.94}$$

Now passing to the momentum space (see Eqs. (4.52) and (4.82)) one obtains

$$\begin{aligned}
D_{11}(0) &= \frac{1}{\beta V} \sum_k \sum_{\omega_n} \frac{\varepsilon_k + X_2}{E_k^2 + \omega_n^2} = \frac{1}{V} \sum_k \frac{W_k(\varepsilon_k + X_2)}{E_k}, \\
D_{22}(0) &= \frac{1}{\beta V} \sum_k \sum_{\omega_n} \frac{\varepsilon_k + X_1}{E_k^2 + \omega_n^2} = \frac{1}{V} \sum_k \frac{W_k(\varepsilon_k + X_1)}{E_k}, \\
D_{12}(0) &= \frac{1}{\beta V} \sum_k \sum_{\omega_n} \frac{\omega_n}{E_k^2 + \omega_n^2} = 0, \qquad D_{21}(0) = -D_{12}(0) = 0,
\end{aligned} \tag{4.95}$$

where we introduced the notation $D_{ij}(0) \equiv D_{ij}(\tau, \mathbf{r}; \tau', \mathbf{r}')|_{\tau' \to \tau, \mathbf{r}' \to \mathbf{r}}$ and have taken Matsubara sums by using following relations

$$\begin{aligned}
\sum_{\omega_n} \frac{1}{a^2 + \omega_n^2} &= \sum_{n=-\infty}^{\infty} \frac{1}{a^2 + (2\pi n T)^2} = \frac{\beta \coth(a/2T)}{2a} \equiv \frac{\beta W_k}{a}, \\
\sum_{\omega_n} \frac{\omega_n}{a^2 + \omega_n^2} &= 0,
\end{aligned} \tag{4.96}$$

which can be easily verified in Mathematica or Maple. Note that here the Bose distribution $1/(\exp(\beta E_k) - 1)$ for quasiparticles (Bogolons) with the dispersion $E_k = \sqrt{(\varepsilon_k + X_1)(\varepsilon_k + X_2)}|_{X_2=0} = \sqrt{\varepsilon_k(\varepsilon_k + X_1)}$ automatically arises from Matsubara summation due to the identity

$$\frac{\coth(E_k/2T)}{2} = \frac{1}{2} + \frac{1}{e^{\beta E_k} - 1} \equiv W_k(E_k). \tag{4.97}$$

The Eqs. (4.93)–(4.97) lead finally to the following expression for the the density of uncondensed fraction:

$$\rho_1 = \frac{D_{11}(0) + D_{22}(0)}{2} = \frac{1}{V} \sum_k \left[\frac{W_k(\varepsilon_k + \Delta)}{E_k} - \frac{1}{2} \right], \qquad (4.98)$$

where $\Delta \equiv X_1/2 = g\rho_0$ and $E_k = \sqrt{\varepsilon_k(\varepsilon_k + 2\Delta)}$, $\varepsilon_k = \mathbf{k}^2/2m$. The origin of the additional term $-1/2$ in (4.98) will be explained further. As to the condensed fraction $n_0 = \rho_0/\rho$ it may be calculated from the normalization condition $\rho_0 = \rho - \rho_1$ where ρ is supposed to be known, while ρ_1 is given by Eq. (4.98). Thus, in the case of contact interaction, the bilinear approximation leads to the following nonlinear algebraic equation with respect to n_0:

$$n_0 = 1 - n_1 = 1 - \frac{4\pi}{(2\pi)^3\rho} \int_0^\infty k^2 dk \left[\frac{W_k(\varepsilon_k + g\rho n_0)}{\sqrt{\varepsilon_k(\varepsilon_k + 2g\rho n_0)}} - \frac{1}{2} \right]. \qquad (4.99)$$

4.8.2   Bogolyubov Approximation

Solving the Eq. (4.99) will be more or less simplified by introducing the dimensionless gas parameter $\gamma = \rho a_s^3$ which is in reality rather small for the weakly interacting degenerated gases: $\gamma \sim 10^{-5}$. In such situations, one may neglect the second term of the equation and feel free to set $n_0 = 1$, i.e. $\rho_0 = \rho$ with the dispersion $E_k = \sqrt{\varepsilon_k(\varepsilon_k + 2g\rho)}$, which is referred in the literature as Bogolyubov dispersion. Note that, the widely used Gross- Pitaevskii equation [5] also assumes Bogolyubov approximation with the 100% condensed fraction. On the other hand, this approximation may serve as a basis of power expansion by γ, such that one will be able to control its validity and accuracy. We shall illustrate this expansion in the next section for $T = 0$.

4.9   Zero Temperature

Below we prove the necessity of the term $-1/2$ in (4.98) and show that the Eq. (4.99) can be solved analytically at $T = 0$. For this purpose lets rewrite the Eq. (4.98) omitting the term $-1/2$ as

$$\rho_1 = \frac{4\pi}{(2\pi)^3} \int_0^\infty k^2 dk \left[\frac{W_k(\varepsilon_k + \Delta)}{\sqrt{\varepsilon_k(\varepsilon_k + 2\Delta)}} \right], \qquad (4.100)$$

where $W_k(T = 0) = 1/2$ and $\Delta = g\rho_0$ is a positive constant (otherwise ρ_1 becomes complex). It can be easily seen that this integral is divergent and hence, needs a renormalization.

4.9.1 Methods of Regularization

To get physical values, we must regularize the otherwise divergent integrals. We know from the textbooks on field theories that there are mainly three widely used methods of regularization:

- Mass or coupling constant renormalization
- Dimensional regularization
- Method of counter terms- subtraction method.

It has been shown [33] that, at least for bosons all the above three methods lead to the same result for physical observables.

For completeness, below we briefly outline each of them. Renormalization of coupling constant is essential in high energy physics, leading to a running coupling constant. In low-energy condensed matter physics, involving scattering theory, this procedure is usually realized by the transformation

$$\frac{m}{4\pi a_s} \to \frac{1}{g} + \frac{1}{V} \sum_k \frac{m}{\mathbf{k}^2}. \tag{4.101}$$

It is almost irreplaceable in the theoretical description of BCS-BEC crossover [34] as well as superconductivity [14].

Dimensional regularization is based on the fundamental 't Hooft and Veltman conjecture stating that, in the context of dimensional regularization integrals over a polynomial identically vanish i.e.

$$\int_0^\infty dq \, q^{D-1}(q^2)^{n-1} = 0, \qquad n = 0, 1, 2\ldots, \tag{4.102}$$

where D may assume non-integer values. To exploit this method one has to make the following transformation

$$\int_0^\infty \frac{d^D q}{(2\pi)^D} f(q^2) \to \frac{S_D}{(2\pi)^D} \left(\frac{\tilde{\mu}^2 e^\gamma}{4\pi}\right)^{-\epsilon} \int_0^\infty dq \, q^{D-1} f(q^2), \tag{4.103}$$

where $\gamma \approx 0.577$ is Euler's constant, $S_D = 2(\pi)^{D/2}/\Gamma(D/2)$, $\epsilon = (D-2)/2$, ($\epsilon \to 0$), and arbitrary constant $\tilde{\mu}$ has been introduced for dimensional reasons. Then typical integrals of quantum field theory can be evaluated analytically by using

$$\int_0^\infty \frac{dt \, t^{x-1}}{(1+t)^{x+y}} = \frac{\Gamma(x)\Gamma(y)}{\Gamma(x+y)}, \qquad Re(x), \quad Re(y) > 0, \tag{4.104}$$

with $\Gamma(x)$ is a gamma function. Some other integrals can be converted to the form (4.104) by using the identities [35]:

$$
\int_0^\infty d\tau \; \tau^{\alpha-1} e^{-\tau a} = \frac{\Gamma(\alpha)}{a^\alpha},
$$
$$
\frac{1}{A^a B^b} = \frac{\Gamma(a+b)}{\Gamma(a)\Gamma(b)} \int_0^1 \frac{x^{a-1}(1-x)^{b-1}}{[Ax + B(1-x)]^{a+b}} dx
\tag{4.105}
$$

We note that, in our knowledge, the applicability of dimensional regularization is limited by the integrals of the kind of (4.104). Nevertheless, it has been successfully used in theoretical models of superfluidity [31,35,36].

In the present work, we exploit the third method. The essence of this method is the following. Let

$$
R = \int_0^\infty f(x)dx \to \infty, \quad \text{since} \quad \lim_{x\to\infty} f(x) = \infty.
\tag{4.106}
$$

Regularization of this integral includes following steps:

- Add to the integrand a polynomial function of x (counter terms)

$$
f(x) \to f(x) + \sum_{i=-k_1}^{k_2} c_i x^i = f_{ren}.
\tag{4.107}
$$

    (you have to guess the integers k_1 and k_2).
- Make the substitution $x = 1/y$. Then you will get $f_{ren}(x) \to \tilde{f}(y, c_i)$.
- Expand $\tilde{f}(y, c_i)$ in power series of y.
- Require that each coefficient of y^i with non-positive i, (say $i = -3, -2, -1, 0$) should vanish
- Find c_i from these requirements and putting them to $\tilde{f}(y, c_i)$, return to the variable x by $y = 1/x$. Check again the limit $\lim_{x\to\infty} f_{ren}(x)$. If now f_{ren} is still divergent, then choose other values of k_1, k_2 (expand their ranges), otherwise, you are lucky and $R_{ren} = \int_0^\infty f_{ren}(x)dx$ will be convergent.

Let's apply this procedure for the integral in (4.100). So, the first step gives

$$
\rho_1 = \frac{1}{2V} \sum_k \frac{\varepsilon_k + \Delta}{E_k} \to \frac{1}{2V} \sum_k \left[\frac{\varepsilon_k + \Delta}{E_k} - c_0 - \frac{c_2}{\varepsilon_k} \right],
\tag{4.108}
$$

where we have chosen $k_1 = 2$, $k_2 = 0$ and used $\varepsilon_k = \mathbf{k}^2/2m$ instead of k^2 for the convenience. Clearly, for a while, c_0 and c_2 are unknown constants. After the transformation $k = 1/y$ we have the following integral:

$$\rho_1^{ren} = \int_0^\infty dy\, R(y),$$

$$R(y) = \frac{2c_2 ms(y)y^2 + c_0 s(y) - 1 - 2\Delta m y^2}{4\pi^2 y^2\, s(y)}, \quad s(y) = \sqrt{1 + 4\Delta m y^2}. \tag{4.109}$$

The power series of $R(y)$ is

$$R(y) = \frac{1 - c_0}{4\pi^2 y^2} - \frac{mc_2}{2\pi^2} + \frac{m^2\Delta^2 y^2}{2\pi^2} + O(y^4). \tag{4.110}$$

Now requiring that the terms with y^{-2} and y^0 should be zero we find

$$1 - c_0 = 0, \quad c_0 = 1, \quad c_2 = 0. \tag{4.111}$$

Using these values and coming back to our variable $k = 1/y$ in (4.109) we can finally represent regularized ρ_1 as[14]

$$\rho_1 = \int_0^\infty k\,dk\, \frac{k^2 + 2\Delta m - k\sqrt{k^2 + 4\Delta m}}{4\pi^2\sqrt{k^2 + 4\Delta m}} = \frac{1}{V}\sum_k \left[\frac{W_k(\varepsilon_k + \Delta)}{E_k} - \frac{1}{2}\right]. \tag{4.112}$$

where $W_k = W_k(T = 0) = 1/2$. Therefore we have proven that the term $-1/2$ in (4.98) actually plays the role of a counter term, serving as a regulator.[15] Finally, one notes that the integral in (4.112) may be evaluated analytically with the net result:

$$\rho_1(T = 0) = \frac{(m\Delta)^{3/2}}{3\pi^2}. \tag{4.113}$$

4.9.2 Power Series in γ for n_0

Now we are on the stage of finding an explicit expression for the condensed fraction as a function $n_0(\gamma)$. For this purpose, we rewrite the normalization condition at $T = 0$

$$\rho_0 = \rho - \rho_1 = \rho - \frac{(m\Delta)^{3/2}}{3\pi^2}, \tag{4.114}$$

in dimensionless form

$$Z^3 + \frac{3\sqrt{\pi}}{8\sqrt{\gamma}}(Z^2 - 1) = 0, \tag{4.115}$$

---

[14] Here we omit the notation "ren".

[15] It is interesting that in the method of Bogolyubov transformations, this term arises automatically.

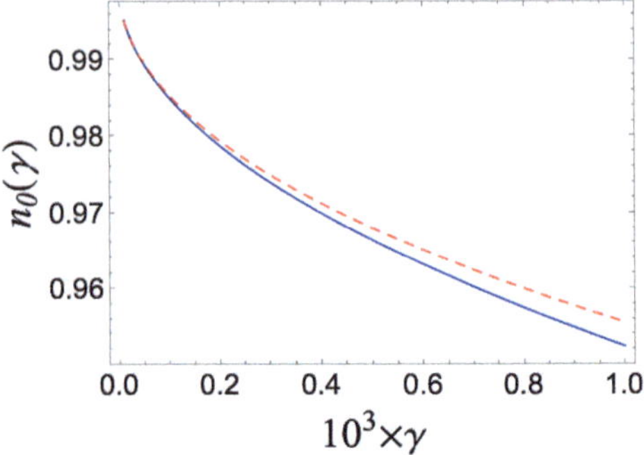

Fig. 4.5 Plot of the condensed fraction at $T = 0$ given by Eqs. (4.119) (solid line) and (4.118) (dashed lines)

where $Z^2 = \Delta/(g\rho)$, $\gamma = \rho a_s^3$, $\Delta = g\rho_0$. That is, we are considering the equation with respect to $\Delta/(g\rho)$ instead of n_0. It can be easily verified that the only physical solution of this cubic equation [37] is given by

$$Z = \frac{\sqrt{\pi}}{8\sqrt{\gamma}} \left[2\sin(\frac{\pi}{6} + \frac{1}{3}\arccos(\pi - \frac{96\gamma}{\pi})) - 1 \right]. \tag{4.116}$$

Thus $\Delta = Z^2 g\rho$ can be expanded in powers of $\sqrt{\gamma}$ as

$$\Delta = g\rho - \frac{8\sqrt{\gamma}g\rho}{3\sqrt{\pi}} + O(\gamma) = g\rho_0, \tag{4.117}$$

which leads for $n_0 = \Delta/(g\rho)$

$$n_0(T = 0, \gamma \ll 1) \approx 1 - \frac{8\sqrt{\gamma}}{3\sqrt{\pi}}, \tag{4.118}$$

This approximation is also referred in the literature as the Bogolyubov approximation with the Bogolyubov depletion $n_1 \approx 8\sqrt{\gamma}/(3\sqrt{\pi})$.[16] In Fig. 4.5 we present the condensed fraction at $T = 0$ in the framework of the bilinear approximation given by the exact expression

$$n_0(\gamma) = \frac{\pi}{64\gamma} \left[1 - 2\sin(\frac{\pi}{6} + \frac{1}{3}\arccos(\pi - \frac{96\gamma}{\pi})) \right]^2, \tag{4.119}$$

and by the Bogolyubov approximation (4.118). It is seen that the latter is good for $\gamma \le 2 \times 10^{-4}$.

---

[16] As far as we know, there is no rigorous definition of Bogolyubov approximation in the literature [38].

4.9.3 Elementary Excitations and Sound Velocity

We know that any quantum mechanical system has its own ground and excited states, which may be discrete or continuous. For example, the energy levels of an atom are discrete, while that of a crystal, consisting of these atoms, are mostly continuous corresponding to the energy spectrum of phonons as $\omega_k \sim ck$. Similarly to crystals, excited states of a condensate can be described in terms of quasiparticles (Bogolons) whose energy depends on their momentum being continuous. This dependence is referred as a dispersion law for the elementary excitations of BEC. In fact, by virtue of the Bogolyubov transformation it can be shown [39] that the Hamiltonian of the system can be reduced to the diagonal form

$$\hat{H} = E_0 + \sum_k E_k \hat{b}_k^+ \hat{b}_k, \tag{4.120}$$

where E_0 is the ground state energy. The second term corresponds to the Hamiltonian of independent (free) quasiparticles having momentum-dependent energy E_k and whose annihilation and creation operators are given by $\hat{b}_k$ and $\hat{b}_k^+$. The ground state then corresponds to the vacuum of Bogolons as $\hat{b}_k |vac\rangle = 0$ for any $\mathbf{k} \neq 0$ (see Eq. (4.23)). Further, we shall calculate E_0 explicitly, but meanwhile, we would like to concentrate on the energy dispersion $E_k = \sqrt{\varepsilon_k(\varepsilon_k + X_1)} = \sqrt{\varepsilon_k(\varepsilon_k + 2g\rho_0)}$. First of all, we note that the condition of dynamical stability in the equilibrium: E_k should be real and positive, requires that $g \geq 0$, that is the contact interaction between the particles (atoms) should be repulsive. Otherwise, the attractive interaction, especially for a one-component BEC, leads to collapse or even "explosion", in which a substantial fraction of the atoms were blown off ("Bosenova" [40]). For this reason, we consider only the system with repulsive interaction . Secondly, it would be interesting to make clear the following question: "It is well known that the ordinary sound propagates in the air with the velocity $c_{sound} \approx 340 m/sec$. So, what is the velocity of bogolons in BEC?" For this purpose, we consider the power series of E_k in the momentum $\mathbf{k}$

$$E_k \approx \frac{\sqrt{X_1}}{\sqrt{2m}}k + \frac{\sqrt{2}}{8\sqrt{X_1 m^3}}k^3. \tag{4.121}$$

This low momentum approximation describes the long-wavelength excitations of an interacting Bose gas which can be interpreted as sound waves with the velocity [5]

$$s = \frac{\sqrt{X_1}}{\sqrt{2m}} = \frac{\sqrt{\Delta}}{\sqrt{m}} = \frac{\sqrt{g\rho_0}}{\sqrt{m}} \approx \frac{\sqrt{g\rho}}{\sqrt{m}}, \tag{4.122}$$

we again note that g characterizes the interaction between the atoms, not between the Bogolons. So, there would be no sound waves for an ideal Bose gas. These excitations can also, be regarded as Goldstone modes associated with the spontaneous breaking

of gauge symmetry caused by BEC (see Eq. (4.124) below). Moreover, it is seen that
the dispersion is gapless and satisfies Landau criterion (3.32):

$$\lim_{k \to 0} \frac{E_k}{k} = s \neq 0, \qquad (4.123)$$

which confirms the fact that three-dimensional ($D = 3$) uniform interacting Bose
has the property of superfluidity. Bogolyubov was the first who derived such gapless
dispersion with $s = \sqrt{g\rho/m}$ explaining the superfluidity in microscopical level [9].

It is interesting that, in the presence of the Bose condensate, the excitation behaves
as a **massless** quasiparticle (Bogolon) traveling with the sound velocity s (the phonon
mode). In fact, its effective mass is given by

$$m^* = \lim_{k \to 0} \frac{k}{\partial E_k / \partial k} = \lim_{k \to 0} \frac{k}{s} = 0. \qquad (4.124)$$

In the opposite limit $k \gg ms$ the dispersion law approaches the free particle law:

$$E_k \approx \varepsilon_k + g\rho_0 \approx \frac{\mathbf{k}^2}{2m} + g\rho. \qquad (4.125)$$

The transition between these two regimes (phonon with $E_k \sim sk$ and particle $E_k \approx
\varepsilon_k + g\rho$) takes place when $\varepsilon_k \sim g\rho \sim ms^2$. By setting $\varepsilon_k = g\rho$ with $k = 1/\xi$ one
can define the characteristic interaction length

$$\xi = \frac{1}{\sqrt{2mg\rho}} = \frac{1}{\sqrt{2}ms}, \qquad (4.126)$$

which can be referred to as the healing length. This is illustrated in Fig. 4.6.

Above we mostly used Bogolyubov approximation: $ms^2 \approx g\rho$. However, by using
the expansion (4.118) one may obtain a little bit more accurate expression:

$$\frac{s^2(T = 0, \gamma \ll 1)}{c^2} \approx \frac{(\hbar c)^2 \pi \gamma}{(mc^2)^2 a_s^2} \left[1 - \frac{8\sqrt{\gamma}}{3\sqrt{\pi}} \right] + O(\gamma^{5/2}), \qquad (4.127)$$

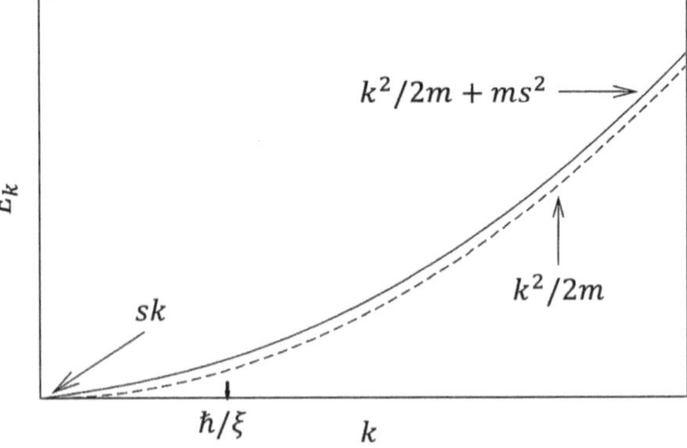

Fig. 4.6 Plot of the energy of the phonon and free particle. The transition between phonon and free particle regimes takes place at $\hbar/\xi$

where we restored the natural units introducing the speed of light c for the convenience of numerical calculations.

Task 4.5: Calculate the velocity of sound in BEC of ^{87}Rb at zero temperature with $\gamma \approx 10^{-5}$ and $a_s \approx 5nm$ [41].

Task 4.6: Make a list of pros and cons of the second quantization in comparison with ordinary quantization in high energy and condensed matter physics.

4.10 Conclusion

In the present chapter, we have stated that the theoretical description of a statistical system of bosons with even simple contact interaction, the analogy of $\lambda\phi^4$ theory in particle physics, cannot be performed analytically as a whole. For this reason, one has to resort to an adequate approximation to explain experimental measurements. One of the most widely used approaches is the bilinear (one loop) approximation which is accurate when the gas parameter γ is enough small: $\gamma \sim 10^{-5}$. However, in reality, there are systems with rather large γ [17] To study such systems one has to seek for a more accurate approximation rather than a bilinear or Bogolyubov one. One of the convenient methods of going beyond these approximations is based on the path integral (Feynman) formulation of statistical quantum mechanics. In the next chapters, we outline this method and present its applications. Here we note that this field of theoretical formalism, although it is not good to study bound states, has been successfully used both in theories of particle physics at high temperatures [23,43] as well as statistical physics at very low temperatures, including polymers [44].

However, before going deep inside the theory it is worth looking at the work of experimenters. In the next chapter we shall make an attempt to look inside laboratories at the bird's-eye level.

References

1. S.K. Adhikari, Vortex-lattice in a uniform Bose-Einstein condensate in a box trap. J. Phys.: Condensed Matter. **31**(27), 275401 (2019). https://dx.doi.org/10.1088/1361-648X/ab14c5
2. H. Hu, X.-J. Liu, Consistent theory of self-bound quantum droplets with bosonic pairing. Phys. Rev. Lett. **125**, 195302 (2020). https://link.aps.org/doi/10.1103/PhysRevLett.125.195302
3. R.A. Aziz, V.P.S. Nain, J.S. Carley, W.L. Taylor, G.T. McConville, An accurate intermolecular potential for helium. J. Chem. Phys. **70**(9), 4330–4342 (1979). https://doi.org/10.1063/1.438007
4. A. Cappellaro, L. Salasnich, Finite-range corrections to the thermodynamics of the one-dimensional Bose gas. Phys. Rev. A **96**, 063610 (2017). https://link.aps.org/doi/10.1103/PhysRevA.96.063610

---

[17] Some of the properties of superfluid ^4He can be understood by treating the potential as a hard-core interaction of diameter $a_s = 2.139\text{Å}$, which, at saturated vapour pressure, corresponds to $\gamma \approx 0.2$ [42].

5. L. Pitaevskii, S. Stringari, *Bose-Einstein Condensation*. Series International Series of Monographs on Physics (Clarendon Press, 2003). https://books.google.co.uz/books?id=rIobbOxC4j4C

6. H. Stoof, D. Dickerscheid, K. Gubbels, *Ultracold Quantum Fields*. Series Theoretical and Mathematical Physics (Springer Netherlands, 2008). https://books.google.co.uz/books?id=-9c2Ns2J5P4C

7. J. Dalibard, Collisional dynamics of ultra-cold atomic gases, in B*ose-Einstein Condensation in Atomic Gases* (IOS Press, 1999), pp. 321–349

8. C. Chin, R. Grimm, P. Julienne, E. Tiesinga, Feshbach resonances in ultracold gases. Rev. Mod. Phys. **82**, 1225–1286 (2010). https://link.aps.org/doi/10.1103/RevModPhys.82.1225

9. N.N. Bogolyubov, On the theory of superfluidity. J. Phys. (USSR) **11**, 23–32 (1947)

10. A. Abrikosov, L. Gorkov, I. Dzyaloshinski, R. Silverman, *Methods of Quantum Field Theory in Statistical Physics*. Series Dover Books on Physics (Dover Publications, 2012). https://books.google.co.uz/books?id=JYTCAgAAQBAJ

11. J. Blaizot, G. Ripka, J.W. Negele, H. Orland, G.E. Brown, Quantum theory of finite systems and quantum many-particle systems. Phys. Today **41**(9), 106–107 (1988). https://doi.org/10.1063/1.2811565

12. S. Chang, *Introduction to Quantum Field Theory*. Series Lecture Notes in Physics Series (World Scientific, 1990). https://books.google.co.uz/books?id=MISUeVE7sJYC

13. L. Landau, E. Lifshitz, *Quantum Mechanics: Non-relativistic Theory*. Series Course of Theoretical Physics, vol. 3 (Elsevier Science, 1991). https://books.google.co.uz/books?id=J9ui6KwC4mMC

14. N. Nagaosa, S. Heusler, *Quantum Field Theory in Condensed Matter Physics*. Series Theoretical and Mathematical Physics (Springer, Berlin, Heidelberg, 2013). https://books.google.co.uz/books?id=dqbvCAAAQBAJ

15. A. Khudoyberdiev, A. Rakhimov, A. Schilling, Bose–Einstein condensation of triplons with a weakly broken U(1) symmetry. New J. Phys. **19**(11), 113002 (2017). https://dx.doi.org/10.1088/1367-2630/aa8a2f

16. D.W. Snoke, A.J. Daley, The question of spontaneous symmetry breaking in condensates. *Universal Themes of Bose-Einstein Condensation*, p. 79 (2017). https://books.google.co.uz/books?id=7tmiDgAAQBAJ

17. V.I. Yukalov, Basics of Bose-Einstein condensation. Phys. Part. Nucl. **42**(3), 460–513 (2011). https://doi.org/10.1134/S1063779611030063

18. V. Ginzburg, L. Landau, On the Heisenberg and Schrödinger pictures. Zh. Eksp. Teor. Fiz. **20**, 1064 (1950)

19. W. Dickhoff, D. Van Neck, *Many-body theory exposed! Propagator Description of Quantum Mechanics in Many-Body Systems* (World Scientific Publishing Company, 2005). https://books.google.co.uz/books?id=gzk8DQAAQBAJ

20. A. Fetter, J. Walecka, *Quantum Theory of Many-particle Systems*. Series Dover Books on Physics (Dover Publications, 2003). https://books.google.co.uz/books?id=0wekf1s83b0C

21. A. Rakhimov, F.C. Khanna, Finite temperature amplitudes and reaction rates in thermofield dynamics. Phys. Rev. C **64**, 064907 (2001). https://link.aps.org/doi/10.1103/PhysRevC.64.064907

22. F.C. Khanna, A.P.C. Malbouisson, J.M.C. Malbouisson, A.E. Santana, *Thermal Quantum Field Theory* (World Scientific, 2009). https://www.worldscientific.com/doi/abs/10.1142/6896

23. M. Bellac, *Thermal Field Theory*. Series Cambridge Monographs on Mathematical Physics (Cambridge University Press, 2000). https://books.google.co.uz/books?id=00_x6GR8GXoC

24. A. Rakhimov, F. Khanna, U. Yakhshiev, M. Musakhanov, Density dependence of meson-nucleon vertices in nuclear matter. Nucl. Phys. A **643**(4), 383–401 (1998). https://www.sciencedirect.com/science/article/pii/S0375947498005612

25. A. Das, *Finite Temperature Field Theory*, 2nd ed. (World Scientific, 2023). https://www.worldscientific.com/doi/abs/10.1142/13308

26. H. Umezawa, H. Matsumoto, M. Tachiki, *Thermo Field Dynamics and Condensed States.* Amsterdam, New York: North-Holland Pub. Co. ; Sole distributors for the USA. and Canada (Elsevier Science Pub. Co. Amsterdam, New York, 1982). https://worldcat.org/title/8114603

27. S. Beliaev, Application of the methods of quantum field theory to a system of bosons. Sov. Phys. JETP **34**(7), 289–299 (1958). http://jetp.ras.ru/cgi-bin/dn/e_007_02_0289.pdf

28. N.M. Hugenholtz, D. Pines, Ground-state energy and excitation spectrum of a system of interacting bosons. Phys. Rev. **116**, 489–506 (1959). https://link.aps.org/doi/10.1103/PhysRev.116.489

29. V. Yukalov, Representative statistical ensembles for Bose systems with broken gauge symmetry. Ann. Phys. **323**(2), 461–499 (2008). https://www.sciencedirect.com/science/article/pii/S0003491607000723

30. S. Watabe, Hugenholtz-pines theorem for multicomponent Bose-Einstein condensates. Phys. Rev. A **103**, 053307 (2021). https://link.aps.org/doi/10.1103/PhysRevA.103.053307

31. J.O. Andersen, Theory of the weakly interacting Bose gas. Rev. Mod. Phys. **76**, 599–639 (2004). https://link.aps.org/doi/10.1103/RevModPhys.76.599

32. E. Braaten, H.-W. Hammer, Universality in the three-body problem for ^{4}He atoms. Phys. Rev. A **67**, 042706 (2003). https://link.aps.org/doi/10.1103/PhysRevA.67.042706

33. L. Salasnich, F. Toigo, Zero-point energy of ultracold atoms. Phys. Rep. **640**, 1–29 (2016). Zero-point energy of ultracold atoms. https://www.sciencedirect.com/science/article/pii/S0370157316301338

34. M. Randeria, W. Zwerger, M. Zwierlein, *The BCS–BEC Crossover and the Unitary Fermi Gas* (Springer Berlin, Heidelberg, 2012), pp. 1–32. https://doi.org/10.1007/978-3-642-21978-8_1

35. H. Kleinert, V. Schulte-Frohlinde, *Critical Properties of Φ^4-theories.* Series G-Reference, Information and Interdisciplinary Subjects Series (World Scientific, 2001). https://books.google.co.uz/books?id=rBfWk4vVi-cC

36. A. Rakhimov, C.K. Kim, S.-H. Kim, J.H. Yee, Stability of the homogeneous Bose-Einstein condensate at large gas parameter. Phys. Rev. A **77**, 033626 (2008). https://link.aps.org/doi/10.1103/PhysRevA.77.033626

37. I.J. Zucker, 92.34 the cubic equation—a new look at the irreducible case. Math. Gazette **92**(524), 264–268 (2008). http://www.jstor.org/stable/27821778

38. C. Vianello, L. Salasnich, Condensate and superfluid fraction of homogeneous Bose gases in a self-consistent Popov approximation. Sci. Rep. **14**(1), 15034 (2024). https://doi.org/10.1038/s41598-024-65897-2

39. V.I. Yukalov, Representative ensembles in statistical mechanics. Int. J. Modern Phys. B **21**(01), 69–86 (2007). https://doi.org/10.1142/S0217979207035893

40. E.A. Donley, N.R. Claussen, S.L. Cornish, J.L. Roberts, E.A. Cornell, C.E. Wieman, Dynamics of collapsing and exploding Bose–Einstein condensates. Nature **412**(6844), 295–299 (2001). https://doi.org/10.1038/35085500

41. E. Henn, J. Seman, G. Seco, E. Olimpio, P. Castilho, G. Roati, D.V. Magalhães, K.M.F. Magalhães, V.S. Bagnato, Bose-Einstein condensation in ^{87}Rb: Characterization of the Brazilian experiment. Brazilian J. Phys. **38**, 279–286 (2008). https://www.redalyc.org/pdf/464/46413553012.pdf

42. V.I. Yukalov, E.P. Yukalova, Bose-Einstein-condensed gases with arbitrary strong interactions. Phys. Rev. A **74**, 063623 (2006). https://link.aps.org/doi/10.1103/PhysRevA.74.063623

43. L.D. Faddeev, A.A. Slavnov, *Gauge Fields: An Introduction To Quantum Theory, Second Edition (Frontiers in Physics)* (CRC Press, Lausanne, Switzerland, 2019)

44. W. Janke, A. Pelster, *Path Integrals: New Trends and Perspectives: Proceedings of the 9th International Conference: Dresden, Germany, 23–28 September 2007* (World Scientific, 2008). https://books.google.co.uz/books?id=o005yAEACAAJ

Experiments in a Nutshell

5.1 Introduction

While these lectures are addressed firstly to theoreticians, anyway, the latters can not feel themselves fully happy without the satisfaction of their curiosity about the work of an experimenter. For this purpose, below we make a little excursion to experimenters.

In the present book, we frequently use the phrase "Weakly interacting ultracold dilute Bose gases". But the question arises, how a thermodynamic system can remain in the gaseous phase at very low temperatures instead of a crystal phase? The question is appropriate, since in accordance with typical $P - T$ phase diagrams from textbooks, a stable equilibrium state of any gas (excluding, maybe hydrogen) should correspond to the crystal phase. This is true, for example, rubidium becomes solid already at $T < 311$K. Actually, it has been observed that any Bose condensed system of atoms has a very short lifetime. Soon after its creation, three-body recombination events (say, forces) lead to the formation of molecules, which eventually bring the BEC gas into the thermodynamic stable solid phase. Thus BEC phase is a metastable state! Moreover, it is not so easy to obtain even such a metastable state, otherwise, BEC could be created nearly 100 years ago, just after its theoretical prediction.

It becomes clear that this can be achieved when at least the following two criteria are satisfied

- The temperature and density should be sufficiently low, otherwise three-body collisions will be inevitable. Actually, typical densities, reached in the gaseous BEC regime correspond to $(10^{13} - 10^{15})$ atoms/cm^3 at temperatures of the order of microkelvins.[1]

---

[1] For comparison, typical density of air in normal conditions, at $T \sim 300$K, is about 10^{19} atoms/cm^3.

© The Author(s), under exclusive license to Springer Nature Switzerland AG 2026
A. Rakhimov and S. Mardonov, *Theory of Quantum Bose Liquids Beyond Bogoliubov Approximation*, Lecture Notes in Physics 1045,
https://doi.org/10.1007/978-3-032-05096-0_5

- The gas should be kept far from any material wall, where the interaction with other atoms would favor the formation of molecules. In other words, the sample should hang like a plasma in the Tokamak.

The first condition can be realized by the appropriate cooling technology (see Appendix B) when excess fast atoms will be expelled. The second one can be achieved by confining the gas in a magnetic or optical trap [1]. Below, we briefly outline these methods of trapping.

5.2   Magnetic Trap

Magnetic trapping for neutral atoms is based on the use of inhomogeneous magnetic fields. The first trap, used to reach the BEC regime in alkalies (Li, Na, K, Rb, Cs gases) was a linear quadrupole trap. This trap is characterized by a magnetic field, which varies linearly in all directions and can be generated by two coils in an "anti-Helmholtz" configuration. The magnetic field is given by the simple law $\mathbf{B} = \mathbf{B}(x,y,-2z)$ satisfying the condition $(\nabla \mathbf{B}) = 0$ and curl $(\mathbf{B}) = 0$. The intensity of the magnetic field has a minimum at the origin, so that atoms occupying low-field-seeking states will be driven towards the center, while atoms occupying high-field-seeking states will be driven out of the trap. Later on, so-called TOP (Time-Orbiting-Potential) traps were constructed [1]. The magnetic field in the TOP trap has the form $\mathbf{B} = (B'x + B_0 \cos(\omega t), B'y + B_0 \sin(\omega t), -2B'z)$ where B' is the radial gradient of the quadrupole trap. Typical values of ω-angular velocity of the rotating field B_0, are a few kHz. It can be shown that, an atom in the TOP trap "feels" as if it is inside an external harmonic potential $U(r) \sim m\omega^2(x^2 + y^2 + 8z^2)$. Nowadays, it is possible to produce effective potentials of different symmetries, including cigar-shaped, triaxial, and even spherical traps designed as a microchip with typical size 2×2 cm.

Theoretical description of a BEC system in a magnetic trap is rather difficult. This is because of the presence of the trapping potential $U(\mathbf{r})$ in the grand canonical many-body Hamiltonian

$$H = \int d\mathbf{r}\psi^\dagger(\mathbf{r})\left[-\frac{\nabla^2}{2m} + U_{\text{trap}}(\mathbf{r}) - \mu + \frac{g}{2}|\psi|^2\right]\psi(\mathbf{r}), \tag{5.1}$$

where the simple contact interaction between atoms (4.1) has been assumed. The spatially varying potential $U_{\text{trap}}(\mathbf{r})$ makes the system inhomogeneous so that the chemical potential and even the number density of atoms will not be constant: $\mu = \mu(r)$, $\rho = \rho(r)$, $\rho_0 = \rho_0(r)$ etc. For this reason, usually one has to use the Thomas-Fermi approximation for the chemical potential

$$\mu \approx \mu_{TF} = \frac{\omega}{2}\left(\frac{15\,Na_s}{a_{\text{ho}}}\right)^{2/5}, \tag{5.2}$$

where $a_{\text{ho}} = \sqrt{1/m\omega}$ is the oscillator length associated with the average frequency ω of the harmonic trap. Moreover, the quantum as well as thermal fluctuations $\tilde{\psi}(\mathbf{r})$ given by the Bogolyubov shift (4.46) have to be neglected. Then one deals with the time-dependent Gross- Pitaevskii equation for the condensate wave function $\psi_0(\mathbf{r}, t)$:

$$\left[-\frac{\nabla^2}{2m} + U_{\text{trap}}(r) + g|\psi_0|^2\right]\psi_0(\mathbf{r}, t) = i\frac{\partial \psi_0(\mathbf{r}, t)}{\partial t}, \tag{5.3}$$

for which several numerical algorithms were developed (see e.g. [2]). Although Gross-Pitaevskii equation is valid only for zero temperature, it has been serving as a good platform to study condensed fractions and especially solitons [3] formed in BEC. Note also that, due to the spatially varying potential $U_{\text{trap}}(r)$, the critical temperature T_c even for an ideal gas in the magnetic trap is significantly altered. Particularly, well known relation $\rho_0 \sim T^{3/2}$ (see Eq. (3.9)) should be replaced by $\rho \sim T^3$ for the harmonic trap [4,5].

5.3   Optical Trapping

This kind of trapping uses a focused laser beam to trap and manipulate small particles. The technique is based on the fact that light exerts a small amount of force on an object, which can be used to confine the atom.

A neutral atom, for instance, of sodium, can be presented as a nuclear with a positive charge, surrounded by the cloud of an electron with a negative charge. Their centers coincide. However, if one places the atom into an electric field, then it acquires a dipole moment, which starts to interact with the external electric field. By choosing the intensity and frequency of the field it is possible to create a potential well, where the atom will be confined. In the dipole approximation this can be written as $V(\mathbf{r}, t) = -\mathbf{d} \cdot \mathbf{E}(\mathbf{r}, t)$ where $\mathbf{d}$ is the electric dipole moment operator for a single atom, and $\mathbf{E}(\mathbf{r}, t) = (\mathbf{E}(\mathbf{r}) \exp(-i\omega t) + c.c.)$ is a time-dependent electric field with the frequency ω. By its turn, the dipole also starts to oscillate as $d \sim E(\mathbf{r}) \exp(-i\omega t)$. As a result, due to the Stark effect, the following effective potential $U \sim - < E^2 >_t$ arises, which will keep the atom in the potential well.

Optical traps and tweezers can be used to build box traps, low-dimensional traps, atom guides, optical lattices [6], rotating traps etc. Among them optical box trap, proposed firstly in Ref. [7] has been successfully used to study properties of BEC of atomic gases [8].

The main advantages of optical trapping, compared with magnetic one are as follows:

- Trapping is not limited to a specific magnetic state and can be employed to investigate the coexistence of multi-spin components with the possible occurrence of new magnetic phases.
- It is rather convenient to tune the value of the scattering length by the Feshbach resonance technique.

- BEC in an optical box trap is homogeneous, while in a magnetic trap is not. Thanks to this property, the application of present mean field approximations directly to BEC in a box trap is straightforward. This is seen also in Fig. 3.3, where the the dots correspond to the experimental data, and the solid line to the formula for the ideal Bose gas given by (3.13).
- In optical trapping, the length of an optical box trap L is much larger than the healing length $\eta = 1/\sqrt{8\pi\rho a_s}$, i.e. $L \approx 250\eta$, so one may use safely the thermodynamic limit: $N \to \infty$, $V \to \infty$, $(N/V) \sim constant$. On the other side, in magnetic trapping one has to take into account finite size effects properly. The radii of the condensate in the harmonic trap can be defined as $r_i \approx \sqrt{2n_0 g/m(2\pi\omega_i)^2}$ where $\omega_i = (\omega_x, \omega_y, \omega_z)$ is the trapping frequency. Inhomogeneous distribution of the condensate is given by $\psi_0^2(x, y, z) = n_0[1 - \sum_i (x_i/r_i)^2]$, where n_0 denotes the peak of density. In spite of this, the excitation spectrum in magnetic trapping remains formally unchanged given by $E_k = \sqrt{\varepsilon_k(\varepsilon_k + 2g\rho)}$ in Bogolyubov approximation.

It should be also underlined that, at finite temperature, two-dimensional BEC cannot exist in optical box traps, while this is possible in the case of magnetic trapping. Now after such a little imaginary excursion to the experimental laboratory, we are on the stage of discussing some measurement procedures for observables.

5.4   Structure Factor

To study the structure of a crystal or a liquid an experimenter in condensed matter physics uses the same method widely used in atomic, nuclear or elementary particle physics: he irradiates the object with a beam of neutrons, electrons, photons etc. For example, to study superfluid helium it has been more convenient to use neutrons or X-rays. On the other hand, for BEC of atomic gases the two-photon Bragg spectroscopy is preferable. Below we discuss briefly both of these methods.

5.4.1   Neutron Scattering

Experiments on the scattering of "cold" neutrons (with the kinetic energy corresponding to the temperature $\sim$50 K) on superfluid helium have a long story, which started from famous experiments by Sweden group more than 65 years ago [9] who used neutron beam from the Stockholm reactor. As it is illustrated in Fig. 5.1 a neutron with initial energy ε_i, and momentum $\mathbf{k}_i$ is directed to the cryostat, filled with liquid helium, scatters on an atom acquiring the final energy ε_f, and momentum $\mathbf{k}_f$, which can be identified by the detector. The scattering angle Θ can be also measured by rotating the position of the detector. The initial energy can be varied by changing the thickness of the Beryllium filters (see Ref. [10] for the details). Then in Born approximation the double differential cross section is given by

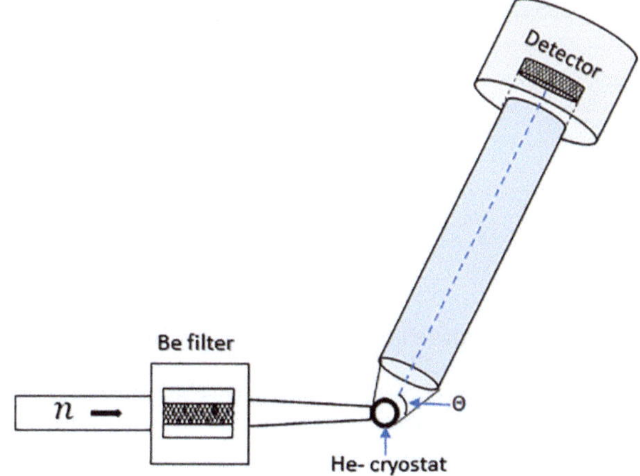

Fig. 5.1 Schematic diagram of apparatures to perform helium experiment [10]

$$\frac{1}{N}\frac{d^2\sigma}{d\Omega_q d\omega} = \frac{k_f}{k_i} b_s^2 S_{\mathrm{dyn}}(\mathbf{q}, \omega), \tag{5.4}$$

where b_s is the scattering length of the neutron on an atom of helium, Ω_q is a solid angle around $\mathbf{q} = \mathbf{k}_i - \mathbf{k}_f$ and $\omega = \varepsilon_i - \varepsilon_f$ is the energy transferred from the neutron to the whole system during the collision. As to the $S_{\mathrm{dyn}}(\mathbf{q}, \omega)$ it is the main quantity under interest, which is referred as the dynamic structure factor, characterizing the structure of the system-liquid ^4He.

Actually, since the cross-section can be measured easily and b_s is supposed to be known, the structure factor is also regarded as a measurable quantity, providing a direct connection between theory and experiment. For this reason, $S_{\mathrm{dyn}}(\mathbf{q}, \omega)$ serves as a nice tool to determine the excitation spectrum E_k of the system. In fact, similarly in accordance with Bohr's postulates, since an atom can change its energy only discretely, liquid helium can disperse incident neutrons inelastically, provided that the transferred energy w and momentum $\mathbf{q}$ corresponds to the excitation spectrum of the liquid, (say, $\omega \sim cq$ for small q). Otherwise, the cross-section will be negligibly small. In other words, $\omega(q)$ should be the average frequency of the dynamic structure factor $S_{\mathrm{dyn}}(\mathbf{q}, \omega)$, which gives the response of the excitation process. In this way, phonon or roton modes of the spectrum, displayed in Fig. 3.7, were observed.

The following explanation can be added to the above statements. Clearly the incident neutron energy ε_i and its momentum $\mathbf{k}_i$ are connected via $\varepsilon_i = \mathbf{k}_i^2/2m$. The same is true for the energy of the outgoing neutron: $\varepsilon_f = \mathbf{k}_f^2/2m$. On the contrary, for the transferred energy $\omega = \varepsilon_i - \varepsilon_f$ the relation $\omega = \mathbf{q}^2/2m$, in general, need not to be true. So, from the first glance ω and $\mathbf{q}$ are seemed to be independent variables. However, their mutual dependence is hidden "inside" the system in terms of the energy spectrum of collective ($T < T_c$) or single particle ($T > T_c$) excitations which we denote by E_k. Thus, in reality, a neutron scattering occurs only providing that $\omega_q \sim E_q$, that is $S(\mathbf{q}, \omega) \sim \delta(\omega_q - E_q)$.

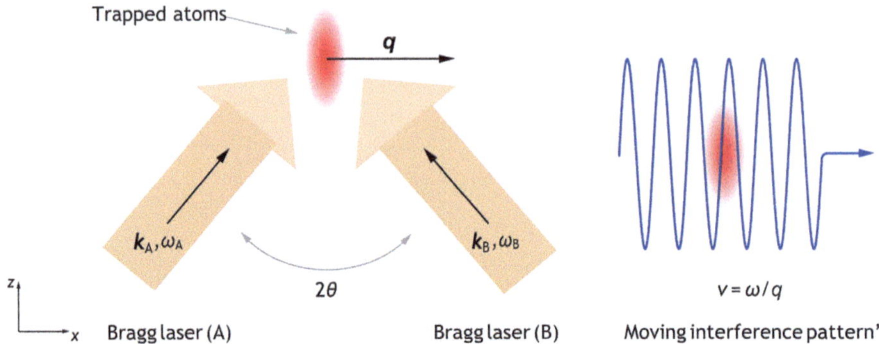

Fig. 5.2 Bragg spectroscopy involves two collimated laser beams, (**A**) and (**B**), with wave vector $\mathbf{q} = \mathbf{k}_A - \mathbf{k}_B$ and frequency detuning $\omega = \omega_A - \omega_B$, where $\omega \ll \omega_{A,B}$, to create a Bragg grating that moves with velocity $v = \omega/q$

5.4.2 Bragg Scattering

Cold and ultracold atoms are highly amenable to optical Bragg spectroscopy, where inelastic light scattering is used to induce weak perturbations to the momentum distribution of an atomic cloud [11,12]. In early two-photon Bragg scattering experiments on single-species Bose gases, phonon- and particle-like Bogolyubov excitations were studied by measuring density-density responses of atomic BECs [13,14].

Two-photon Bragg scattering is a Raman scattering process involving the stimulated absorption and emission of a photon between an atom and two optical fields thereby transferring momentum $\mathbf{k}$ and energy ω. For clarity, we present the probing scheme in Fig. 5.2.   Here Two collimated laser beams, intersecting at an angle of 2θ, illuminate a trapped atomic cloud. The difference in the wave vectors defines the momentum transfer, $\mathbf{q} = \mathbf{k}_A - \mathbf{k}_B$, while their frequency difference defines the amount of energy transferred, $\omega = \omega_A - \omega_B$. From previous chapters, we know that, the excitation spectrum E_k of BEC, or the superfluid helium is different in low and high momentum, as illustrated in Fig. 3.7. Two-photon Bragg scattering is a good tool to study both these cases: experiments with low momentum transfer reveals the collective excitations, while with the high momentum transfer reveals the single-particle excitations.

The probe particles are photons provided by the Bragg laser beams. The total momentum and the energy of the photons have to be conserved in a photon absorption-emission process, where an atom compensates for the momentum and energy transfer. By measuring the change in the center of mass or the energy of the atomic distribution after a short Bragg pulse, the spectral response becomes evident. This leads to the dynamic structure factor and the static structure factor, which is linked to the two-body correlation function (see below).

In the Bragg scattering the two beams form a moving periodic potential. Absorption of N photons from one laser beam and subsequent stimulated emission into the other imparts a momentum to a number of atoms where the order of the scattering process is N. The momentum transfer may be written as

$$q = 2Nk_L \sin(\theta), \tag{5.5}$$

where $|k_A| \approx |k_B| = k_L$ has been assumed. The energy transfer to an atom with initial momentum k_i is

$$\omega = \frac{\mathbf{q}^2}{2\,m} + \frac{\mathbf{k}_i \cdot \mathbf{q}}{m}. \tag{5.6}$$

These two equations fulfil energy and momentum conservation. A resonance will occur when the momentum of the moving periodic potential, $2q$, matches the momentum of the atom mv with the mass of an atom m. For a single atom, this simplifies to the usual Bragg resonance frequency, $\omega_R = 2q^2/m$. The second term in equation (5.6) is related to the Doppler shift of the resonance and shows that the Bragg scattering is a velocity-selective process. Energy and momentum can be decoupled if the angle between k_i and q is small.

It should be underlined that, unlike the textbook Bragg scattering (or diffraction) of a light in crystals, present Bragg scattering process is a two-photon process where fluctuations are caused by the time-varying electric field of the Bragg laser beams. The potential seen by the atoms oscillates with time and induces transitions of energy ω. The spatial period of the Bragg potential sets the momentum range being probed. In two-photon Bragg spectroscopy, two momentum states of the same internal atomic state are connected by a stimulated two-photon process: The stimulated process implies higher resolution and sensitivity in the measurement as the signal is background-free. Compared to radio-frequency spectroscopy or Raman spectroscopy, Bragg scattering does not involve a third atomic state or an intermediate metastable state and thus final state interaction effects are avoided.

Similarly to the neutron scattering, the Bragg scattering also provides access to the dynamic structure factor, $S(\mathbf{q}, \omega)$. The momentum and energy transferred to the cloud lifts the ground state, $|0\rangle$, into an excited state, $|n\rangle$, according to equation (5.6), and the particle gains a momentum $\mathbf{q}$. As mentioned above, this is a stimulated process. The many-body system is perturbed by the periodic potential, V_r, formed by the electric fields of the Bragg laser beams. The Fourier transform of the perturbing potential, $V_q = 2\Omega_R$, can be expressed in terms of the two-photon Rabi frequency, Ω_R, which expresses the coupling strength between the atom and light field. Implicitly, it is assumed that at $t \to -\infty$ the system is governed by the unperturbed Hamiltonian. The interaction between the atoms and the probe, $H_{\text{int}} = H_{Br}$, may be written as

$$H_{\text{int}} = \frac{V_{\mathbf{q}}}{2} \left(\delta\rho_{\mathbf{q}}^{\dagger} e^{-i\omega t} + \delta\rho_{\mathbf{q}} e^{i\omega t} \right), \tag{5.7}$$

where $\delta\rho_{\mathbf{q}}$ is the fluctuation of the particle density, generated by the external field. This interaction causes the transition $|0, \mathbf{p}\rangle \to |n, \mathbf{p} - \mathbf{q}\rangle$. The probability of such a transition $W(\mathbf{q}, \omega)$ per unit of time can be measured in a standard way. And then by using Fermi's golden rule

$$W(\mathbf{q}, \omega) = 2\pi |V_{\mathbf{q}}|^2 \, S(\mathbf{q}, \omega), \tag{5.8}$$

one can extract desired quantity-$S(\mathbf{q}, \omega)$ which contains information, particularly on the energy dispersion E_k.

5.4.3   Static Structure Factor

Moreover, the structure factor contains information on the correlation between atoms, critical temperature, as well as on the condensate fraction in the BEC phase even for dilute Bose gases. So for example, following algebraic equation for the condensate fraction at zero temperature was proposed in Ref. [15]

$$
n_0 = \frac{\rho_0}{\rho} = \frac{N_0}{N} = \exp\left\{ -\int \frac{d\mathbf{q}}{(2\pi)^3} \frac{[1 - S(\mathbf{q})]^\alpha}{4\rho S(\mathbf{q})} \right\}, \quad \alpha = \frac{2}{2 - n_0}. \tag{5.9}
$$

In (5.9) $S(\mathbf{q})$ is referred as a static structure factor, defined as follows. Suppose that an experimenter has measured $S_{\mathrm{dyn}}(\mathbf{q}, \omega)$ in several discrete points and obtained a set

$$
\{S_{\mathrm{dyn}}(\mathbf{q}_1, \omega_1), S_{\mathrm{dyn}}(\mathbf{q}_2, \omega_2) \ldots S_{\mathrm{dyn}}(\mathbf{q}_n, \omega_n)\}.
$$

Then he is able to determine the whole continuous $S_{\mathrm{dyn}}(\mathbf{q}, \omega)$ at any physically possible point, using an interpolation technique to exploit it as an analytical function of ω. In this way, he can verify some theoretical properties of the dynamic structure factor, such as sum rules [4], which include integration over ω. One of such integrals namely

$$
S(\mathbf{q}) \equiv \int_{-\infty}^{\infty} S_{\mathrm{dyn}}(\mathbf{q}, \omega) \frac{d\omega}{2\pi}, \tag{5.10}
$$

defines a static structure factor $S(\mathbf{q})$. Clearly, $S(\mathbf{q})$ is also an observable quantity.

In Fig. 5.3a, b we present an experimental static structure factor of ^4He at $T = 1.38K < T_c$ (a) and $T = 4.18K > T_c$ (b). It is seen that in the BEC phase it has maximum at $q \sim 2\text{\AA}^{-1}$, while in the normal phase $S(\mathbf{q})$ exhibits minimum near $q \sim 0.5\text{\AA}^{-1}$. The peak exceeding the unity $S(\mathbf{q}) > 1$ near $q \sim 2\text{\AA}^{-1}$ occurs for the excitations whose wavelength $2\pi/q$ is equal to the characteristic wavelength of density fluctuations in the ground state wave function of the quantum liquid. Superfluid helium exhibits such a peak, which can be find via the Feynman-Bijl relation at zero temperature $T = 0$ [17]

$$
S(\mathbf{q}) = \frac{\mathbf{q}^2}{2\, m\omega(\mathbf{q})}. \tag{5.11}
$$

It is remarkable that, having measured $S(\mathbf{q})$ in some region of the small momentum transfer and making extrapolation it to $q = 0$, one can determine the sound velocity s [18]

$$
s = \lim_{q \to 0} \left[\frac{T}{m\, S(\mathbf{q})} \right]^{1/2}, \tag{5.12}
$$

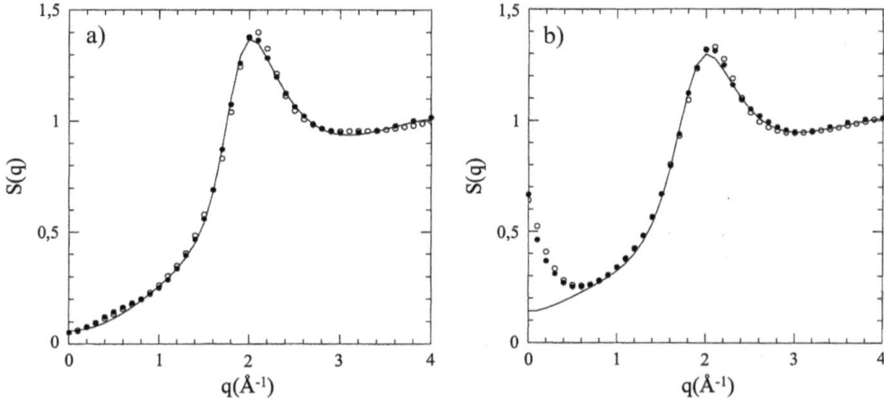

Fig. 5.3 a The structure factor of ^4He at $1.38K$ (**a**), $4.25K$ **b** temperatures and saturated vapor pressures. Reprinted with permission from [16]. © 1995, American Physical Society. All rights reserved

as well as the isothermal compressibility

$$\kappa_T \equiv -\frac{1}{V}\left(\frac{\partial V}{\partial P}\right)_{TN} = \frac{1}{\rho}\left(\frac{\partial \rho}{\partial P}\right)_{TN} = \lim_{q\to 0}\frac{S(\mathbf{q})}{\rho T}. \qquad (5.13)$$

Therefore, experimental data on the static structure factor provides with large amount of useful information on the system.

5.5   The Sound Velocity

In principle, the speed of the sound propagating inside BEC can be measured in different ways [19]: Via observation of static structure factor (5.12); by the spectrum of collective excitations $E_k \sim ks$; or by measuring the compressibility (5.13).

However, the first measurement was realized by the MIT group in 1997 [20], who obtained the speed of sound vs the density directly. The idea is the following. One may generate sound waves in BEC gases by focusing a laser pulse in the cigar-type trap. A wave packet can be formed in this way, and propagates outwards, which will be imaged at different times giving the value of the sound velocity.

The experimenters used sodium atoms in $|F = 1, m_F = -1\rangle$ ground state. Atoms have been optically cooled and trapped and transferred into a magnetic cigar trap where they were further cooled by radio frequency (RF)-induced evaporation. Thus condensates typically containing 5×10^6 atoms were produced every $30\,s$. The atom clouds were cigar-shaped with long axis horizontal. Then the laser beam was focused into the center of the trap and could be switched on and off in less than milliseconds. Localized increases in density were created by suddenly switching on the argon ion laser beam after the condensate had been formed. This repulsive optical dipole force expelled atoms from the center of the condensate, creating two density peaks which propagate symmetrically outward. Then images were taken by dividing a

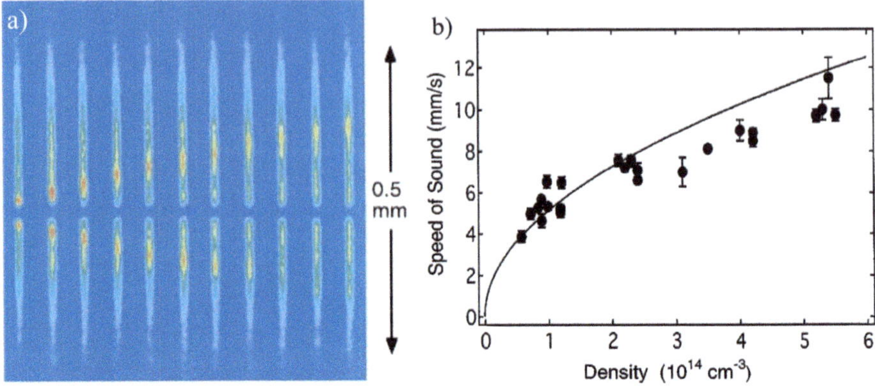

Fig. 5.4 a Observation of sound propagation in cigar-type BEC. Images were taken every $1.3ms$. **b** Speed of sound versus condensate peak density $n(0)$. The solid line corresponds to the Bogolyubov approximation, $ms^2 = gn(0)$. Reprinted with permission from [20]. © 1997, American Physical Society.

CCD camera chip into eleven strips (see Fig. 5.4a) every 1.3 ms, beginning 1 ms after switching on the argon ion laser. In such conditions, the speed of the travelling pulse (perturbation), can be measured and identified as the speed of the sound. Thus performing measurements with BECs with various densities the MIT group have obtained the speed of the sound versus the condensate peak density (see Fig. 5.4b) for the first time.

5.6   Pair Correlation Function

There is another measurable important quantity to study internal properties of a many-body system-pair correlation function $g(\mathbf{r}_1, \mathbf{r}_2)$. Physically, it is defined as follows:

$[g(\mathbf{r}_1 - \mathbf{r}_2) - 1]d\mathbf{r}_1 d\mathbf{r}2=$ *Probability of simultaneously finding a particle in the volume element $d\mathbf{r}_1$ about $\mathbf{r}_1$ and another particle in the volume element $d\mathbf{r}_2$ about $\mathbf{r}_2$.*

For a uniform system, it clearly depends on the distance between the two particles, $g(\mathbf{r}_1, \mathbf{r}_2) = g(|\mathbf{r}_1 - \mathbf{r}_2|) \equiv g(|\mathbf{r}|) = g(r)$ and, obviously decreases with increasing the distance r. Especially, for an ideal Bose gas, it vanishes very fast as $g(r) - 1 \sim \exp(-2\pi r^2/\lambda_T^2)$ [21]. In general, the function $g(r)$ varies between one and two: $1 \leq g(r) \leq 2$, reaching its maximum for a totally coherent system. Namely,

$$g(r) = \begin{cases} 1, & \text{chaos} \\ 2, & \text{coherence} . \end{cases} \tag{5.14}$$

How to measure the pair correlation function? Fortunately, $g(r)$ can be extracted from experimental data on static structure factor via following equation [4]

$$g(r) = 1 + \frac{1}{(2\pi)^3 \rho} \int d\mathbf{q}\, e^{-i(\mathbf{qr})}[S(\mathbf{q}) - 1] = 1 + \frac{1}{2\pi^2 \rho r} \int dq\, q \sin(qr)[S(\mathbf{q}) - 1], \quad (5.15)$$

where we used the integral

$$\int d\Omega_{\mathbf{q}} e^{-i(\mathbf{qr})} = \int_0^{2\pi} d\phi \int_{-1}^{1} d\cos\theta\, e^{iqr\cos\theta} = 4\pi j_0(qr) = \frac{4\pi \sin(qr)}{qr}. \quad (5.16)$$

Actually, the correlation function gives an opportunity to identify the critical temperature and even the condensate fraction.

5.6.1 T_c via $g(r)$

In their pioneering experimental investigation of the pair correlation function in liquid ^4He, Svensson et al. [22] obtained experimental data on the static structure factor, and hence for also $g(\mathbf{r})$ at temperatures $1K \leq T \leq 4.27K$ covering a wide range region of wave vectors $0.8\text{Å}^{-1} \leq q \leq 10.8\text{Å}^{-1}$. They have observed that, in the superfluid phase $(T < T_c)$ the correlation function $g(r)$ has a peak near $r \sim 3.5\text{Å}$ (see Fig. 5.5a). Further, analyzing the temperature dependence of this peak, they found that it becomes extremely large close to a temperature T_c which should be clearly identified with the transition temperature, namely, $T_c \sim 2.17K$ for the liquid ^4He (see Fig. 5.5b). It is understood that such a method can be used as an additional tool to fix the critical temperature of BEC phase transition in alkali Bose gases [23,24]. Meanwhile, the questions arise:*What about the condensate fraction? How the data e.g. in Fig. 3.3 were obtained?*

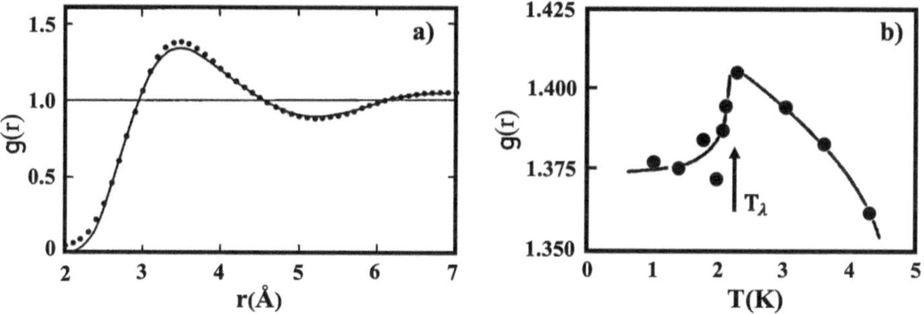

Fig. 5.5 a The correlation function for ^4He at $T = 1K$. **b** Comparison of the results for $g(r)$ at 1 K (circles) with the results of Monte Carlo calculations [24]. Reprinted with permission from [22].

5.7   Measurement of Condensed Fraction

Actually, the method based on the static structure factor (see Eq. (5.9)) is convenient for quantum liquids. On the other hand, there is also another method of measuring $n_0 = N_0/N$ in dilute Bose gases which does not involve neutron scattering. The method is based on analyzing a series of images of ultracold clouds of atoms. Such images, displaying the velocity distribution of rubidium atoms, were first taken by Anderson et al. [25] (see Fig. 5.6). Further, the JILA group [26] obtained $n_0(T)$ presented in Fig. 3.3 in the following way.

They optically trapped precooled ^{87}Rb atoms and loaded them into a purely magnetic trap. They further cooled atoms in the time-averaged optical potential (TOP) with forced evaporation. After the emergence of the atom cloud, typical to the image of BEC (middle image in Fig. 5.6), the confining TOP is **suddenly turned off** and as a result, the cloud started to expand freely. The velocity distributions of atoms in the cloud were obtained by probing the cloud with the pulse of light of proper frequency. These distributions contain a wealth of thermodynamic information. For instance, the integrated area under the distribution is proportional to the total number of atoms N in the sample. The condensate appears as a narrow feature centered on zero velocity [25]; the number of atoms in the ground state, N_0 is then proportional to the integrated area under this feature. From the mean square radius of the expanded cloud and the expansion time, they got the mean square velocity, or average energy, of the cloud. Finally, the temperature is extracted from the images, even though the temperature is not merely proportional to the mean energy in a degenerate Bose cloud.

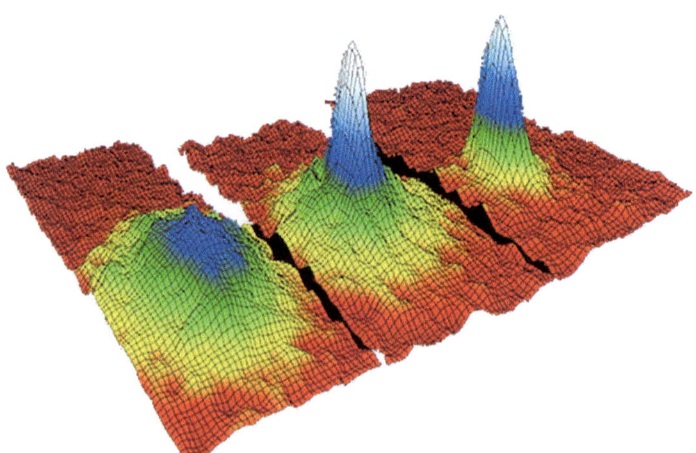

Fig. 5.6 Images of the velocity distribution of rubidium atoms in the experiment by Anderson et al. [25] taken by means of the expansion method. The left frame corresponds to a gas at a temperature just above condensation; the centre frame, just after the appearance of the condensate, the right frame, after further evaporation, leaves a sample of nearly pure condensate. Reprinted with permission from [27]. © 2016, C. Barenghi and N.G. Parker. All rights reserved

On the other hand, two-photon Bragg spectroscopy also has been successfully used to investigate BEC of dilute gases. The condensate fraction as well as quasiparticle energy inhomogeneous optically trapped atoms of ^{39}K have been measured in Refs [28,29].

5.8   Tan's Contact

There is one more measurable thermodynamic variable, C, introduced recently by Shina Tan and referred by its name as Tan's contact. Some years ago by using rigorous mathematical methods to study the system of fermions with contact interaction, he obtained [30–32] exact universal relations involving C. He proved that this quantity measures the density of pairs at short distances and determines the exact large momentum or high-frequency behavior of various physical observables. Further, Tan's ideas have been expanded for bosonic systems also [33]. It is remarkable that Tan's relations hold for any state of the system, few-body or many-body, homogeneous or inhomogeneous (magnetic trapped), superfluid or normal, zero or non-zero temperatures. The only basic requirement is that binary interaction between particles should be described by a single parameter, the s- wave scattering length a_s i.e. the particles are allowed to interact with each other only via contact interaction (4.1).[2] Then, considering a_s as a thermodynamic variable, one can define its thermodynamic conjugate, the so-called Tan's contact for bosons as follows:

$$C \equiv 8\pi ma_s^2 \left(\frac{\partial F}{\partial a_s}\right), \tag{5.17}$$

where the derivative of the free energy of the whole system F with respect to the scattering length is taken at constant particle number, volume and entropy. So, the variation of the free energy can be written as [4]

$$dF = -SdT - PdV + \mu dN - \frac{VC}{8\pi m}d\left(\frac{1}{a_s}\right). \tag{5.18}$$

In our units ($\hbar = 1$, $k_B = 1$) the contact has a dimensionality $(lenght)^{-4}$, e.g. cm^{-4}.

In general, C may be measured and evaluated by using its definition (5.17) as well as by any of the following relations or methods:

1. By the asymptotic behavior of the momentum distribution n_k:

$$C = \lim_{k\to\infty} k^4 n_k. \tag{5.19}$$

---

[2] For three-body interactions a similar quantity C_3 has been introduced by Braaten et al. [34].

2. By the operator product expansion of the expectation value of the interaction term

$$C = \frac{m^2 g^2}{V} \int d\mathbf{r} \langle \psi^\dagger(\mathbf{r}) \psi^\dagger(\mathbf{r}) \psi(\mathbf{r}) \psi(\mathbf{r}) \rangle, \tag{5.20}$$

where $g = 4\pi a_s/m$

3. By the rate of transferring atoms from one state to another state (see Eq. (5.22) below);

4. By the static structure factor through the relation

$$S(q) = 1 + \frac{C}{8\rho} \left(\frac{1}{q} - \frac{4}{\pi a_s q^2} \right). \tag{5.21}$$

Now we are on the state of answering the question about the physical meaning of Tan's contact. This can be understood from Eqs. (5.17)-(5.21). In fact, Tan's contact characterizes various thermodynamic quantities, like the dependence of the energy on the scattering length, the relation between the energy and the momentum distribution, the rate for the transferring atoms from one state to another state and the behavior of the static factor at large wave vectors.

For bosons it was measured firstly by the JILA group in 2012 [35] by using the third-radio frequency (RF) spectroscopy method. Below we outline the main scope of this experiment and further, in Chap. 9, we shall make an attempt to evaluate it. So, the experimenters used Bose condensed ^{85}Rb atoms, which were initially in the $|i\rangle = |F = 2, m_F = -2\rangle$ state, where F is the atomic spin and m_F is the spin projection. Then they imposed RF pulse with the frequency ω to drive the $|i\rangle = |F = 2, m_F = -2\rangle \rightarrow |f\rangle = |F = 2, m_F = -1\rangle$ transition, and measured its rate $\Gamma(\omega)$ by "counting" the number of atoms transferred to the final $|f\rangle$ state. Clearly, due to the difference in spin projections, the scattering lengths of atoms in both states need not to be the same, $a_s^i \neq a_s^f$. Now the contact may be determined by using Braaten's formula [34]

$$C = \frac{4\pi \omega^{3/2} \sqrt{m}(1 + \omega/E_f)}{\Omega^2 (a_s^f/a_s^i - 1)^2} \lim_{\omega \to \infty} \Gamma(\omega), \tag{5.22}$$

where $E_f = 1/m(a_s^f)^2$, and Ω is the Rabi frequency.   The results are displayed in Fig. 5.7, where following parameters were used: $\omega \approx 2\pi \times 40$ kHz, $\rho \approx 5.8 \times 10^{12}/cm^3$, $\sqrt{\gamma} = \sqrt{\rho(a_s^i)^3} \approx 0.043$, $E_f \approx 133$kHz, $T \sim 0$K. It is seen that C slowly increases with increasing scattering length. Here it should be underlined that, unfortunately, the temperature dependence of Tan's contact for bosons, $C(T)$, has not been experimentally studied yet.

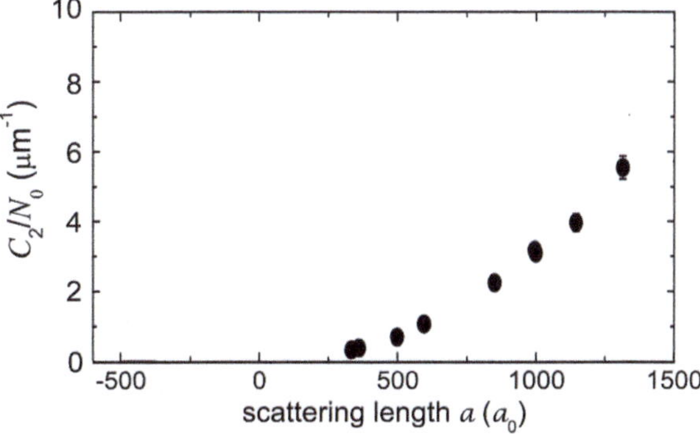

Fig. 5.7 The contact versus the s-wave scattering length for ^{87}Rb. Experimental points are taken from Ref. [35]. (a_0 is the Bohr radius). Reprinted with permission from [35]. ©2012, American Physical Society. All rights reserved

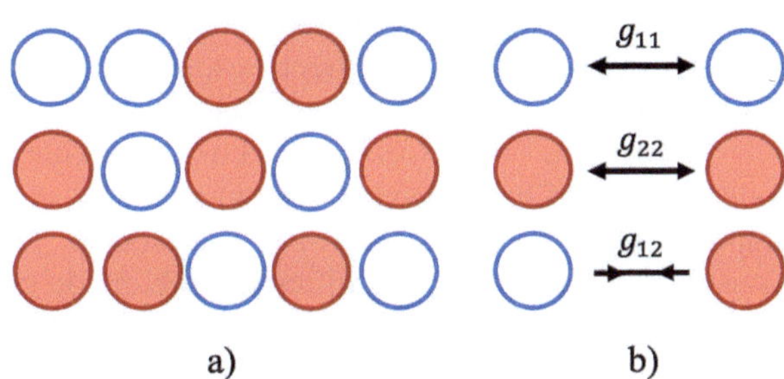

Fig. 5.8 A two-component BECs. g_{ij} are coupling constants of the contact interactions

5.9   Quantum Liquid Droplets

In Chap. 2 we have made a little introduction to quantum liquid droplets. Below we bring a sketch of their experimental observation in a two-component mixture of Bose condensates.

Actually, experimental observation of the droplets is based on the following idea, developed by Petrov [36]. Let's consider a system composing of two different Bosonic atoms, interacting via contact potential (2.2). Naturally, at temperatures close to zero both components undergo Bose-Einstein condensation, so the whole system can be considered as a two-component BEC. Let intracomponent interactions are repulsive, i.e. $g_{11} > 0$ and $g_{22} > 0$, while the intercomponent interaction is attractive ($g_{12} < 0$) as illustrated in Fig. 5.8.

It is expected that, each condensate is stable, but the whole system may collapse due to attractive intercomponent interaction $V_{12} = -|g_{12}|\delta(\mathbf{r}_1 - \mathbf{r}_2)$. Petrov predicted that, by altering coupling constants, (or equivalently the s-wave scattering lengths) one can reach a balance between these constructive and destructive forces. In the balance, an atom feels itself as if it is localized in an effective potential, similar to the van der Waals one, which creates a nice condition for the formation of a quantum droplet.[3]

Soon experimenters have confirmed this idea in reality [37–41] They have demonstrated the self-bound character of mixture droplets and directly measured their ultralow densities and micrometer-scaled sizes. Moreover, through comparison with a single-component condensate with only contact interactions, they confirmed that their stability stems from quantum fluctuations.

The experiments have been performed with two ^{39}K Bose-Einstein condensates in states $|1\rangle \equiv |F = 1, m_F = -1\rangle$ and $|2\rangle \equiv |F = 1, m_F = 0\rangle$, where F is the total angular momentum and m_F is its projection. The scattering lengths a_{11}, a_{22}, and a_{12} were controlled and manipulated by an external magnetic field B (see Fig. 1 of Ref. [37]). To characterize various regimes of the system following widely used parameter $\delta a = a_{12} + \sqrt{a_{11}a_{22}}$ has been introduced, such that the condition $\delta a = 0$ separates the repulsive ($\delta a > 0$) and attractive ($\delta a < 0$) regimes. The spin composition of the mixture is verified via Stern-Gerlach separation during time-of-flight expansion. Starting with $\delta a \approx 7a_0$ (a_0-Bohr radius), they slowly ramped down the magnetic field and passed to the attractive regime with $\delta a < 0$. Then they switched off the trap, allowing the atoms to evolve freely out of the trap. The integrated atomic density was imaged in situ at different evolving times.

Typical images of the mixture time evolution in the repulsive and attractive regimes have displayed that, e.g. for $\delta a \approx 1.2a_0 > 0$ the cloud expands progressively in the plane. In contrast, in the attractive regime, $\delta a \approx -3.2a_0 < 0$, the dynamics of the system are remarkably different, and atoms reorganize in an isotropic self-bound liquid droplet. For completeness and to make sure in reliability, they also made a single-component attractive condensate and observed the collapse of such a BEC.

Besides, by performing quantitative analysis of the images fitting the integrated atomic density profiles, the experimenters were able to extract the number of atoms N as well as radial size σ_r of droplets. It has been shown that a critical minimal number of atoms N_c is essential for droplet formation, as it was predicted by Petrov [36]. This critical number is about $N_c \sim 5000$ at moderate values of the external magnetic field ($B \sim 56G$). They also observed the fate of a droplet: It decays in several milliseconds, exhibiting a liquid-to-gas transition. A more detailed description of the experiments can be found, e.g. in the PhD thesis by Cabrera [42].

References

1. W.D. Phillips, Nobel lecture: laser cooling and trapping of neutral atoms. Rev. Mod. Phys. **70**, 721–741 (1998). https://doi.org/10.1103/RevModPhys.70.721
2. D. Vudragović, I. Vidanović, A. Balaž, P. Muruganandam, S.K. Adhikari, C programs for solving the time-dependent Gross–Pitaevskii equation in a fully anisotropic trap. Comput.

---

[3] This picture will be explained quantitatively in Chaps. 11–12.

Phys. Commun. **183**(9), 2021–2025 (2012). https://www.sciencedirect.com/science/article/pii/S0010465512001270

3. F. Abdullaev, J. Engelbrecht, S. Darmanyan, P. Khabibullaev, *Optical Solitons*, Series Springer Series in Nonlinear Dynamics (Springer, Berlin, Heidelberg, 2014). https://books.google.co.uz/books?id=wFXGoAEACAAJ

4. L. Pitaevskii, S. Stringari, *Bose-Einstein Condensation*, Series International Series of Monographs on Physics (Clarendon Press, 2003). https://books.google.co.uz/books?id=rIobbOxC4j4C

5. P.W. Courteille, V.S. Bagnato, V. Yukalov, Bose-Einstein condensation of trapped atomic gases. Laser Phys. **11**(6), 659–800 (2001)

6. M. Lewenstein, A. Sanpera, V. Ahufinger, *Ultracold Atoms in Optical Lattices: Simulating Quantum Many-Body Physics* (Oxford University Press, Oxford, UK, 2017). https://global.oup.com/academic/product/ultracold-atoms-in-optical-lattices-9780198785804

7. T.P. Meyrath, F. Schreck, J.L. Hanssen, C.-S. Chuu, M.G. Raizen, Bose-Einstein condensate in a box. Phys. Rev. A **71**, 041604 (2005). https://doi.org/10.1103/PhysRevA.71.041604

8. N. Navon, R.P. Smith, Z. Hadzibabic, Quantum gases in optical boxes. Nat. Phys. **17**(12), 1334–1341 (2021). https://doi.org/10.1038/s41567-021-01403-z

9. H. Palevsky, K. Otnes, K.E. Larsson, R. Pauli, R. Stedman, Excitation of rotons in helium ii by cold neutrons. Phys. Rev. **108**, 1346–1347 (1957). https://doi.org/10.1103/PhysRev.108.1346

10. H. Palevsky, K. Otnes, K.E. Larsson, Excitation of rotons in helium ii by cold neutrons. Phys. Rev. **112**, 11–18 (1958). https://doi.org/10.1103/PhysRev.112.11

11. P.J. Martin, B.G. Oldaker, A.H. Miklich, D.E. Pritchard, Bragg scattering of atoms from a standing light wave. Phys. Rev. Lett. **60**, 515–518 (1988). https://doi.org/10.1103/PhysRevLett.60.515

12. M. Kozuma, L. Deng, E. W. Hagley, J. Wen, R. Lutwak, K. Helmerson, S.L. Rolston, W.D. Phillips, Coherent splitting of Bose-Einstein condensed atoms with optically induced bragg diffraction. Phys. Rev. Lett. **82**, 871–875 (1999). https://doi.org/10.1103/PhysRevLett.82.871

13. D.M. Stamper-Kurn, A.P. Chikkatur, A. Görlitz, S. Inouye, S. Gupta, D.E. Pritchard, W. Ketterle, Excitation of phonons in a Bose-Einstein condensate by light scattering. Phys. Rev. Lett. **83**, 2876–2879 (1999). https://doi.org/10.1103/PhysRevLett.83.2876

14. J. Stenger, S. Inouye, A.P. Chikkatur, D.M. Stamper-Kurn, D.E. Pritchard, W. Ketterle, Bragg spectroscopy of a Bose-Einstein condensate. Phys. Rev. Lett. **82**, 4569–4573 (1999). https://doi.org/10.1103/PhysRevLett.82.4569

15. Y.E. Lozovik, I.L. Kurbakov, G.E. Astrakharchik, J. Boronat, Estimation of the condensate fraction from the static structure factor. Phys. Rev. B **103**, 094511 (2021). https://doi.org/10.1103/PhysRevB.103.094511

16. D.M. Ceperley, Path integrals in the theory of condensed helium. Rev. Mod. Phys. **67**, 279–355 (1995). [Online]. https://doi.org/10.1103/RevModPhys.67.279

17. J. Steinhauer, R. Ozeri, N. Katz, N. Davidson, Excitation spectrum of a Bose-Einstein condensate. Phys. Rev. Lett. **88**, 120407 (2002). https://doi.org/10.1103/PhysRevLett.88.120407

18. V. Yukalov, Structure factor of Bose-condensed systems. J. Phys. Stud. **11**, 55–62 (2007)

19. A.R. Fritsch, P.E.S. Tavares, F.A.J. Vivanco, G.D. Telles, V.S. Bagnato, E.A.L. Henn, Thermodynamic measurement of the sound velocity of a Bose gas across the transition to Bose–Einstein condensation. J. Stat. Mech.: Theory Exp. **2018**(5), 053108 (2018). https://doi.org/10.1088/1742-5468/aabbcf

20. M.R. Andrews, D.M. Kurn, H.-J. Miesner, D.S. Durfee, C.G. Townsend, S. Inouye, W. Ketterle, Propagation of sound in a Bose-Einstein condensate. Phys. Rev. Lett. **79**, 553–556 (1997). https://doi.org/10.1103/PhysRevLett.79.553

21. K. Huang, *Introduction to Statistical Physics*. Series Civil and Environmental Engineering (CRC Press, 2009). https://books.google.co.uz/books?id=rKmC3bMEVxIC

22. E.C. Svensson, V.F. Sears, A.D.B. Woods, P. Martel, Neutron-diffraction study of the static structure factor and pair correlations in liquid ^{4}He. Phys. Rev. B **21**, 3638–3651 (1980). https://doi.org/10.1103/PhysRevB.21.3638

23. J. Steinhauer, R. Ozeri, N. Katz, N. Davidson, Peak in the static structure factor of a Bose-Einstein condensate. Phys. Rev. A **72**, 023608 (2005). https://doi.org/10.1103/PhysRevA.72.023608
24. P.A. Whitlock, D.M. Ceperley, G.V. Chester, M.H. Kalos, Properties of liquid and solid ^{4}He. Phys. Rev. B **19**, 5598–5633 (1979). https://doi.org/10.1103/PhysRevB.19.5598
25. M.H. Anderson, J.R. Ensher, M.R. Matthews, C.E. Wieman, E.A. Cornell, Observation of Bose-Einstein condensation in a dilute atomic vapor. Science **269**(5221), 198–201 (1995). https://doi.org/10.1126/science.269.5221.198
26. J.R. Ensher, D.S. Jin, M.R. Matthews, C.E. Wieman, E.A. Cornell, Bose-Einstein condensation in a dilute gas: Measurement of energy and ground-state occupation. Phys. Rev. Lett. **77**, 4984–4987 (1996). https://doi.org/10.1103/PhysRevLett.77.4984
27. C. Barenghi, N.G. Parker, *A Primer on Quantum Fluids*. Series Springer Briefs in Physics (Springer Cham, 2016). https://doi.org/10.1007/978-3-319-42476-7
28. R. Lopes, C. Eigen, N. Navon, D. Clément, R.P. Smith, Z. Hadzibabic, Quantum depletion of a homogeneous Bose-Einstein condensate. Phys. Rev. Lett. **119**, 190404 (2017). https://doi.org/10.1103/PhysRevLett.119.190404
29. R. Lopes, C. Eigen, A. Barker, K.G.H. Viebahn, M. Robert-de Saint-Vincent, N. Navon, Z. Hadzibabic, R.P. Smith, Quasiparticle energy in a strongly interacting homogeneous Bose-Einstein condensate. Phys. Rev. Lett. **118**, 210401 (2017). https://doi.org/10.1103/PhysRevLett.118.210401
30. S. Tan, Energetics of a strongly correlated Fermi gas. Ann. Phys. **323**(12), 2952–2970 (2008). https://www.sciencedirect.com/science/article/pii/S0003491608000456
31. S. Tan, Large momentum part of a strongly correlated Fermi gas. Ann. Phys. 323(12), 2971–2986 (2008). https://www.sciencedirect.com/science/article/pii/S0003491608000432
32. S. Tan, Generalized virial theorem and pressure relation for a strongly correlated Fermi gas. Ann. Phys. **323**(12), 2987–2990 (2008). https://www.sciencedirect.com/science/article/pii/S0003491608000420
33. R. Combescot, F. Alzetto, X. Leyronas, Particle distribution tail and related energy formula. Phys. Rev. A **79**, 053640 (2009). https://doi.org/10.1103/PhysRevA.79.053640
34. E. Braaten, D. Kang, L. Platter, Universal relations for identical bosons from three-body physics. Phys. Rev. Lett. **106**, 153005 (2011). https://doi.org/10.1103/PhysRevLett.106.153005
35. R.J. Wild, P. Makotyn, J.M. Pino, E.A. Cornell, D.S. Jin, Measurements of Tan's contact in an atomic Bose-Einstein condensate. Phys. Rev. Lett. **108**, 145305 (2012). https://doi.org/10.1103/PhysRevLett.108.145305
36. D.S. Petrov, Quantum mechanical stabilization of a collapsing Bose-Bose mixture. Phys. Rev. Lett. **115**, 155302 (2015). https://doi.org/10.1103/PhysRevLett.115.155302
37. C.R. Cabrera, L. Tanzi, J. Sanz, B. Naylor, P. Thomas, P. Cheiney, L. Tarruell, Quantum liquid droplets in a mixture of Bose-Einstein condensates. Science **359**(6373), 301–304 (2018). https://doi.org/10.1126/science.aao5686
38. G. Semeghini, G. Ferioli, L. Masi, C. Mazzinghi, L. Wolswijk, F. Minardi, M. Modugno, G. Modugno, M. Inguscio, M. Fattori, Self-bound quantum droplets of atomic mixtures in free space. Phys. Rev. Lett. **120**, 235301 (2018). https://doi.org/10.1103/PhysRevLett.120.235301
39. C. D'Errico, A. Burchianti, M. Prevedelli, L. Salasnich, F. Ancilotto, M. Modugno, F. Minardi, C. Fort, Observation of quantum droplets in a heteronuclear bosonic mixture. Phys. Rev. Res. vol. 1, p. 033155 (2019). https://doi.org/10.1103/PhysRevResearch.1.033155
40. A. Burchianti, C. D'Errico, M. Prevedelli, L. Salasnich, F. Ancilotto, M. Modugno, F. Minardi, C. Fort, A dual-species Bose-Einstein condensate with attractive interspecies interactions. Condensed Matt. **5**(1) (2020). https://www.mdpi.com/2410-3896/5/1/21
41. Z. Guo, F. Jia, L. Li, Y. Ma, J.M. Hutson, X. Cui, D. Wang, Lee-Huang-Yang effects in the ultracold mixture of ^{23}Na and ^{87}Rb with attractive interspecies interactions. Phys. Rev. Res. **3**, 033247 (2021). https://doi.org/10.1103/PhysRevResearch.3.033247
42. C.R.C. Cabrera, *PhD Thesis: Quantum liquid droplets in a mixture of Bose-Einstein condensates* (Polytechnic University of Catalonia. Institute of Photonic Sciences, Catalonia, 2018)

Path Integrals for Bosons 6

The intersection of condensed matter and high-energy physics, for which the path integral method was originally invented, is based on the fundamental formula

$$\Omega = -T \ln Z = -T \ln \sum_n e^{-\beta E_n} \tag{6.1}$$

where Z is the partition function - statistical sum,

$$Z = \sum_n \langle n | \exp(-\beta H) | n \rangle = \mathrm{Tr}\, \exp(-\beta H) \tag{6.2}$$

and E_n are the eigenvalues of the Hamiltonian H. It is well known that the grand canonical thermodynamic potential $\Omega(T, V, \mu)$ contains all the information on the equilibrium system confined in the volume V with the chemical potential μ [1,2]. The primary object in the evaluation of Z is the effective action S, which is supposed to be known analytically. So, we start from a brief review of its properties.

6.1 The Principle of Least Action and Gross-Pitaevskii Equation

First, we consider the case of classical and quantum mechanics for a single particle and then by means of second quantization pass to quantum field theory.

A. Rakhimov and S. Mardonov, *Theory of Quantum Bose Liquids Beyond Bogoliubov Approximation*, Lecture Notes in Physics 1045, https://doi.org/10.1007/978-3-032-05096-0_6

6.1.1   Classical Mechanics

Lets assume that the Hamiltonian or Lagrangian is given as a function of momentum $\mathbf{k}$ and position $\mathbf{r}$. Anyway, by using the Legendre transformation:

$$\mathcal{L}(\mathbf{r}, \dot{\mathbf{r}}, t) = \mathbf{k}\dot{\mathbf{r}} - H(\mathbf{r}, \mathbf{k}) \tag{6.3}$$

we define the action as

$$S = \int dt \mathcal{L}(\mathbf{r}, \dot{\mathbf{r}}, t) \tag{6.4}$$

where $\dot{\mathbf{r}} = d\mathbf{r}/dt$. In this sense, strictly speaking, S should be referred not to the function but the functional of the variable $\mathbf{r}(t)$ as $S = S[\mathbf{r}(t)]$. (The reader unfamiliar with the functionals and functional derivatives is referred to the Appendix C). Let's determine the classical trajectory $r_{cl}(t)$ where the end points remain fixed as (r_i, t_i) and (r_f, t_f). The principle of least action states that classical trajectory $r_{cl}(t)$ minimizes the functional:

$$\left.\frac{\delta S[\mathbf{r}]}{\delta \mathbf{r}(t)}\right|_{r=r_{cl}} = 0. \tag{6.5}$$

So, by using the rules given in the Appendix C we find that

$$\frac{\delta S[\mathbf{r}]}{\delta \mathbf{r}(t)} = \int_{t_i}^{t_f} dt' \left[\frac{\partial \mathcal{L}}{\partial \mathbf{r}}\delta(t - t') + \frac{d}{dt}\left(\frac{\partial \mathcal{L}}{\partial \dot{\mathbf{r}}}\right)\delta(t - t') \right] = 0 \tag{6.6}$$

Performing the integration, we arrive at

$$\frac{\partial \mathcal{L}}{\partial \mathbf{r}} = \frac{d}{dt}\frac{\partial \mathcal{L}}{\partial \dot{\mathbf{r}}}, \tag{6.7}$$

which can be solved using additional conditions that the end points are fixed as $r_i(t_i) \equiv r_i$ and $r_f(t_f) \equiv r_f$. The above equation is known as the Euler-Lagrange equation. Particularly, for a particle with mass m and moving in the potential $U(\mathbf{r}, t)$ with the Lagrangian given by

$$\mathcal{L} = \frac{m\dot{\mathbf{r}}^2}{2} - U(\mathbf{r}, t), \tag{6.8}$$

we find Newton's familiar equation

$$m\frac{d^2\mathbf{r}}{dt^2} = -\frac{\partial U(\mathbf{r}, t)}{\partial \mathbf{r}}. \tag{6.9}$$

6.1.2   Classical Statistical Mechanics

In classical statistical mechanics, where we want to determine the canonical partition function $Z \sim \int d\mathbf{r} d\mathbf{k} \exp(-\beta H(\mathbf{r}, \mathbf{k}))/(2\pi)^3$, the action will be the functional of variables $\mathbf{k}(t)$ and $\mathbf{r}(t)$:

$$S[\mathbf{k}(t), \mathbf{r}(t)] = \int_{t_i}^{t_f} dt \, [\mathbf{k}(t)\dot{\mathbf{r}}(t) - H(\mathbf{k}(t), \mathbf{r}(t), t)] , \qquad (6.10)$$

The variation of the action has to vanish for the classical path $r_{cl}(t)$ and $k_{cl}(t)$, leading to the canonical Hamilton equations by demanding that, the functional derivatives with respect to position and momentum are zero, giving

$$\frac{d\mathbf{k}(t)}{dt} = -\frac{\partial H}{\partial \mathbf{r}}, \qquad \frac{d\mathbf{r}(t)}{dt} = \frac{\partial H}{\partial \mathbf{k}}, \qquad (6.11)$$

which serve as the basic equations in nonequilibrium thermodynamics. Particularly, in the case of classical mechanics with the Lagrangian (6.8), corresponding to the Hamiltonian

$$H = \frac{\mathbf{k}^2}{2m} + U(r, t), \qquad (6.12)$$

the canonical equations have the form:

$$m\frac{d\mathbf{r}(t)}{dt} = \mathbf{k}(t), \qquad \frac{d\mathbf{k}(t)}{dt} = -\frac{\partial U}{\partial r}, \qquad (6.13)$$

which are familiar from textbooks [3].

6.1.3   Quantum Statistical Mechanics

The ideas of classical physics, outlined above, can be easily transferred to problems of many- particle systems by means of second quantization method in terms of field operators $\hat{\psi}^{\dagger}$ and $\hat{\psi}$ outlined in Chap. 4. This may be realized by following transformation [4]

$$r \to \hat{\psi}, \qquad k \to i\hat{\psi}^{\dagger}(r), \qquad (6.14)$$

where $(\mathbf{r}, \mathbf{k})$ are considered as generalized coordinates. Thus, applying this procedure to the classical action (6.10) and passing to the Euclidean space we derive the field theoretical effective action

$$S[\psi, \psi^{\dagger}] = \int_0^{\beta} d\tau \int d\mathbf{r} \left[\psi^{\dagger}(\mathbf{r}, \tau)\partial_{\tau}\psi(\mathbf{r}, \tau) + \frac{(\nabla \psi^{\dagger})(\nabla \psi)}{2m} - \mu\psi^{\dagger}(\mathbf{r}, \tau)\psi(\mathbf{r}, \tau) \right]$$

$$+ \frac{1}{2}\int_0^{\beta} d\tau \int d\mathbf{r} \int d\mathbf{r}' \left[\psi^{\dagger}(\mathbf{r}, \tau)\psi^{\dagger}(\mathbf{r}', \tau)U(\mathbf{r} - \mathbf{r}')\psi(\mathbf{r}', \tau)\psi(\mathbf{r}, \tau) \right],$$

$$(6.15)$$

which was partly exploited in Chap. 4. One may understood that, in Eq. (6.15), the first term is originated, actually, from the Legendre transformation (6.3), the second and fourth terms from the Eq. (6.12), while the third one with the chemical potential has been added by hand in the spirit of the grand canonical ensemble.

6.1.4 Gross-Pitaevskii Equation

Going further, performing the transformation (6.14) in Eq. (6.11) leads to

$$i\frac{\partial \psi(\mathbf{r}, t)}{\partial t} = \frac{\delta H[\psi^{\dagger}(\mathbf{r}, t), \psi(\mathbf{r}, t)]}{\delta \psi^{\dagger}(\mathbf{r}, t)}, \tag{6.16}$$

where, this time, the Hamiltonian given previously in Eq. (4.18), should be considered as a functional. Now using the following formulas (see Appendix C)

$$\frac{\delta \int dx f(x)}{\delta f(y)} = \delta(x - y), \qquad \int d\mathbf{r}(\nabla \psi^{\dagger})(\nabla \psi) = -\int d\mathbf{r} \psi^{\dagger}(\nabla^2 \psi), \tag{6.17}$$

we obtain nonlinear Schrodinger equation [5] :

$$\begin{aligned}
i\frac{\partial \psi(\mathbf{r}, t)}{\partial t} &= -\frac{\nabla^2 \psi(\mathbf{r}, t)}{2m} + U_{ext}(\mathbf{r}, t)\psi(\mathbf{r}, t) \\
&+ \int d\mathbf{r}' \psi^{\dagger}(\mathbf{r}', t)U(\mathbf{r} - \mathbf{r}')\psi(\mathbf{r}', t)\psi(\mathbf{r}, t),
\end{aligned} \tag{6.18}$$

which is good even for the finite range potential. Particularly, for the zero-range potential, $U(\mathbf{r} - \mathbf{r}') = g\delta(\mathbf{r} - \mathbf{r}')$ this gives a widely used version of the Gross-Pitaevskii equation:

$$i\frac{\partial \psi(\mathbf{r}, t)}{\partial t} = \left[-\frac{\nabla^2}{2m} + U_{ext}(\mathbf{r}, t) + g|\psi(\mathbf{r}, t)|^2 \right] \psi(\mathbf{r}, t). \tag{6.19}$$

Note that, in most of the works this equation has been exploited by the following assumptions:

(1) $T = 0$, i.e. the temperature depletion (thermal fluctuations) is neglected;
(2) $\psi(\mathbf{r}, t) \approx \sqrt{\rho_0(\mathbf{r}, t)}$, i.e. the quantum fluctuations are also neglected. However, due to the trapping potential $U_{ext}(\mathbf{r}, t)$, the condensate fraction is not constant, $\rho_0 = \rho_0(\mathbf{r}, t) \neq const$, in contrast to the uniform case.
  Nevertheless, the Gross-Pitaevskii equation has been successfully used as a good tool to study the properties of a trapped BEC, skyrmions, solitons in one dimension [6], Josephson effects, etc. [7].

6.2   Path Integrals in Quantum Mechanics: Physical Interpretation

Before going into the complicated mathematics of path integrals, we would like to explain their physical origin in the spirit of the Copenhagen interpretation [8].

Let's concentrate on the double slit experiment, illustrated in Fig. 6.1. First, imagine that the slits are enough wide and we are shooting with ordinary bullets from the gun G. If you aim to the slit A then the bullet will hit point C, lying on a straight line GAC, which we call as a classical trajectory. Now, replace the gun with an electronic cannon and "shoot" with a single electron. You may observe that the electron also hits point C like a bullet (particle). Now we diminish the experimental equipment many times until its average size is comparable with the Broglie wavelength $\lambda \sim h/p$ of the electron, and repeat shooting with a single electron over a long period, aiming at the slit A. This time you will see that some electrons hit the points on the screen lying out of the classical trajectory. In other words, the observed distribution of electrons at all points of the screen becomes the interference pattern of a wave. Here a philosophical question arises: "Can a single electron interfere with itself?". Our answer is negative, since to observe such a pattern one has to take single shots many times or to use a whole stream of electrons. The point is that, in the microcosm, the concept of trajectory loses its meaning. Or the same thing, as, an electron can move along any trajectory, connecting two points G and P. In particular, when both slits, A and B are open, some part of the stream of electrons come to the same point P via the slit A and some part through B to make an interference pattern near the point P.

Next, we consider the case, when the number of slits is large (see Fig. 6.2a). Now there are many possibilities for the electron, starting from the point G, to reach the point P. After its detection at point P, nobody knows which trajectory it has chosen to proceed from G to P! What will happen if we "drill" more and more slits? Well, by increasing the number of screening layers and slits more and more, and finally reaching an infinite number, each screening layer will have a lot of slits and finally disappears. The interval between the gun and the screen will become a continuous

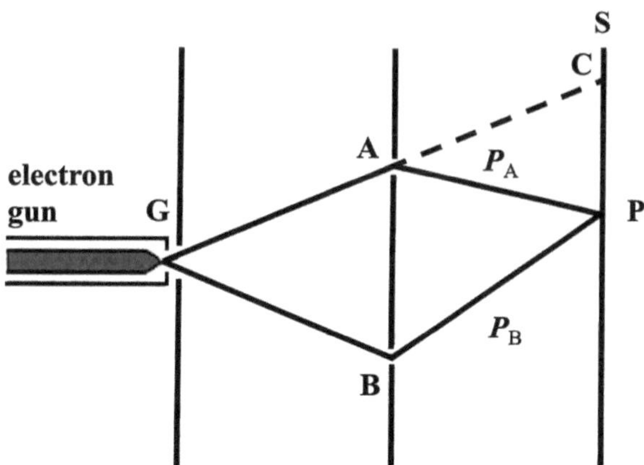

Fig. 6.1 Interference experiment with electron waves. The waves that passed through slits A and B interfere on the screen S

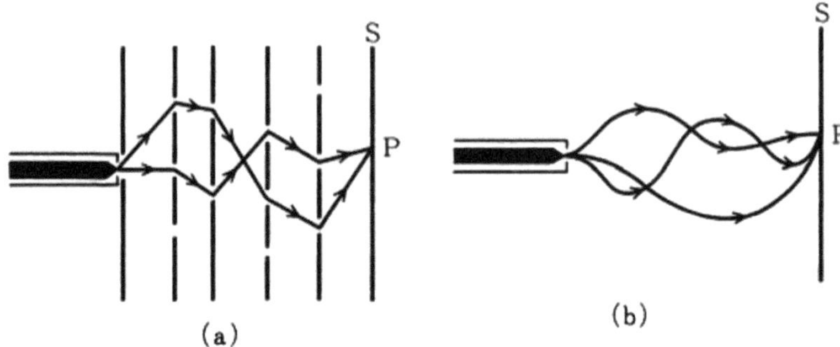

Fig. 6.2 Generalization of the interference experiment of Fig. 6.1. The number of screens and the number of slits in the screen is increased

space (Fig. 6.2b). The path will mutate to an arbitrary path from G to P, depending on the "combination of slits", and the sum of all amplitudes of possible ways will mutate to an integral over the paths, that is the path integral. That is to say, the principle that "the amplitude corresponding to a transition from a starting point to an end point, corresponding to an integral over the amplitudes of all possible paths linking these two points" can be regarded as the principle of quantum mechanics in Feynman's formulation. Thus, the path integral formulation replaces the classical notion of a single, unique trajectory with a sum over all possible trajectories. The trajectories are added together by using functional integration. Certainly, calculations made in this formulation give the same numerical results as they were obtained by using Schrödinger equations. The point is in the convenience: Schrödinger formalism is indispensable when one deals with tasks on boundary states.

6.3   Path Integrals in Quantum Mechanics: Mathematics

We now turn to mathematics to establish a connection between the partition function Z and the action S. For this purpose, we start from the Schrödinger equation

$$i\frac{\partial \psi(x,t)}{\partial t} = -\frac{1}{2m}\frac{d^2\psi(x,t)}{dx^2} + U(x)\psi(x,t) = H\psi(x,t). \tag{6.20}$$

For a moment, we restrict the motion of the particle to a one-dimensional space with space coordinate x. The Eq. (6.20) can be solved formally as:

$$|\psi(t')\rangle = W(t',t)|\psi(t)\rangle = e^{-iH(t'-t)}|\psi(t)\rangle, \qquad t' > t, \tag{6.21}$$

which describes the time evaluation of the state $|\psi(t)\rangle$ occurring at t through $W(t',t)$ during the time $t'-t$, leading to the state $|\psi(t')\rangle$. Here we omit the label x, intentionally, because $|\psi(t)\rangle$ should be regarded as a continuous infinite dimensional vector with components $\psi(x,t)$ labeled by x, and $W(t',t)$ should be regarded as a matrix with components $W(x',t',x,t)$.

In the coordinate representation, we obtain

$$\psi(x', t') = \int dx dt\, W(x', t', x, t)\psi(x, t). \tag{6.22}$$

Now making an analogy with the Fig. 6.2b, where we divided the space into many "invisible screens and slits", we decompose $t' - t$ into small time intervals. Let $t' - t \equiv N\Delta t$, and $t_n = t + n\Delta t$. The time evaluation corresponding to these steps can be expressed as [4]

$$W(t', t) = W(t', t_{N-1})W(t_{N-1}, t_{N-2})\ldots W(t_2, t_1)W(t_1, t), \tag{6.23}$$

using matrix multiplication of $W(t_{n+1}, t_n)$. In components, we obtain

$$W(x', t', x, t) = \int dx(N-1)dx(N-2)\ldots dx(1) \times W(x', t', x(N-1), t_{N-1})$$
$$\times W(x(N-1), t_{N-1}, x(N-2), t_{N-2}) \times \ldots W(x(1), t_1, x, t). \tag{6.24}$$

Here, the matrix components are given by the "bra-and ket-sandwich"

$$W(x(n+1), t_{n+1}, x(n), t_n) = \langle x(n+1)|e^{-iH\Delta t}|x(n)\rangle. \tag{6.25}$$

When Δt is small enough, the exponential can be expanded as

$$\langle x(n+1)|e^{-iH\Delta t}|x(n)\rangle \approx \langle x(n+1)|\,[1 - i\Delta t H]\,|x(n)\rangle. \tag{6.26}$$

Passing to the momentum space we obtain

$$\langle x(n+1)|H|x(n)\rangle = \int dk(n)\langle x(n+1)|k(n)\rangle\langle k(n)|H|x(n)\rangle. \tag{6.27}$$

Since the Hamiltonian is expressed in terms of $\hat{k}$ and $\hat{x}$ we have

$$\langle k(n)|H(\hat{k}, \hat{x})|x(n)\rangle = H(k(n), x(n))\langle k(n)|x(n)\rangle. \tag{6.28}$$

Here $\hat{k} = -id/dx$. Using this, the Eq. (6.25) can be written as

$$W(x(n+1), t_{n+1}, x(n), t_n) \approx \int dp(n)\langle x(n+1)|p(n)\rangle\langle p(n)|x(n)\rangle$$
$$\times\, [1 - i\Delta t H(p(n), x(n))]$$
$$\approx \int \frac{dp(n)}{2\pi} \exp\left[ip(n)(x(n+1) - x(n))\right.$$
$$\left. -i\Delta t H(p(n), x(n))\right]. \tag{6.29}$$

Inserting this for each term in (6.24) leads for $W(x', t', x, t)$

$$W(x', t', x, t) = \int \frac{dp(N-1)}{2\pi} \cdots \int \frac{dp(0)}{2\pi} \int dx(N-1) \ldots \int dx(1)$$

$$\times \exp\left[i \sum_{n=0}^{N-1} (p(n)(x(n+1) - x(n)) - \Delta t H(p(n), x(n))) \right].$$
(6.30)

By performing the limit $N \to \infty$ and $\Delta t \to 0$ one obtains

$$x(n+1) - x(n) \to \dot{x}(t)\Delta t, \qquad \sum_{n=0}^{N-1} \Delta t \to \int_t^{t'} dt. \tag{6.31}$$

Thus we can represent (6.30) as

$$W(x', t', x, t) = \int_{x(t')=x'}^{x(t)=x} \mathcal{D}p(t'')\mathcal{D}x(t'')$$

$$\times \exp\left[i \int_t^{t'} dt''(p(t'')\dot{x}(t'') - H(p(t''), x(t''))) \right],$$
(6.32)

where we denoted multiple integrals over $p(n)$ and $x(n)$ as $\mathcal{D}p(t)$ and $\mathcal{D}x(t)$, respectively. Further, they will be referred to as path integrals. The Eq. (6.32) can be simplified for the concrete Hamiltonian (6.12) by performing integration over the momentum and rewritten as [4]

$$W(x', t', x, t) = \int_{x(t')=x'}^{x(t)=x} \mathcal{D}x(t'')$$

$$\times \exp\left[i \int_t^{t'} dt''(\frac{m\dot{x}(t'')^2}{2} - U(x(t''))) \right].$$
(6.33)

Here we discover that the argument of the exponent in (6.33) is similar to the action, namely, to iS. Thus we can finally express the amplitude $W(x', t', x, t)$ in terms of the action:

$$W(x', t', x, t) = \int_{x(t')=x'}^{x(t)=x} \mathcal{D}x(t'') \exp\left(i S[x(t'')]\right) \tag{6.34}$$

where

$$S = \int_{t_1}^{t_2} dt \left[\frac{m\dot{x}(t)^2}{2} - U(x(t)) \right] = \int dt \mathcal{L} \tag{6.35}$$

In the above case, it was possible to integrate out the momentum, however, in general, this is not always possible.

6.4   Imaginary Time and Field Formulation

By making replacement

$$\int dt'' \to -i \int d\tau', \qquad \dot{x}(t'') \to i\dot{x}(\tau') \tag{6.36}$$

and setting $\tau = \beta$, $x' = x$ one may obtain from Eqs. (6.2) and (6.34)

$$Z = \mathrm{Tr}\, e^{-\beta H} = \int dx \langle x|e^{-\beta H}|x\rangle, \tag{6.37}$$

where

$$\langle x'|e^{-\beta H}|x\rangle = \int \mathcal{D}x(\tau') \int \mathcal{D}p(\tau') \exp\left(-S[x(\tau'), p(\tau')]\right). \tag{6.38}$$

Finally, making in (6.38) the transformation $x \to \psi(\mathbf{r})$ and $p \to i\psi^\dagger(\mathbf{r})$ we obtain

$$Z = \int \mathcal{D}\psi^\dagger(\mathbf{r}, \tau)\mathcal{D}\psi(\mathbf{r}, \tau)e^{-S[\psi^\dagger, \psi]}, \tag{6.39}$$

which together with $\Omega = -T \ln(Z)$ will be our working equation for the parturition function for bosons in the path integral formalism with the action given by (6.15).

6.5   Evaluation of Path Integrals

In contrast to usual integrals, for which there are a lot of handbooks [9] and codes in Mathematica, evaluation of path integrals, even numerically, is a rather troublesome business. This is why there is no table book of path integrals. The reason is that, actually, a path integral is a multiple integral with, in general, infinite multiplicity. From the previous section, we understood that physically this multiplicity is caused by the number of possible trajectories along which a quantum mechanical particle can move to reach the end point (t_f, x_f) starting from the point (t_i, x_i).

The meager list of path integrals, taken analytically, is headed by the following equality:

$$\int \mathcal{D}\phi e^{-\frac{1}{2}\int dx dy \phi(x)A(x,y)\phi(y)} = (Det\, A)^{-1/2}, \tag{6.40}$$

where $A(x, y)$ is a symmetric function $A(x, y) = A(y, x)$. Integral in the exponent, strictly speaking, is supposed to be taken in the whole range $(-\infty.. + \infty)^1$. As to the determinant $(Det\, A)$ it will be clarified later.

---

[1] To facilitate the explanation, here we consider the case of one component path integral in $D = 1$.

The derivation or justification of Eq. (6.40) is based on the following reasoning. Let's start from the well-known Gaussian integral given by

$$\int dx\, e^{-\frac{1}{2}\alpha x^2} = \left(\frac{2\pi}{\alpha}\right)^{1/2}. \tag{6.41}$$

The product of such integrals, being a multiple integral is

$$\int dx_1 dx_2 \ldots dx_n \exp\left[-\frac{1}{2}\sum_n \alpha_n x_n^2\right] = \frac{(2\pi)^{n/2}}{\sqrt{\alpha_1 \alpha_2 \ldots \alpha_n}}. \tag{6.42}$$

Now, suppose that A is a diagonal matrix with elements $\alpha_1, \alpha_2, \ldots \alpha_n$ and x is n-dimensional vector $(x_1, x_2 \ldots x_n)$. Then we may rewrite the sum in (6.42) as the following scalar product:

$$\sum_n \alpha_n x_n^2 = (x, Ax). \tag{6.43}$$

Since A is diagonal, then $det\, A = \alpha_1 \alpha_2 \ldots \alpha_n$, and hence (6.42) can be exactly rewritten as

$$\int d^n x \exp\left(-\frac{1}{2}(x, Ax)\right) = (2\pi)^{n/2}(det\, A)^{-1/2}. \tag{6.44}$$

Now redefinition of the measure of the integral as $(dx) = d^n x (2\pi)^{-n/2}$ converts the Eq. (6.44) into the form, similar to the path integral (6.40).

The list of "handbook of path integrals" has been extended by Julian Schwinger, who put forward more extended version of (6.40) as [10]

$$\int \mathcal{D}\phi\, e^{-\frac{1}{2}\int dx dy \phi(x) A(x,y)\phi(y) + \int dx\, J(x)\phi(x)}$$

$$= (Det\, A)^{-1/2} e^{\frac{1}{2}\int dx dy\, J(x) A^{-1}(x,y) J(y)}, \tag{6.45}$$

where $J(x)$ may be considered as an arbitrary function. A beginner can understood this formula on the base of the following table integral:

$$\int dx\, e^{-\alpha x^2 + \beta x} = \frac{\sqrt{\pi}}{\sqrt{\alpha}} e^{\frac{\beta^2}{4\alpha}}. \tag{6.46}$$

Note that, a more rigorous proof of (6.40) and (6.45) can be found e.g. in the textbook by Ramon [11].

In physical tasks, we deal with the case when $A(x, y)$ represents the inverse of the Green function. In this case, the path integral in (6.45) corresponds to the amplitude of vacuum-vacuum transition in the presence of the source $J(x)$ and referred as a generating functional:

$$Z_0[J] = \int \mathcal{D}\phi\, e^{-\frac{1}{2}\int dx dy \phi(x) G^{-1}(x,y)\phi(y) + \int dx\, J(x)\phi(x)}$$

$$= (Det\, G^{-1})^{-1/2} e^{\frac{1}{2}\int dx dy\, J(x) G(x,y) J(y)}. \tag{6.47}$$

It is remarkable that, by using this equation and rules for functional derivatives, outlined in the Appendix C, one will be able to calculate analytically more complicated versions of the Gaussian path integrals, such as

$$I_n = \int \mathcal{D}\phi \; \phi^n \; e^{-\frac{1}{2}\int dx dy \phi(x) G^{-1}(x,y)\phi(y)}. \tag{6.48}$$

In fact, it is easy to make sure that,

$$I_n = \frac{\delta^n}{\delta J^n} Z_0[J]|_{J=0}, \tag{6.49}$$

and hence, the path integral in (6.49) with an odd power of ϕ equals to zero: $I_{2n+1} = 0$ (n- integer). This is similar to the ordinary integral $\int_{-\infty}^{\infty} dx x^{2n+1} \exp(-x^2) = 0$.

Task 6.1: By performing functional integration, prove that e.g. $I_3 = 0$.

So, only the integrals with even powers of ϕ will survive. Particularly, we obtain

$$\begin{aligned}
I_2 &= \int \mathcal{D}\phi \; \phi^2 \; e^{-\frac{1}{2}\int dx dy \phi(x) G^{-1}(x,y)\phi(y)} \\
&= \frac{\delta^2}{\delta J^2} Z_0[J]|_{J=0} = (Det G^{-1})^{-1/2} G(x, x')|_{x' \to x}.
\end{aligned} \tag{6.50}$$

In general, the Eqs. (6.48) and (6.49) can be extended for an arbitrary composite function $V(\phi)$ as

$$\int \mathcal{D}\phi \; V(\phi) \; e^{-\frac{1}{2}\int dx dy \phi(x) G^{-1}(x,y)\phi(y)} = \left[V(\frac{\delta}{\delta J}) Z_0[J] \right]\Big|_{J=0}, \tag{6.51}$$

where $V(\phi)$ is supposed to be expanded in power series of ϕ. Moreover, it can be shown that the sign of the derivative can be moved out from the functional integral: $\mathcal{D}\phi \nabla \to \nabla \mathcal{D}\phi$. For example, one may obtain,

$$\begin{aligned}
&\int dx dx' \int \mathcal{D}\phi \; (\nabla_x \phi)(\nabla_{x'} \phi) \; e^{-\frac{1}{2}\int dx dy \phi(x) G^{-1}(x,y)\phi(y)} \\
&= \int dx dx' \nabla_x \nabla_x' \int \mathcal{D}\phi \; \phi(x)\phi(x') \; e^{-\frac{1}{2}\int dx dy \phi(x) G^{-1}(x,y)\phi(y)} \\
&= (Det G^{-1})^{-1/2} \int dx dy \nabla_x \nabla_x' G(x, x')
\end{aligned} \tag{6.52}$$

For practical calculations, we bring the identity for the two-component case: [12]:

$$\int \mathcal{D}\psi_1 \mathcal{D}\psi_2 \exp\left[-\frac{1}{2} \sum_{a,b=1,2} \int \psi_a(x) G_{ab}^{-1}(x, y)\psi_b(y)dxdy \right.$$

$$\left. + \sum_{a=1,2} \int J_a(x)\psi_a(x)dx \right]$$

$$\equiv Z_0[J_1, J_2] = (Det G^{-1})^{-1/2} \exp\left[\sum_{a,b=1,2} \int dxdy\, J_a(x)G_{ab}(x, y)J_b(y) \right]$$

$$(6.53)$$

which is just the extension of (6.47). Particularly, setting here $J_1 = J_2 = 0$ gives

$$Z_0[0, 0] = \int \mathcal{D}\psi_1 \mathcal{D}\psi_2 \exp\left[-\frac{1}{2} \sum_{a,b=1,2} \int \psi_a(x) G_{ab}^{-1}(x, y)\psi_b(y)dxdy \right]$$

$$= (Det G^{-1})^{-1/2}.$$

$$(6.54)$$

The extension of the other above equations for two (or more component) cases is straightforward. More detailed information on the properties of path integrals such as integration by parts, replacing variables etc. are rigorously presented in the book by Slavnov and Faddeev [12].

Thus, in our knowledge, the list of functional integrals, taken analytically, is being exhausted. Then, the question arises how one may calculate other kinds of path integrals, say

$$\int \mathcal{D}\phi \exp\left[-\frac{1}{2} \int dxdy\phi(x)A(x, y)\phi(y) + \int dx\phi^4(x) \right], \qquad (6.55)$$

which is typical for the $\lambda\phi^4$ theory? Well, there are two ways: Numerical and analytical. In numerical calculations, one has to deal with integrals with many multiplicity, which are referred as Monte-Carlo lattice calculations. Although these business requires rather large machine time and generation of special algorithms [13], nowadays its results are considered as very reliable, as if they were obtained by experimental observations.

On the other hand, if you prefer to go on with analytics, then you have to choose one of the well-known approximations, or, in the best way to generate your own one. Among these approximations, the loop expansion is rather popular [14]. Actually, in the framework of filed theories, this corresponds to the expansion in powers of $\hbar$. In particular, the first stage of loop expansion ($\hbar^0$) corresponds to the classical trajectory, outlined in the Sect. 6.3, while the next stage $\sim (\hbar^1)$ corresponds to the bilinear (Gaussian) approximation, which we have studied in Sect. 4.7. There are also perturbative methods, mostly used in Quantum Electrodynamics, as well as non-perturbative methods, which we will consider later.

6.5.1 Determinant G^{-1} and Lee-Huang Yang Term

To evaluate the expression $(Det\ D^{-1})^{-1/2}$ we start from the well-known formula

$$(Det\,\hat{A})^{1/2} = \exp\left(\frac{1}{2}\mathrm{Tr}\,\ln\,\hat{A}\right), \tag{6.56}$$

which is valid for any diagonal matrix A. Now let the Green function be, in general, given by the Eq. (4.82), whose determinant equals to $Det(D^{-1}) = (E_k^2 + \omega_n^2)$ with $E_k^2 = (\varepsilon_k + X_1)(\varepsilon_k + X_2)$. Since our final goal is not the calculation of the determinant by itself, but its logarithm, $\Omega \sim \ln(Z)$, we consider the expression $\ln(Det\ D^{-1})^{-1/2} = (-1/2)\mathrm{Tr}\,\ln(D^{-1})$, where the trace will correspond to the summation $\sum_n \sum_k$.

It can be shown that[2], for $\omega_n = 2\pi n T$

$$L_s(a) = \sum_{n=-\infty}^{\infty} \ln(\omega_n^2 + a^2) = \beta a + 2\ln(1 - e^{-\beta a}). \tag{6.57}$$

In fact, differentiating $L_s(a)$ with respect to a gives

$$\frac{dL_s(a)}{da} = \frac{d}{da}\sum_{n=-\infty}^{\infty} \ln(\omega_n^2 + a^2) = 2a\sum_{n=-\infty}^{\infty}\frac{1}{\omega_n^2 + a^2} = 2\beta\left(\frac{1}{2} + \frac{1}{e^{\beta a} - 1}\right), \tag{6.58}$$

where we used the Eqs. (4.96) and (4.97). Now, after we get rid off Matsubara summation, we integrate both parts of the last equation by a and using the identity

$$\frac{1}{\beta}\ln\left(1 - e^{-\beta a}\right) = \int da\,\frac{1}{e^{\beta a} - 1}, \tag{6.59}$$

obtain the Eq. (6.57). Thus, for the inverse Green function D^{-1} in Eq. (4.81) we finally get

$$\ln(Det\ D^{-1})^{-1/2} = -\frac{1}{2}\beta\sum_k E_k - \sum_k \ln\left(1 - e^{-\beta E_k}\right). \tag{6.60}$$

Note that, we still have not escaped from divergences. In fact, the first term in Eq. (6.60) is, clearly, divergent: $\sum_k E_k \sim \int_0^\infty dk\,k^2\sqrt{(k^2)(k^2 + 2m\Delta)} = \infty$. The situation may be cured by introducing appropriate counter terms, which leads to a regularized expression:

$$\frac{1}{2}\sum_k E_k \rightarrow \frac{1}{2}\sum_k\left[E_k - \varepsilon_k - \Delta + \frac{\Delta^2}{2\varepsilon_k}\right], \tag{6.61}$$

---

[2] More rigorous derivation and the way of "making this infinite sum as finite" is presented in the book by Capusta [15]. See also Refs. [16,17].

where $E_k = \sqrt{(\varepsilon_k)(\varepsilon_k + 2\Delta)}$ and $\varepsilon_k = \mathbf{k}^2/2m$. Further, for $D = 3$ the integral in (6.61) can be taken analytically leading to

$$\frac{1}{2}\sum_k \left[E_k - \varepsilon_k - \Delta + \frac{\Delta^2}{2\varepsilon_k} \right] = \frac{8Vm^{3/2}\Delta^{5/2}}{15\pi^2}. \tag{6.62}$$

Therefore, we can present (6.60) as

$$\ln(Det\ D^{-1})^{-1/2} = -\frac{8V\beta m^{3/2}\Delta^{5/2}}{15\pi^2} - \sum_k \ln\left(1 - e^{-\beta E_k}\right). \tag{6.63}$$

Now, one may exploit this equation to evaluate Ω by means of Eqs. (6.1) and (6.54). It is seen that, for the thermodynamic potential at zero temperature only the first term of (6.63) survives. In thermodynamics, this term with $\Delta = \mu = g\rho$ corresponds to the famous Lee, Huang and Yang correction [18] being referred to a zero point energy of Bogolyubov excitations, while the second term of (6.63) describes thermal fluctuations. Above, we used the method of counter terms. It has been proven that, [19] the Eq. (6.63) remains true for any type of regularization scheme!

The present chapter would not be complete without an explanation of one more method of evaluation of path integrals. It is based on the Hubbard-Stratanovich transformation.

6.5.2   Hubbard-Stratanovich Transformation for Bosons

So, we have established that only the path integral of the Gaussian type can be evaluated analytically. However, life is more complicated. Usually, the action of real physical systems involves a quartic term in fields also. For instance, the action of spinless Bosons, described by the Lagrangian (4.80) has quadratic as well as quartic parts:

$$S = S_0 + S_2 + S_4, \qquad S_0 = \int dx(-\mu\rho_0 + \frac{U_x\rho_0^2}{2}),$$

$$S_2 = \int dxdy\tilde{\psi}^\dagger(x)\left[(\partial_\tau - \frac{\nabla^2}{2m} - \mu + \rho_0 U_x)\delta(x-y) + \rho_0 U(x-y) \right]\tilde{\psi}(y)$$

$$+ \frac{\rho_0}{2}\int dxdy\, U(x-y)\left[\tilde{\psi}^\dagger(x)\tilde{\psi}^\dagger(y) + \tilde{\psi}(x)\tilde{\psi}(y)\right],$$

$$S_4 = \frac{1}{2}\int dxdy\,\tilde{\psi}^\dagger(x)\tilde{\psi}(x)U(x-y)\tilde{\psi}^\dagger(y)\tilde{\psi}(y),$$

$$\tag{6.64}$$

where $\int dx \equiv \int_0^\beta \int d\mathbf{r}$, $U_z = \int dt\, U(z-t)$, and $\tilde{\psi}(x)$ is a fluctuating field given by the Bogolyubov shift.[3] Particularly, for the contact interaction $U(x-x') = g\delta(\tau - \tau')\delta(\mathbf{r}-\mathbf{r}')$.

The idea of the application of the Hubbard-Stratanovich transformation is to, somehow, totally eliminate the S_4 term from the partition function

$$Z = \int \mathcal{D}\tilde{\psi}^\dagger \mathcal{D}\tilde{\psi}\, e^{-S_0 - S_2 - S_4}. \tag{6.65}$$

Such an idea was firstly proposed by Stratanovich [20] on the base of the simple identity

$$e^{-a^2} \int_{-\infty}^{\infty} dt\, e^{-\pi t^2 - 2\sqrt{\pi}at} = 1, \tag{6.66}$$

and popularized by Hubbard [21] later. Below we illustrate it for bosons described by the action (6.64).

Let's start with a similar identity

$$1 = e^{-\frac{1}{2}\mathrm{Tr}\,\ln U^{-1}} \int \mathcal{D}\phi\, e^{W[\phi]}, \tag{6.67}$$

with

$$W = \frac{1}{2}\int dx\,dy\{\phi(x) - U_x\tilde{\psi}^\dagger(x)\tilde{\psi}(x)\}U^{-1}(x-y)\{\phi(y) - U_y\tilde{\psi}^\dagger(y)\tilde{\psi}(y)\}, \tag{6.68}$$

which follows from Eqs. (6.45) and (6.56). In (6.67) trace of any operator $\hat{A}(x,y)$ and its inverse are defined as

$$\mathrm{Tr}\,\hat{A} = \int dx\, A(x,y)|_{y\to x}, \qquad \int dy\, A^{-1}(x-y)A(y-z) = \delta(x-z) \tag{6.69}$$

The functional in (6.68) has following expanded form:

$$\begin{aligned} W = &\frac{1}{2}\int dx\,dy\,\phi(x)U^{-1}(x-y)\phi(y) - \int dx\,\phi(x)\tilde{\psi}^\dagger(x)\tilde{\psi}(x) \\ &+ \frac{1}{2}\int dx\,dy\,\tilde{\psi}^\dagger(x)\tilde{\psi}(x)U(x-y)\tilde{\psi}^\dagger(y)\tilde{\psi}(y). \end{aligned} \tag{6.70}$$

Now when we multiply both sides of (6.65) to unity by means of (6.67) the quartic term totally cancels, leaving us with the Gaussian (quadratic) path integral in $\tilde{\psi}$ fields, which can be integrated out.

---

[3] See Eqs. (8.1)–(8.3) in Chap. 8.

In fact, we have

$$Z \cdot 1 \equiv Z = e^{-\frac{1}{2}\mathrm{Tr}\,\ln U^{-1}} e^{-S_0}$$
$$\times \int \mathcal{D}\phi \exp\left[\frac{1}{2}\int dxdy\phi(x)U^{-1}(x-y)\phi(x)\right]\mathcal{D}\tilde{\psi}^\dagger \mathcal{D}\tilde{\psi}\, \exp[-\tilde{S}_2],$$

$$\tilde{S}_2 = \int dxdy\,\tilde{\psi}^\dagger(x)\left[\partial_\tau - \frac{\nabla^2}{2m} - \mu + \rho_0 U_x - \phi(x)\right]\tilde{\psi}(y)\delta(x-y)$$
$$+ \frac{\rho_0}{2}\int dxdy\,U(x-y)\left[\tilde{\psi}^\dagger(x)\tilde{\psi}^\dagger(y) + \tilde{\psi}(x)\tilde{\psi}(y)\right]$$
$$+ \rho_0 \int dxdy\,\tilde{\psi}^\dagger(x)U(x-y)\tilde{\psi}(x).$$

$$(6.71)$$

It is easily understood that, the factor $\exp\{-\frac{1}{2}\mathrm{Tr}\,\ln U^{-1}\}$ will give a trivial additional constant term to Ω, and hence, can be further omitted.

Now let's concentrate only on the particular case of contact interaction with $U(x - x') = g\delta(\mathbf{r} - \mathbf{r}')\delta(\tau - \tau')$, so that $U_x = \int dz\,U(x-z) = g$. Then for a homogeneous system, the partition function Z will be simplified as

$$Z = e^{-V\beta[-\mu\rho_0 + \frac{g\rho_0^2}{2}]}\int \mathcal{D}\phi \exp\left[\frac{1}{2g}\int dx(\phi(x))^2\right]\mathcal{D}\tilde{\psi}^\dagger \mathcal{D}\tilde{\psi}$$
$$\times \exp\left\{-\int dxdy\,\tilde{\psi}^\dagger(x)\left[\partial_\tau - \frac{\nabla^2}{2m} - \mu + 2g\rho_0 - \phi(x)\right]\tilde{\psi}(y)\delta(x-y)\right\}$$
$$\times \exp\left\{-\frac{g\rho_0}{2}\int dx \int dy\left[\tilde{\psi}^\dagger(x)\tilde{\psi}^\dagger(y) + \tilde{\psi}(x)\tilde{\psi}(y)\right]\delta(x-y)\right\}.$$

$$(6.72)$$

Now, passing to the real field representation by Eq. (4.55) and rewriting the integrals in momentum space one can present (6.72) as follows

$$Z = e^{-V\beta[-\mu\rho_0 + \frac{g\rho_0^2}{2}]}\int \mathcal{D}\phi \exp\left[\frac{1}{2g}\int dx(\phi(x))^2\right]\mathcal{D}\psi_1\mathcal{D}\psi_2$$
$$\times \exp\left\{-\frac{1}{2}\sum_{i,j}\sum_{\omega_n,\mathbf{k}}\psi_i(\omega_n,\mathbf{k})(D_\phi^{-1})_{i,j}(\omega_n,\mathbf{k})\psi_j(\omega_n,\mathbf{k})\right\},$$

$$(6.73)$$

where $(i, j) = (1, 2)$, ϕ_k is the Fourier transform of the auxiliary field, $\phi_k = \int d\mathbf{r}\exp(i\mathbf{k}\cdot\mathbf{r})\phi(\mathbf{r})$ and the inverse Green function is similar to that in the Bilinear approximation (4.81):

$$D_\phi^{-1}(\omega_n,\mathbf{k}) = \begin{pmatrix} \varepsilon_k - \mu + 3g\rho_0 - \phi_k & \omega_n \\ -\omega_n & \varepsilon_k - \mu + g\rho_0 - \phi_k \end{pmatrix}$$
$$\equiv \begin{pmatrix} \varepsilon_k + X_1(k) & \omega_n \\ -\omega_n & \varepsilon_k + X_2(k) \end{pmatrix}.$$

$$(6.74)$$

Now, the path integral over atomic fields is Gaussian and can be evaluated directly by using the formula (6.53), resulting in

$$Z = e^{-V\beta[-\mu\rho_0 + \frac{g\rho_0^2}{2}]} \int \mathcal{D}\phi \, e^{\frac{1}{2g}\int dx (\phi(x))^2} (Det\, D_\phi^{-1})^{-1/2}$$
$$= e^{-V\beta[-\mu\rho_0 + \frac{g\rho_0^2}{2}]} \int \mathcal{D}\phi \, e^{\frac{1}{2g}\int dx (\phi(x))^2} e^{-\frac{1}{2}Tr\, \ln D_\phi^{-1}}, \tag{6.75}$$

where we used the Eq. (6.56). So, by using the Hubbard-Stratanovich transformation we have successfully got rid of S_4 term and easily obtained <u>exact</u> simple expression for the partition function Z, which is related to the thermodynamic potential $\Omega = -T \ln Z$.

Unfortunately, there is a law of conservation of difficulties: now we have to overcome the difficulties with the calculation of the path integral over the auxiliary scalar field ϕ. This may be realized in the spirit of saddle point approximation, which may convert the functional integral in (6.75) to the sum of integrals of kind (6.48). In practical calculations, one introduces a fluctuating field $\tilde{\phi}$ by

$$\phi(x) = \phi_0 + \tilde{\phi}(x), \tag{6.76}$$

where ϕ_0 is constant, and makes an expansion the integrand in (6.75) in powers of $\tilde{\phi}$. In fact, after the shift (6.76) the functional integration proceeds by $\tilde{\phi}$, i.e. $\int \mathcal{D}\phi \to \int \mathcal{D}\tilde{\phi}$ which is still complicated due to the logarithmic factor. The latter may be expanded by using Dyson equation (4.51) for D_ϕ^{-1}:

$$D_\phi^{-1} = D_{\phi_0}^{-1} - \Sigma_{\tilde{\phi}}, \tag{6.77}$$

where we have introduced "free" Green function and the self energy $\Sigma_{\tilde{\phi}}$ as follows

$$D_{\phi_0}^{-1}(\omega_n, \mathbf{k}) = \begin{pmatrix} \varepsilon_k - \mu + 3g\rho_0 - \phi_0 & \omega_n \\ -\omega_n & \varepsilon_k - \mu + g\rho_0 - \phi_0 \end{pmatrix}, \tag{6.78}$$

$$\Sigma_{\tilde{\phi}} = \begin{pmatrix} \tilde{\phi}_k & 0 \\ 0 & \tilde{\phi}_k \end{pmatrix}. \tag{6.79}$$

Then for the logarithm, we shall have

$$\ln D_\phi^{-1} = \ln D_{\phi_0}^{-1}[1 - D_{\phi_0}\Sigma_{\tilde{\phi}}] = \ln D_{\phi_0}^{-1} + \ln[1 - D_{\phi_0}\Sigma_{\tilde{\phi}}], \tag{6.80}$$

where the first term does not depend on $\tilde{\phi}$, and the second one can be expanded in powers of fluctuating field $\tilde{\phi}$ elementary by using

$$\ln(1 - \epsilon) = -\sum_{n=1}^{\infty} \frac{\epsilon^n}{n}. \tag{6.81}$$

As to the classical field ϕ_0 it will be fixed after the integration by the requirement of minimization of the thermodynamic potential as $\partial\Omega/\partial\phi_0 = 0$.

In our knowledge, the trick described above, although it is not widely used in the literature for bosons, has been a very convenient tool in the microscopical description of superconductivity [4,22] as well as BCS-BEC crossover [23].

6.6   Expected Values

In quantum mechanics, the expected value is the probabilistic expected (mean) value of the result (measurement) of an experiment. It is a fundamental concept in all areas of physics. We know that each physical observable has its own operator, say, $\hat{O}$, working on the Hilbert space of eigenfunctions ψ of the Hamiltonian H. In quantum mechanics, its expected value, i.e. average value is defined as [24]

$$\bar{O} = <\hat{O}> = \frac{\int \psi^\dagger (\hat{O}\psi)d\mathbf{r}}{\int \psi^\dagger \psi d\mathbf{r}}. \tag{6.82}$$

In quantum statistical mechanics, where in the equilibrium the temperature $T = 1/\beta$ and the chemical potential μ are involved, the definition of $\bar{O}$ is similar to (6.82) [25]

$$\bar{O} = <\hat{O}> = \frac{\mathrm{Tr}\,[\hat{O}e^{-\beta(H-\mu N)}]}{\mathrm{Tr}\,[e^{-\beta(H-\mu N)}]}. \tag{6.83}$$

This can be directly extended to quantum field theory in terms of functional integrals by means of secondary quantization:

$$\bar{O} = <\hat{O}(\psi^\dagger, \psi)> = \frac{1}{Z_0} \int \mathcal{D}\psi^\dagger \mathcal{D}\psi \; \hat{O}(\psi^\dagger, \psi)e^{-S[\psi^\dagger, \psi]}, \tag{6.84}$$

where

$$Z_0 = \int \mathcal{D}\psi^\dagger \mathcal{D}\psi \; e^{-S[\psi^\dagger, \psi]}. \tag{6.85}$$

For the uniform Bose system in the condensed phase it is convenient to introduce Bogolyubov shift $\psi = \sqrt{\rho_0} + \tilde{\psi}$ and real fields by Eq. (4.32). Then, due to the identical transformation $\mathcal{D}\psi^\dagger \mathcal{D}\psi \rightarrow \mathcal{D}\tilde{\psi}^\dagger \mathcal{D}\tilde{\psi} \rightarrow \mathcal{D}\psi_1 \mathcal{D}\psi_2$ one may perform integration in variables (ψ_1, ψ_2).[4]

---

[4] Minor constants coming from the Jacobean of such transformation will be canceled in (6.84).

6.6.1 Expected Values in One Loop Approximation for Uniform Systems

For further application, below we consider expectation values for some operators in bilinear (Gaussian) approximation outlined in Chap. 4. So, the action, corresponding to the Lagrangian (4.85) with the contact potential is given by

$$
S = S_0 + S_1 + S_2,
$$

$$
S_0 = \int_0^\beta d\tau \int d\mathbf{r} \left(-\mu\rho_0 + \frac{g\rho_0^2}{2} \right),
$$

$$
S_1 = \sqrt{2\rho_0} \int_0^\beta d\tau \int d\mathbf{r} \left(\frac{\partial}{\partial\tau} - \frac{\nabla^2}{2m} + g\rho_0 - \mu \right) \psi_1
$$

$$
= \sqrt{2\rho_0} \int_0^\beta d\tau \int d\mathbf{r} \, (g\rho_0 - \mu) \, \psi_1,
$$

$$
S_2 = \frac{1}{2} \int_0^\beta d\tau \int d\mathbf{r} \sum_{i,j} \left[i\varepsilon_{ij}\psi_i\partial_\tau\psi_j + \psi_i(-\frac{\nabla^2}{2m} + X_i)\psi_j\delta_{ij} \right]
$$

$$
= \frac{1}{2} \sum_{\omega_n,\mathbf{k}} \sum_{i,j=1,2} \psi_i(\omega_n,\mathbf{k})(D^{-1})_{ij}(\omega_n,\mathbf{k})\psi_j(\omega_n,\mathbf{k}),
$$

(6.86)

D^{-1} is given by (4.81). Above in the linear term, S_1 we used the identities

$$
\int_0^\beta d\tau \sqrt{\rho_0}\frac{\partial}{\partial\tau}\psi_1(\tau,\mathbf{r}) = 0, \qquad \int d\mathbf{r}\sqrt{\rho_0}\nabla^2\psi_1(\tau,\mathbf{r}) = 0, \tag{6.87}
$$

which holds for the homogeneous system with $\rho_0 = const$ and can be proven by partial integration.

Performing functional integration in (6.85) (see (6.54)) we obtain

$$
Z_0 = \int \mathcal{D}\tilde{\psi}^\dagger \mathcal{D}\tilde{\psi}\, e^{-S_0-S_1-S_2} = e^{-S_0} \int \mathcal{D}\psi_1 \mathcal{D}\psi_2\, e^{-S_2[\psi_1,\psi_2]}
$$

$$
= e^{-S_0}(Det\, D^{-1})^{-1/2}.
$$

(6.88)

Here, a curious reader may ask "Where is the contribution from S_1?". Actually, here not only its expectation value $< S_1 >$ but the linear term S_1 by itself should vanish. In fact, by definition of the order parameter in Bogolyubov shift $\psi = \psi_0 + \tilde{\psi} = \sqrt{\rho_0} + \tilde{\psi} = \sqrt{\rho_0} + (\psi_1 + \psi_2)/2$ the expectation value of ψ must be equal to ψ_0, i.e. $\langle\psi\rangle = \psi_0 = \sqrt{\rho_0}$. And hence, we require that $\langle\tilde{\psi}\rangle = \langle\psi_1\rangle = \langle\psi_2\rangle = 0$. From the definition of the expectation value (6.84) it is understood that

$$
\langle\psi_1\rangle \sim \int \mathcal{D}\psi_1 \mathcal{D}\psi_2\, e^{-S_0-S_1-S_2}\psi_1, \tag{6.89}
$$

vanishes if and only if $S_1 = 0$, since $S_2[\psi_1, \psi_2]$ is a quadratic form in (ψ_1, ψ_2). The requirement $S_1 = 0$ leads to

$$- \mu + g\rho_0 = 0, \tag{6.90}$$

which is nothing but the Hugenholtz—Pines relation discussed in Sect. 4.5.

Now we consider the numerator of Eq. (6.84). It can be rewritten as:

$$\int \mathcal{D}\tilde{\psi}^\dagger \mathcal{D}\tilde{\psi}\, \hat{O}(\tilde{\psi}^\dagger, \tilde{\psi})\, e^{-S[\tilde{\psi}^\dagger, \tilde{\psi}]} = e^{-S_0} \int \mathcal{D}\psi_1 \mathcal{D}\psi_2 \hat{O}(\psi_1, \psi_2)$$

$$\times\, e^{-S_2[\psi_1, \psi_2]} e^{\int dx[J_1(x)\psi_1(x) + J_2(x)\psi_2(x)]}\Big|_{J_1=0, J_2=0}$$

$$= e^{-S_0}(\mathrm{Det}D^{-1})^{-1/2}\hat{O}\left(\frac{\delta}{\delta J_1}, \frac{\delta}{\delta J_2}\right)$$

$$\times\, \exp\left[\sum_{a,b=1,2}\int dx\, dy\, J_a(x)D_{a,b}(x, y)J_b(y)\right]\Bigg|_{J_1=0, J_2=0}, \tag{6.91}$$

where we used Eqs. (6.51)–(6.54). Thus from (6.84), (6.88) and (6.91) we see that the factor $\exp(-S_0)(Det D^{-1})^{-1/2}$ remarkably cancels to give finally

$$\bar{O} = \langle \hat{O}(\psi_1, \psi_2)\rangle$$

$$= \hat{O}\left(\frac{\delta}{\delta J_1}, \frac{\delta}{\delta J_2}\right) \exp\left[\sum_{a,b=1,2}\int dx\, dy\, J_a(x)D_{a,b}(x, y)J_b(y)\right]\Bigg|_{J_1=0, J_2=0}, \tag{6.92}$$

where $x \equiv (\tau, \mathbf{r})$ and $D_{a,b}(x, y)$ is given by

$$D_{ab}(\tau, \mathbf{r}; \tau', \mathbf{r}') = \frac{1}{V\beta}\sum_k \sum_{\omega_n} \exp[i\omega_n(\tau - \tau')]\exp[i\mathbf{k}(\mathbf{r} - \mathbf{r}')]D_{ab}(\omega_n, \mathbf{k}), \tag{6.93}$$

with $D_{ab}(\omega_n, \mathbf{k})$ in Eq. (4.82). For further applications, we present the expectation values of some operators:

$$\langle \psi_a(\mathbf{r}, \tau)\psi_b(\mathbf{r}', \tau')\rangle = D_{ab}(\mathbf{r}, \tau; \mathbf{r}', \tau'),$$

$$\langle \psi_1(\mathbf{r})\psi_2(\mathbf{r})\rangle = D_{12}(0) = \frac{1}{\beta}\sum_n D_{12}(\mathbf{k}, \omega_n) = \frac{1}{\beta}\sum_{n=-\infty}^{\infty}\frac{\omega_n}{\omega_n^2 + E_k^2} = 0, \tag{6.94}$$

$$\langle \psi_a^4(\mathbf{r})\rangle = 3D_{aa}^2(0), \qquad \langle \psi_1^2(\mathbf{r})\psi_2^2(\mathbf{r})\rangle = D_{11}(0)D_{22}(0),$$

$$\langle \psi_{a_1}, \psi_{a_2}\ldots\psi_{a_n}\rangle = 0, \qquad n = 1, 3, 5\ldots.$$

where $D_{ab}(0) \equiv D_{ab}(\tau - \tau', \mathbf{r} - \mathbf{r}')|_{\tau \to \tau', \mathbf{r} \to \mathbf{r}'}$. These relations may be obtained by using the Eq. (6.92) and the rules presented in the Appendix C.

Task 6.2

Show that anomalous density, defined as

$$\sigma = \frac{1}{2V} \int d\mathbf{r}[\langle \tilde{\psi}^\dagger(\mathbf{r})\tilde{\psi}^\dagger(\mathbf{r})\rangle + \langle \tilde{\psi}(\mathbf{r})\tilde{\psi}(\mathbf{r})\rangle], \tag{6.95}$$

equals to

$$\sigma = \frac{X_2 - X_1}{2V} \sum_k \frac{W_k}{E_k} = -\frac{\Delta}{V} \sum_k \frac{W_k}{E_k}, \tag{6.96}$$

where the contact potential has been assumed and other notations are introduced in Chap 4. Note that the absolute value of σ has the meaning of the density of pair-correlated particles.

Task 6.3 Bearing in mind that, at zero temperature for $D = 3$, the momentum integral in (6.96) is divergent, find an appropriate counterterm to prove

$$\sigma(T = 0) = 3\rho_1(T = 0). \tag{6.97}$$

6.6.2 Grand Thermodynamic Potential Ω

Now we are on the stage of derivation an explicit expression for the Grand Thermodynamic potential for Bose system in the BEC phase in the Bilinear approximation. In fact one obtains

$$\Omega = -T \ln(Z_0) = -T \ln[e^{-S_0}(Det D^{-1})^{-1/2}] = T S_0 - T \ln[(Det D^{-1})^{-1/2}]. \tag{6.98}$$

For homogeneous system $\rho_0 = const$ and hence $S_0 = \beta V(-\mu\rho_0 + g\rho_0^2/2)$. As to the second term in (6.98) it has been evaluated in Sect. 6.5. Therefore, finally we obtain following final expression for Ω

$$\Omega(T < T_c) = -\mu V \rho_0 + \frac{V g \rho_0^2}{2} + \frac{1}{2} \sum_k \left[E_k - \varepsilon_k - \Delta + \frac{\Delta^2}{2\varepsilon_k} \right] + T \sum_k \ln\left(1 - e^{-\beta E_k}\right), \tag{6.99}$$

where we have still $E_k^2 = (\varepsilon_k + X_1)(\varepsilon_k + X_2)$, $X_1 = 2\Delta = 2g\rho_0$ and $X_2 = 0$. Note that, when the gas parameter is enough small, $\gamma << 1$, one may set $\rho_0 \approx \rho$ and $\Delta = g\rho$, coming back to Bogolyubov approximation with Lee-Huang-Yang correction i.e. the second term in (6.99).

It is well known that Ω contains all information about the system in the equilibrium. What does it say about the condensed fraction? One of the main principles of the physics is that any isolated systems tends to minimize its total energy when it is passing to its stationary "eternal" state, which is referred as a ground state. For example, if you throw an apple up, then it starts to fall down, since the closer it to the earth, the less it's potential energy. Another example from quantum mechanics, is that, if you somehow excite an atom then it tends to come back to its ground state with minimal energy by emitting photons. However, if you mix two portions of water with different temperatures then the whole system goes to its equilibrium (ground state) not only by minimization of its energy E, but also by increasing the entropy S-in accordance with the second law of thermodynamics. Mathematically this corresponds to the minimum of the free energy: $F = E - TS = \Omega + \mu N$. Clearly, at zero temperature F coincides with E. In grand canonical ensemble one should minimize $\Omega = -PV$, where P is the pressure. So, considering ρ_0 as a variational parameter, and minimizing Ω with respect to ρ_0, we get from the condition $\partial\Omega/\partial\rho_0 = 0$ following equation:

$$- \mu + g\rho_0 = 0 \tag{6.100}$$

which exactly coincides with (6.90)!

It is seen that in the framework of one loop (Bilinear) approximation all three needed conditions, namely

- Relevance of the Hugenholtz-Pines relation, which is equivalent to the emergence of Goldstone mode with massless particles (bogolons in this case);
- Vanishing of expectation value of fluctuating fields: $\langle \tilde{\psi} \rangle = 0$, which is equivalent to the condition $S_1 = 0$;
- Stability of the stationary state corresponding to the minimum of the thermodynamic potential,

are successfully fulfilled. Therefore, in this sense, the Bilinear approximation is self-consistent. This is nice. However, on the other hand, there is another fact that leads to a little disappointment: it is not complete. In fact, in the spirit of Bilinear approximation, only quadratic terms in the action are taken into account while a possible S_4 term has been neglected. Actually, due to the shift $\psi = \psi_0 + \tilde{\psi}$ bilinear approximation is equivalent to the famous saddle point approximation, being a first order of the expansion in powers of the Planck constant $\hbar$.

At the end of the present section, we bring explicit expressions for other thermodynamic quantities for further reference. Namely,

$$F = \Omega + \mu N = \frac{Vg\rho^2}{2}(1 - n_1^2) + \frac{1}{2}\sum_k \left[E_k - \varepsilon_k - \Delta + \frac{\Delta^2}{2\varepsilon_k} \right]$$

$$+ T \sum_k \ln\left(1 - e^{-\beta E_k}\right),$$

$$S = -\left(\frac{\partial F}{\partial T}\right)_V = -\sum_k \ln(1 - e^{-\beta E_k}) + \beta \sum_k E_k f_B(E_k), \tag{6.101}$$

$$E = F + TS = \frac{Vg\rho^2}{2}(1 - n_1^2) + \frac{1}{2}\sum_k \left[E_k - \varepsilon_k - \Delta + \frac{\Delta^2}{2\varepsilon_k} \right]$$

$$+ \sum_k E_k f_B(E_k),$$

where $f_B(x) = 1/(\exp(\beta x) - 1)$, $n_1 = \rho_1/\rho = N_1/N$. Note that, in the literature [19,26], following the ideology of the Bilinear approach, the fluctuations higher than the first-order are usually neglected by setting $(1 - n_1^2) \approx 1$ in Eqs. (6.99) and (6.101).

6.7 Conclusion

Therefore the first approximation in theoretical description of BEC was the Bogolyubov approximation, proposed by him nearly eighty years ago [27]. Although this approximation assumes $\rho_0 \approx \rho$, i.e. $n_0 \approx 1$, neglecting condensate depletion, it was able to describe the deep mechanism of superfluidity: emergence of quasiparticles with the dispersion that satisfies Landau criterium. The one loop approximation, outlined above, goes beyond Bogolyubov approximation by taking into account depletions also e.g. through the Eq. (4.98). The condensed fraction in this case is not hundred percent, $n_0 \neq 1$, but determined from the Eq. (4.99). Moreover, being applied in the formalism of functional integrals, it leads to the explicit expression for the grand thermodynamic potential, given by Eq. (6.99), which involves also the LHY term, so that, further thermodynamic characteristics of the BEC system can be found by using well known thermodynamic relations (see e.g. Ref. [2]). For example, $F = \Omega + \mu N$, $S = -(\partial F/\partial T)$, $E = F + TS$ etc.

How can we go beyond one-loop approximation? Well, immediately what comes to mind is the two-loop approximation. However, it is rather complicated due to the fact that, in this case, one will have to face a problem of evaluation of two-loop integrals and double Matsubara summations [28]. In our opinion one of the more convenient approximations, taking into account higher order terms rather than quadratic ones, is optimized perturbation theory (OPT). The rest part of the present work is devoted to OPT and its applications. Here we note that a reader can find more information about field theoretical approaches in the review by Andersen [26].

References

1. L. Landau, E. Lifshitz, *Statistical Physics*, vol. 5 (Elsevier Science, 2013). https://books.google.co.uz/books?id=VzgJN-XPTRsC
2. V.I. Yukalov, Theory of cold atoms: basics of quantum statistics. Laser Phys. **23**(6), 062001 (2013). https://doi.org/10.1088/1054-660X/23/6/062001
3. L. Landau, E. Lifshitz, J. Sykes, *Mechanics*, Course of Theoretical Physics, vol. 1 (Butterworth-Heinemann, 1976). https://books.google.ru/books?id=e-xASAehg1sC
4. N. Nagaosa, S. Heusler, *Quantum Field Theory in Condensed Matter Physics*, Theoretical and Mathematical Physics (Springer Berlin Heidelberg, 2013). https://books.google.co.uz/books?id=dqbvCAAAQBAJ
5. N. Bogoliubov, *Lectures on Quantum Statistics*, Lectures on Quantum Statistics, vol. 1 (Gordon and Breach Science Publishers, 1967). https://books.google.co.uz/books?id=sZsOAAAAQAAJ

6. F. Abdullaev, J. Engelbrecht, S. Darmanyan, P. Khabibullaev, *Optical Solitons*, Springer Series in Nonlinear Dynamics (Springer Berlin Heidelberg, 2014). https://books.google.co.uz/books?id=wFXGoAEACAAJ

7. L. Pitaevskii, S. Stringari, *Bose-Einstein Condensation*, International Series of Monographs on Physics (Clarendon Press, 2003). https://books.google.co.uz/books?id=rIobbOxC4j4C

8. R. Feynman, *The Feynman Lectures on Physics*, vol. 7–8 (Basic Books, 2006)

9. I. Gradshteyn, I. Ryzhik, *Table of Integrals, Series, and Products* (Elsevier Science, 2014). https://books.google.co.uz/books?id=F7jiBQAAQBAJ

10. J. Schwinger, A theory of the fundamental interactions. Ann. Phys. **2**(5), 407–434 (1957). https://www.sciencedirect.com/science/article/pii/0003491657900155

11. P. Ramond, *Field Theory: A Modern Primer*, Frontiers in Physics: A Lecture Note and Reprint Series (Benjamin/Cummings Publishing Company, Advanced Book Program, 1981). https://books.google.co.uz/books?id=Mviy7nPGy1MC

12. L.D. Faddeev, A.A. Slavnov, *Gauge Fields: An Introduction To Quantum Theory, (Frontiers in Physics)*, 2nd edn. (CRC Press, Lausanne, Switzerland, 2019)

13. B. Svistunov, E. Babaev, N. Prokof'ev, *Superfluid States of Matter*, 1st edn. (CRC Press, Boca Raton, 2015). https://doi.org/10.1201/b18346

14. C. Itzykson, J. Zuber, *Quantum Field Theory*, Dover Books on Physics (Dover Publications, 2012). https://books.google.co.uz/books?id=CxYCMNrUnTEC

15. J. Kapusta, *Finite-Temperature Field Theory*, Cambridge Monographs on Mathematical Physics (Cambridge University Press, 1989). https://books.google.co.uz/books?id=nT2kbcDtMl8C

16. H. Kleinert, Z. Narzikulov, A. Rakhimov, Phase transitions in three-dimensional bosonic systems in optical lattices. J. Stat. Mech.: Theory Exp. **2014**(1), P01003 (2014). https://doi.org/10.1088/1742-5468/2014/01/P01003

17. H. Kleinert, Z. Narzikulov, A. Rakhimov, Quantum phase transitions in optical lattices beyond the Bogoliubov approximation. Phys. Rev. A **85**, 063602 (2012). https://doi.org/10.1103/PhysRevA.85.063602

18. T.D. Lee, K. Huang, C.N. Yang, Eigenvalues and eigenfunctions of a Bose system of hard spheres and its low-temperature properties. Phys. Rev. **106**, 1135–1145 (1957). https://doi.org/10.1103/PhysRev.106.1135

19. L. Salasnich, F. Toigo, Zero-point energy of ultracold atoms. Phys. Rep. **640**, 1–29 (2016), zero-point energy of ultracold atoms. https://www.sciencedirect.com/science/article/pii/S0370157316301338

20. R.L. Stratonovich, On a method of calculating quantum distribution functions. Doklady Akad. Nauk S.S.S.R. **115**, 1097 (1957)

21. J. Hubbard, Calculation of partition functions. Phys. Rev. Lett. **3**, 77–78 (1959). https://link.aps.org/doi/10.1103/PhysRevLett.3.77

22. H. Stoof, D. Dickerscheid, K. Gubbels, *Ultracold Quantum Fields*, Theoretical and Mathematical Physics (Springer Netherlands, 2008). https://books.google.co.uz/books?id=-9c2Ns2J5P4C

23. E. Taylor, A. Griffin, N. Fukushima, Y. Ohashi, Pairing fluctuations and the superfluid density through the BCS-BEC crossover. Phys. Rev. A **74**, 063626 (2006). https://link.aps.org/doi/10.1103/PhysRevA.74.063626

24. L. Landau, E. Lifshitz, *Quantum Mechanics: Non-relativistic Theory*, Course of Theoretical Physics, vol. 3 (Elsevier Science, 1991). https://books.google.co.uz/books?id=J9ui6KwC4mMC

25. A. Fetter, J. Walecka, *Quantum Theory of Many-particle Systems*, Dover Books on Physics (Dover Publications, 2003). https://books.google.co.uz/books?id=0wekf1s83b0C

26. J.O. Andersen, Theory of the weakly interacting Bose gas. Rev. Mod. Phys. **76**, 599–639 (2004). https://link.aps.org/doi/10.1103/RevModPhys.76.599

27. N.N. Bogolyubov, On the theory of superfluidity. J. Phys. (USSR) **11**, 23–32 (1947)

28. E. Braaten, A. Nieto, Quantum corrections to the ground state of a trapped Bose-Einstein condensate. Phys. Rev. B **56**, 14 745–14 765 (1997). https://link.aps.org/doi/10.1103/PhysRevB.56.14745

Optimized Perturbation Theory 7

7.1 Introduction

From the previous chapter, it is understood that the main thermodynamic potential $\Omega = -T \ln Z$ of the many-particle Bose system can not be calculated exactly even in the case of simple contact interaction. In the framework of the path integral formalism the reason for such failure is that, the functional integral

$$Z = \int \mathcal{D}\psi^\dagger \mathcal{D}\psi \, e^{-S[\psi^\dagger, \psi]} \tag{7.1}$$

with the action S given by (6.15) can not be evaluated analytically. In this regard, we remember that a path integral only the type of (6.48), i.e. the Gaussian one, can be taken exactly. Thus, in the language of path integrals the goal of any approach is somehow converting (7.1) to the form of e.g. sum of simple Gaussian integrals.

Here, the first idea that comes to mind is the standard perturbation theory (SPT), which assumes using the power series in the coupling constant. The best example is Quantum Electrodynamics, where the power series in the small parameter, fine-structure constant $e^2/4\pi\epsilon_0\hbar c \approx 1/137$ is used. As to our task, there is indeed a good parameter $\gamma = \rho a_s^3$, which is small enough for a uniform weakly interacting dilute Bose gases. Hence, at zero temperature, one may construct a perturbative scheme in a power series of γ. However, this would not work for finite temperatures! The reason is that, as it has been shown, e.g. in the book by Le Bellac [1], when the temperature is involved, the perturbative theory becomes unreliable, because powers of the coupling constant (or γ) became surmounted by the powers of the temperature. Problems with SPT arise especially near the critical temperature since large fluctuations can emerge

© The Author(s), under exclusive license to Springer Nature Switzerland AG 2026   105
A. Rakhimov and S. Mardonov, *Theory of Quantum Bose Liquids Beyond Bogoliubov Approximation*, Lecture Notes in Physics 1045,
https://doi.org/10.1007/978-3-032-05096-0_7

in the system due to infrared divergences.[1] Unfortunately, IR divergences create a serious problem also in the approach, referred to as loop expansion, which is, actually, the expansion in power series of $\hbar$ [2].

Thus, to calculate the partition function Z beyond the bilinear approximation, outlined in previous chapters, an analytic nonperturbative method should be developed. It must obey at least two basic requirements. First, it should be self-consistent, and second, it should produce useful results already at the lowest orders without the need for going to higher orders. That is, it should produce results that quickly converge to some order, where calculations are still feasible analytically or semianalytically. In our opinion Optimized Perturbation Theory (OPT) is the best one, satisfying both these requirements. The present chapter is devoted to describe the basic features of OPT. In the next chapters, we shall apply it to study BEC more accurately beyond the bilinear approximation.

In the last year's Optimized Perturbation Theory, first proposed by Yukalov [3], and developed by many authors (see review article [4]) become a powerful tool in theoretical physics. This theory sometimes appears in the literature under different names, such as modified perturbation theory, renormalized perturbation theory, variational Gaussian approximation [5], variational perturbation theory, linear delta expansion (LDE) and so on. It has been successfully applied to treating anharmonic quantum crystals, the theory of melting, physics of polymers, quantum magnets [6] and many other systems and phenomena [4]. In general, OPT includes at least one variational parameter as well as a unique artificial parameter-δ, which serves as a "small parameter" for expansion.

7.2   OPT: How Does It Work?

To facilitate the readers understanding, we start with a simple example, which is interpreted in the literature as the partition function of the zero-dimensional ϕ^4 theory. So, imagine that the following integral

$$I(g) = \int_{-\infty}^{\infty} dx\, e^{-x^2 - gx^4},\tag{7.2}$$

has to be calculated, where g is an arbitrary positive constant. We shall make calculations in three ways:

---

[1] Infrared divergence (IR) corresponds to the divergence of momentum integrals at small momentum: $\int_0^\infty f(p)dp \to \infty$ due to $\lim_{p\to 0} f(p) = \infty$, while an ultraviolet divergence (UV) is related to the behavior of $f(p)$ at large momentum: $\lim_{p\to\infty} f(p) = \infty$.

(1) Exactly[2];
(2) By expanding in the power series of g;
(3) By using OPT, that is delta expansion method.

The result of exact computations is presented in Fig. 7.1 (solid line). The second method gives

$$I(g) = \int_{-\infty}^{\infty} dx e^{-x^2 - gx^4} \approx \int_{-\infty}^{\infty} dx e^{-x^2}[1 - gx^4 + \frac{g^2 x^8}{2} + \cdots]$$
$$= \sqrt{\pi}[1 - \frac{3g}{4} + \frac{105g^2}{32} + O(g^3)], \tag{7.3}$$

where we used the tabular integral

$$\int_{-\infty}^{\infty} dx x^n e^{-x^2} = \frac{1}{2}\Gamma\left(\frac{1}{2} + \frac{n}{2}\right)[1 + (-1)^n], \tag{7.4}$$

with $\Gamma(x)$ is the Gamma function. The result is presented in Fig. 7.1 by the dashed line.

For the third case, following the ideology of OPT, we rewrite the integrand as

$$e^{-x^2 - gx^4} = e^{-\lambda^2 x^2} e^{-\delta[gx^4 + (1 - \lambda^2)x^2]}, \tag{7.5}$$

which turns into an identity for $\delta = 1$. As to λ, we shall consider it as a trial parameter. Now, making expansion in the powers of δ, we have

$$I(g) = \int_{-\infty}^{\infty} dx e^{-\lambda^2 x^2} + \delta \int_{-\infty}^{\infty} dx e^{-\lambda^2 x^2}[-gx^4 - x^2(1 - \lambda^2)] + O(\delta^2)$$
$$\equiv I_\delta(\lambda). \tag{7.6}$$

Only at this step we have right to set $\delta = 1$ to obtain

$$I(g) \approx I_\delta(\lambda)|_{\delta=1} = \frac{\sqrt{\pi}}{4\lambda^5}[6\lambda^4 - 2\lambda^2 - 3g]. \tag{7.7}$$

What about λ? We fix this trial parameter by using Principal of Minimal Sensitivity (PMS) [7]:

$$\frac{dI(\lambda, g)}{dg} = 0, \tag{7.8}$$

---

[2] One may find explicit expression in terms of Bessel functions by using MAPLE or MATHEMATICA.

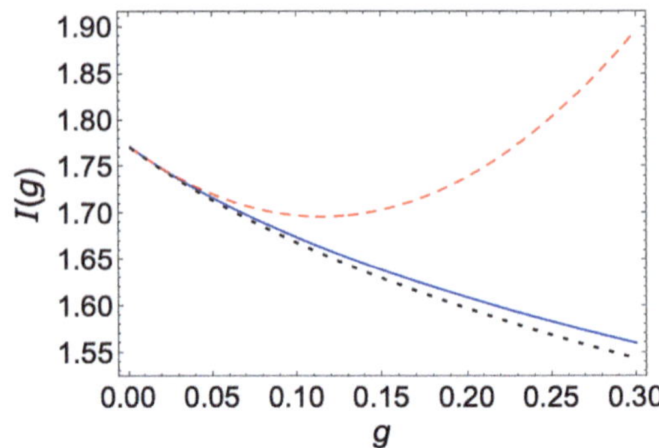

Fig. 7.1 The integral in Eq. (7.2) evaluated by three methods: solid line by exact numerical calculation, dashed line by SPT, Eq. (7.3) and dotted line by OPT, Eq. (7.10)

which has the solution $\lambda = \bar{\lambda}$ as

$$\bar{\lambda} = \frac{\sqrt{2 + 2\sqrt{1 + 10g}}}{2} \tag{7.9}$$

Using this solution in (7.7) we finally get

$$I(\lambda = \bar{\lambda}, g) \approx \frac{4\sqrt{\pi}(1 + 6g + \sqrt{1 + 10g})}{(1 + \sqrt{1 + 10g})^2\sqrt{2 + 2\sqrt{1 + 10g}}}. \tag{7.10}$$

This result is plotted in Fig. 7.1 by dotted curve. It is seen that the standard perturbation scheme, that is the direct expansion in powers of g leads to an unexpected increase of the integral $I(g)$ (dashed line). Meanwhile, the result of OPT (dotted line) which seems rather artificial, almost coincides with the exact one (solid line) even in the first step of the approximation, when only the linear term in the delta has been taken into account.

7.3   OPT: General Ideology

By definition, optimised perturbation theory is a systematic method defining a sequence of successive approximations, whose convergence is governed by control functions [4].

Suppose that we are looking for a real function $f(g)$ of a real variable g. When the sought function is defined by complicated problem, that cannot be solved exactly, one invokes a kind of perturbation theory yielding a sequence of approximations $f_k(g)$, with $k = 0, 1, 2, \ldots$. As a rule, the sequence $\{f_k(g)\}$ is divergent, similar to $I(g)$, approximated by a power series of g in previous section. The main idea of OPT is to rearrange the sequence $\{f_k(g)\}$ to another form $\{F_k(g, \lambda_k)\}$ by introducing trial parameters λ_k that make the rearranged sequence convergent. The functions $\{F_k(g, \lambda_k)\}$ are referred as control functions. Such a rearrangement can be realized in many ways [4, 8–10]. Below we outline a version of OPT, which is called as linear delta expansion or variational Gaussian approximation [11, 12].

What is the difference between a usual variational approximation and a variational Gaussian approximation? In the former one presents the sought function, say, (a solution of Schrödinger or Gross-Pitaevskii equation), in an explicit form intuitively, and makes an attempt to vary its trial parameters in order to describe experimental data. Instead, in OPT, the variational parameters are introduced directly into the action or the Hamiltonian from the very beginning.

Let's consider a system characterized by a Hamiltonian $H(g)$, where g is, in general, a coupling constant. If the system is complicated and the corresponding equations cannot be solved, and hence, the Hamiltonian can not be diagonalized exactly, we can take as a zero approximation a simpler Hamiltonian, say $H_0(\lambda)$ with λ will be considered as a variational parameter. Then the whole Hamiltonian can be presented identically as

$$
\begin{aligned}
H(g) &\to H^\delta = H_{\text{free}}(\lambda) + H_{\text{int}}(\delta, \lambda, g), \\
H_{\text{int}} &= \delta(H - H_{\text{free}}),
\end{aligned}
\tag{7.11}
$$

where δ is a dummy parameter to be set to unity ($\delta = 1$) at the end of calculations. The Hamiltonian H^δ interpolates between the solvable $H_{\text{free}}(\lambda)$ (when $\delta = 0$) and the original one (when $\delta = 1$). This is why we have the right to declare it as a parameter of perturbation theory, making an expansion in power series of δ. Here we have to underline that linear delta expansion does not mean taking only the linear term in the expansion! It just means that the expansion parameter δ is introduced to the Hamiltonian linearly. Further, suppose that, by using such expansion to order δ^k we obtain an explicit expression for a physical quantity, say, $\Phi^{(k)}(\lambda)$. Then the trial parameter λ can be fixed by using one of the following methods:

(1) Principal of minimal sensitivity (PMS) [7]:

$$
\frac{d\Phi^{(k)}(\lambda, \delta = 1)}{d\lambda} = 0.
\tag{7.12}
$$

This naturally means that any physical observable should be less sensitive to the artificially introduced parameter.[3] The optimum value of $\bar{\lambda}$, which satisfies the last equation, is a function of the original parameters of the theory. In particular, it is a nontrivial function of the coupling constants and because of this, nonperturbative results are generated. In some cases, it will be hard to solve the Eq. (7.12). In such cases, one may use another method of optimization-Fastest Apparent Convergence (FAC) criteria [7,13]. This procedure requires from the k-th coefficient of the perturbative expansion

$$
\Phi^{(k)} = \sum_{i=0}^{k} c_i \delta_i
\tag{7.13}
$$

that $[\Phi^{(k)} - \Phi^{(k-1)}]|_{\delta=1} \to 0$.

---

[3] If there are several trial parameters, then PMS should be imposed for each of them separately.

Thus there are at least two methods of fixation of trial parameters: PMS and FAC. Moreover, if we take into account that, in general, PMS equation (7.12) is nonlinear and may have several solutions, then we are struck with the problem of making a preference. In practice such a choice is made in the following way: The trivial solution and those that are not dependent on the coupling constant (and thus cannot lead to nonperturbative results) are not considered. In addition to these, at first order, there is only one nontrivial solution (first family) consistent with all other reliable approximations. In particular, if there is a quadratic term in the Hamiltonian, then it would be desirable when some part of the results coincide with that of the bilinear approximation. This family is then followed in the next order and used in all calculations. It was shown that [14] this is the most consistent and unambiguous way for choosing the optimal value of the trial parameter. As to the physical quantity Φ to be used for the optimization procedure, here, in general, any physical observable may be considered. So, for example, it is customary to use the energy spectrum E_n in quantum mechanics or the thermodynamic potential Ω in the equilibrium statistical mechanics.

It is also worth underlining that, in OPT, the selection and evaluation of Feynman diagrams proceed in the same fashion as in ordinary perturbation theory, including the renormalization procedure [11, 15]. In the next sections, we illustrate the application of OPT in quantum mechanics and in $\lambda\phi^4$ field theory.

7.4   Eigenvalue Problem: Anharmonic Oscillator

In the present section, we shall study the energy spectrum of a one-dimensional anharmonic oscillator with the Hamiltonian[4]

$$H = H_0 + V,$$

$$H_0 = -\frac{d^2}{2mdx^2} + \frac{m\omega^2 x^2}{2},$$

$$V = \frac{m^2\omega^2 g x^4}{\hbar^2},$$

(7.14)

where oscillator parameters, m, ω and g are supposed to be given. The energy spectrum of such system, i.e. the diagonal matrix elements of H, correspond to the solution of stationary Schrödinger equation $H\psi = E_n\psi$. Note that, the solutions for the harmonic oscillator ($g = 0$ in (7.14)) are well known and could be find elsewhere [16], while for the anharmonic oscillator, they do not exist analytically. We shall estimate E_n both in standard and optimized perturbation theories for comparison, based on the solutions of the harmonic oscillator. Before passing to the concrete calculations, we refresh in our mind the main ideas of any perturbation theory.

---

[4] Below we write $\hbar$ explicitly for convenience.

A perturbation theory in quantum mechanics is in fact an iterative procedure. Similarly to the scattering problem, where one may solve the Lippmann-Schwinger (4.39) iteratively

$$T = V + V G_0 V + V G_0 V G_0 V + \cdots , \tag{7.15}$$

the energy levels in the boundary problem can also be presented in the iterative scheme. More exactly, let's suppose that the whole Hamiltonian can be presented as $H = \tilde{H}_0 + \tilde{H}_{\text{int}}$ and the eigenvalues of its part $\tilde{H}_0$ are known as $\tilde{E}_n^{(0)}$. Then the spectrum of H, where $\tilde{H}_{\text{int}}$ is considered as a perturbation, can be found as:

$$\tilde{E}_n = \tilde{E}_n^{(0)} + \Delta E_n \equiv \tilde{E}_n^{(0)} + \Delta E_n^{(1)} + \Delta E_n^{(2)} + \cdots ,$$

$$\Delta E_n = \langle n | \tilde{H}_{\text{int}} | n \rangle + \sum_{m \neq n} \langle n | \tilde{H}_{\text{int}} | m \rangle \langle m | G_0 | m' \rangle \langle m' | \tilde{H}_{\text{int}} | n \rangle$$

$$+ \sum_{m \neq n, m'' \neq n} \langle n | \tilde{H}_{\text{int}} | m \rangle \langle m | G_0 | m' \rangle \langle m' | \tilde{H}_{\text{int}} | m'' \rangle \tag{7.16}$$

$$\times \langle m'' | G_0 | m''' \rangle \langle m''' | \tilde{H}_{\text{int}} | n \rangle + \cdots ,$$

where the "free" propagator $G_0 = 1/(\tilde{H}_0 - \tilde{E}_n^{(0)})$ is diagonal in the space of eigenfunctions $|m\rangle$ of the operator $\tilde{H}_0$: $\tilde{H}_0 | m \rangle = \tilde{E}_m^{(0)} | m \rangle$, so, $\langle m | G_0 | m' \rangle = \delta_{mm'}/(\tilde{E}_m^{(0)} - \tilde{E}_n^{(0)})$. Thus, we obtain the textbook results for the first and second-order corrections as follows

$$\Delta E_n^{(1)} = \langle n | \tilde{H}_{\text{int}} | n \rangle ,$$

$$\Delta E_n^{(2)} = \sum_{m \neq n} \frac{(\langle n | \tilde{H}_{\text{int}} | m \rangle)^2}{\tilde{E}_m^{(0)} - \tilde{E}_n^{(0)}} , \tag{7.17}$$

where the summation includes all possible values of the unperturbed energy levels, but $\tilde{E}_n^{(0)}$. Here, one may ask, "Why is the index n allocated?". This is just because we would like to find the shift of the unperturbated energy level "n" due to the $\tilde{H}_{\text{int}}$.

7.4.1 Standart Perturbation Theory

Following the book by Flugge [16] we rewrite the Hamiltonian in (7.14) as

$$H = H_0 + H_{\text{int}},$$

$$H_0 = -\frac{d^2}{2m d x^2} + \frac{m \omega^2 x^2}{2} ,$$

$$H_{\text{int}} = \frac{m^2 \omega^2 g x^4}{\hbar^2} . \tag{7.18}$$

The nonperturbative harmonic oscillator Hamiltonian has well-known eigenvalues $E_n^{(0)} = \hbar\omega(n + 1/2)$, $n = 0, 1, 2, \ldots$ and eigenfunctions

$$|n\rangle = u_n(x) = (2^n n! l \sqrt{\pi})^{-1/2} e^{-\eta^2/2} H_n(\eta), \qquad \eta = x/l, \quad l^2 = \hbar/m\omega, \tag{7.19}$$

where $H_n(\eta)$ are Hermite polynomials: $H_0(\eta) = 1$, $H_1(\eta) = 2\eta$, $H_2(\eta) = 4\eta^2 - 1 \cdots$. Now using Eqs. (7.17), (7.18) and (7.19) we obtain the first order correction to the energy level n as

$$\Delta E_n^{(1)} = g \int_{-\infty}^{\infty} \eta^4 u_n^2(\eta) dx = \frac{3g(2n^2 + 2n + 1)}{4}. \tag{7.20}$$

For the second order, one has to perform summation in (7.17), which includes the following integrals

$$\langle n|\eta^4|m\rangle = \frac{1}{l} \int_{-\infty}^{\infty} \eta^4 u_n(\eta) u_m(\eta) d\eta. \tag{7.21}$$

It can be shown that not all of such integrals are finite. Direct integration by using properties of $H_n(\eta)$ leads to following results

$$\begin{aligned}
\langle n|\eta^4|n + 2\rangle &= \sqrt{(n + 2)(n + 1)}(n + 3/2), \\
\langle n|\eta^4|n - 2\rangle &= \sqrt{n(n - 1)}(n - 1/2), \\
\langle n|\eta^4|n + 4\rangle &= \frac{1}{4}\sqrt{(n + 4)(n + 3)(n + 2)(n + 1)}, \\
\langle n|\eta^4|n - 4\rangle &= \frac{1}{4}\sqrt{n(n - 1)(n - 2)(n - 3)}
\end{aligned} \tag{7.22}$$

while all other nondiagonal matrix elements of η^4 vanish. Thus, the summation overall states in (7.17) will be simplified, resulting in

$$\Delta E_n^{(2)} = -\frac{g^2}{8\hbar\omega}[34n^3 + 51n^2 + 59n + 21]. \tag{7.23}$$

Particularly, the ground state energy ($n = 0$) of anharmonic oscillator in the second order of SPT will be

$$E_0 = \hbar\omega \left[\frac{1}{2} + \frac{3g}{4\hbar\omega} - \frac{21}{8}\left(\frac{g}{\hbar\omega}\right)^2\right]. \tag{7.24}$$

7.4.2 OPT Calculations

Now we proceed with similar calculations in the linear delta expansion, which is one of the popular versions of the optimized perturbation theory. For this purpose, we rewrite the Hamiltonian (7.14) as $H = \tilde{H}_0 + \tilde{H}_{\text{int}}$ but this time we have

$$\begin{aligned}
\tilde{H}_0 &= -\frac{d^2}{2mdx^2} + \frac{m\lambda^2 x^2}{2}, \\
\tilde{H}_{\text{int}} &= \delta\left[\frac{m^2 x^2(\omega^2 - \lambda^2)}{2} + \frac{m^2\omega^2 g x^4}{\hbar^2}\right],
\end{aligned} \tag{7.25}$$

where λ is a trial parameter. Clearly, the nonperturbative harmonic oscillator Hamiltonian $\tilde{H}_0$ has the eigenfunctions given by (7.19) with the eigenvalues $E_n^0(\lambda) = \hbar\lambda(n + 1/2)$. However, this time we have the following difference: $\eta^2 = x^2 m\lambda/\hbar$, $l = x/\eta$.

In the zero order of δ we have

$$E_n^{LDE} = \langle n|\tilde{H}_0|n\rangle = E_n^0(\lambda), \tag{7.26}$$

which corresponds to the harmonic oscillator with the "unknown" frequency λ.

The first order in delta gives

$$E_n^{LDE} = E_n^0(\lambda) + \langle n|\tilde{H}_{\text{int}}|n\rangle \equiv E_n^0(\lambda) + E_n^{(1)}. \tag{7.27}$$

It is convenient to calculate diagonal matrix elements $\langle n|x^2|n\rangle$ and $\langle n|x^4|n\rangle$ in Fock space where the creation and annihilation operators [17]

$$\hat{a}^+ = \frac{1}{\sqrt{2}}\left(\eta - \frac{\partial}{\partial\eta}\right), \qquad \hat{a} = \frac{1}{\sqrt{2}}\left(\eta + \frac{\partial}{\partial\eta}\right), \tag{7.28}$$

are introduced. They are normalized as $\langle n|\hat{a}\hat{a}^+|\bar{m}\rangle = \delta_{\bar{m}n}(n + 1)$, $\langle n|\hat{a}^+\hat{a}|\bar{m}\rangle = \delta_{\bar{m}n}n$. So, one obtains

$$\langle n|x^2|n\rangle = (n + 1/2)\frac{\hbar}{m\lambda},$$
$$\langle n|x^4|n\rangle = (1 + 2n + 2n^2)\frac{3\hbar^2}{4\,m^2\lambda^2}. \tag{7.29}$$

Thus using these in (7.27) gives

$$E_n^{LDE} = E_n^0(\lambda) + c_1\langle n|x^2|n\rangle + c_2\langle n|x^4|n\rangle \equiv \hbar(n + \frac{1}{2})F(\lambda, \omega, g),$$
$$F(\lambda, \omega, g) = \lambda + \frac{\omega^2 - \lambda^2}{2\lambda} + \frac{3\hbar g\omega^2 a_n}{4\lambda^2}, \tag{7.30}$$

where $a_n = (1 + 2n^2 + 2n)/(n + 1/2)$. Now we are on the stage of fixing λ by PMS condition:

$$\frac{dE_n^{LDE}}{d\lambda} = 0, \tag{7.31}$$

which leads to the following cubic equation with respect to λ

$$\lambda^3 - \lambda\omega^2 + 3\hbar g\omega^2 a_n = 0, \tag{7.32}$$

whose real positive solution $\bar{\lambda}$ is given by

$$\bar{\lambda}_n(g, \omega) = \frac{\alpha}{6} + \frac{2\omega^2}{\alpha},$$
$$\alpha^3 = 324\hbar g\omega^2 a_n + 12\sqrt{729\hbar^2 g^2 a_n^2\omega^4 - 12\omega^6}. \tag{7.33}$$

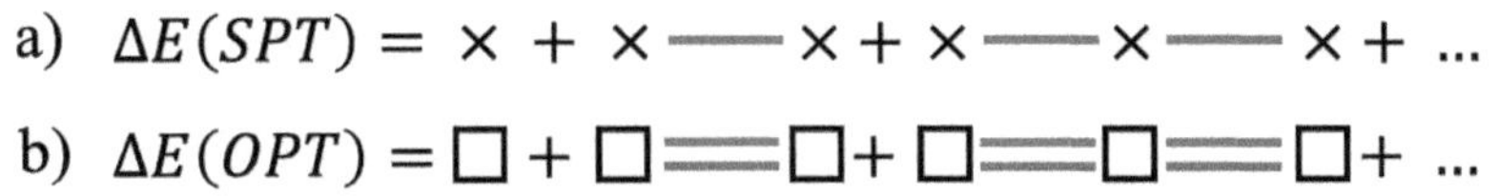

a) $\Delta E(SPT) = \times + \times \!\!-\!\!-\!\! \times + \times \!\!-\!\!-\!\! \times \!\!-\!\!-\!\! \times + \ldots$

b) $\Delta E(OPT) = \square + \square =\!\!=\!\!= \square + \square =\!\!=\!\!= \square =\!\!=\!\!= \square + \ldots$

Fig. 7.2 Perturbation series for the shift of the energy due to the perturbative interaction in SPT (**a**) and OPT (**b**). Here $\times$ vertex corresponds V in Eq. (7.14) and $\square$ vertex corresponds to $\tilde{H}_{\text{int}}$ in Eq. (7.25). Single (—) and double (=) lines correspond to the propagators $1/(H_0 - E_n^0)$ and $1/(\tilde{H}_0 - E_n^0)$, respectively

Putting this solution to (7.30) we finally get

$$E_n^{LDE} = \hbar(n + \frac{1}{2})\frac{\alpha^6 + 72\alpha^4\omega^2 + 864\alpha^2\omega^4 + 1728\omega^6 + 324\hbar g\omega^2\alpha^3 a_n}{12\alpha(\alpha^2 + 12\omega^2)^2},$$

$$(7.34)$$

which is the energy spectrum of the anharmonic oscillator in the first order of delta expansion.

Therefore, we have studied eigenvalues of the Hamiltonian (7.18) by using two different approximations. The question arises, "How much are they reliable?" To discuss their accuracy, we shall compare them with the exact numerical results [18, 19] by introducing the percentage error:

$$\epsilon_n^{SPT,OPT} = \left| \frac{E_n^{SPT,OPT} - E_n^{exact}}{E_n^{exact}} \right| \times 100\%. \qquad (7.35)$$

Our numerical analyses, performed for the ground state ($n = 0$) show that $\epsilon_0^{SPT} \approx 20\%$, while $\epsilon_0^{OPT} \approx 2\%$. That is, linear delta expansion gives a nice accuracy even in its first order. Moreover, high-order corrections to the spectrum in OPT fast converge: $\epsilon_0^{OPT}(\delta^2) \approx 0.8\%$, $\epsilon_0^{OPT}(\delta^3) \approx 0.04\%$ [4], while those in SPT start to diverge even in the third order of g [18]. The perturbation series for standard and optimized theories are illustrated in Fig. 7.2a and b, respectively.

OPT has been successfully applied not only to the anharmonic oscillator but also to other quantum mechanical systems such as hydrogen atoms, helium atoms, crystals etc. [4, 20].

7.5　Relativistic Finite Temperature Field Theory

OPT is a good tool in high-energy physics, also. Below we outline the application of the delta expansion for the relativistic $\lambda\phi^4$ theory, which is closely related to the problem of the early universe. Above, we underlined the breakdown of standard perturbation theory at high temperatures and its poor convergence property. However, OPT is free of such failures.

The procedure of application of LDE to a theory described by the Lagrangian density $\mathcal{L}$ starts with an interpolation defined by

$$\mathcal{L} \to \mathcal{L}^\delta = (1 - \delta)\mathcal{L}_0(\eta) + \delta\mathcal{L} = \mathcal{L}_0(\eta) + \delta[\mathcal{L} - \mathcal{L}_0(\eta)] \qquad (7.36)$$

where $\mathcal{L}_0$ is the Lagrangian density of a solvable theory (similar to the "harmonic oscillator"), which is modified by the introduction of an arbitrary mass parameter η. The Lagrangian density $\mathcal{L}^\delta$ interpolates between the solvable $\mathcal{L}_0(\eta)$ (when $\delta = 0$) and the original $\mathcal{L}$ (when $\delta = 1$). The procedure defined by (7.36) leads to modified Feynman vertices that become multiplied by δ, and modified propagator that now depends on the variational parameter η. Actually, due to the freedom in fixing η "nonperturbative" results will be generated.

So, for the original Lagrangian of the $\lambda\phi^4$ theory

$$\mathcal{L} = \frac{(\partial_\mu\phi)^2}{2} - \frac{m_0^2\phi^2}{2} - \frac{\lambda\phi^4}{4!},\tag{7.37}$$

the procedure (7.36) gives

$$\mathcal{L}^\delta = \mathcal{L}_0(\eta) - \frac{\delta\lambda\phi^4}{4!} + \frac{\delta\eta^2\phi^2}{2} + \mathcal{L}_{ct}^\delta,\tag{7.38}$$

where

$$\mathcal{L}_0(\eta) = \frac{(\partial_\mu\phi)^2}{2} - \frac{m_0^2\phi^2}{2} - \frac{\eta^2\phi^2}{2},\tag{7.39}$$

and $\mathcal{L}_{ct}^\delta$ is the part of the Lagrangian density carrying the renormalization terms needed to render the model finite. This procedure introduces a new quadratic interaction term, with the Feynman rule $i\delta\eta^2$ corresponding to the vertex. Besides, the original propagator

$$G_0(k) = i(k^2 - m_0^2 + i\epsilon)^{-1},\tag{7.40}$$

becomes[5]

$$G_0^\delta(k) = i(k^2 - m_0^2 - \eta^2 + i\epsilon)^{-1},\tag{7.41}$$

while the original quartic vertex is changed as $-i\lambda \to -i\lambda\delta$. Now making the usual shift of the scalar field $\phi = \phi_0 + \tilde{\phi}$ the Lagrangian density can be rewritten as

$$\mathcal{L} = \mathcal{L}_{free} + \delta\mathcal{L}_{int},$$
$$\mathcal{L}_{free} = \frac{(\partial_\mu\tilde{\phi})^2}{2} - \frac{\omega^2\tilde{\phi}^2}{2},\tag{7.42}$$
$$\mathcal{L}_{int} = -\frac{\lambda\phi_0\tilde{\phi}^3}{6} - \frac{\lambda\tilde{\phi}^4}{4!},$$

where

$$\omega^2 = \pm m_0^2 + \frac{\delta\lambda\phi_0^2}{2} + (1 - \delta)\eta^2.\tag{7.43}$$

---

[5] Here k is a four-vector.

Here we introduced the $\pm$ sign in purpose:$+$ sign corresponds to the symmetric case, while $-$ sign leads to the SSB regime. In both regimes, the desired quantity-thermodynamic potential can be evaluated after calculation of the partition function as $\Omega = -T \ln Z$.

Since $\mathcal{L}_{free}$ is quadratic in $\tilde{\phi}$, the relevant part of the partition function

$$Z^{(2)} = \int \mathcal{D}\tilde{\phi}\, e^{i \int dx^4 \mathcal{L}_{free}(\tilde{\phi})}, \tag{7.44}$$

can be obtained directly by using the rules outlined in the Chap. 6. However, when one also introduces $\mathcal{L}_{int}$ part, the path integral will be such complicated, as there would be no way but the expansion in powers of δ:

$$e^{i \int dx^4 \mathcal{L}_{free}(\tilde{\phi})} e^{i\delta \int dx^4 \mathcal{L}_{int}(\tilde{\phi})} = e^{i \int dx^4 \mathcal{L}_{free}(\tilde{\phi})}$$

$$\times \left[1 + i\delta \int dx^4 \mathcal{L}_{int}(\tilde{\phi}(x)) + \frac{i^2\delta^2}{2!} \int dx^4 dy^4 \mathcal{L}_{int}(\tilde{\phi}(x))\mathcal{L}_{int}(\tilde{\phi}(y)) + O(\delta^3) \right]. \tag{7.45}$$

Then one may use formulas (6.47)–(6.49) to evaluate functional integrals. For the finite temperature effects, one may perform calculations in the imaginary time formalism outlined in Chap. 4 and find the thermodynamic potential in a power series of δ, which will then be set to unity. The only complication may be caused by regularization, which will be cured by a proper choice of $\mathcal{L}_{ct}$.

Since in the present lectures our aim is not to go deep into the problems of high energy field theory, we limit ourselves to the illustration of the power of optimal perturbation theory.[6] In Fig. 7.3, the pressure of the system of self-interacting scalar bosons, $P = -\Omega/V$, vs the renormalized coupling constant is presented to orders of δ and δ^2. For comparison, the results of the first and second orders of standard perturbation theory are also plotted. It is seen that the SPT predictions diverge already from the second order, while the OPT ones converge and are physically reliable. Note that OPT has also been applied to study the critical temperature [15] and critical exponents of phase transitions [4, 11].

7.6   Conclusion

In the present chapter, we have made an introduction to the optimized perturbation theory at the student level to show that it is rather a powerful tool for theorists in a wide range of physics. OPT makes it possible to obtain quite accurate approximations for those complicated physical problems for which only a few terms of a perturbative algorithm are available. This approach does not require any small parameters of the physical problem considered. For instance, the coupling parameters can be arbitrarily

---

[6] A reader can find the details, e.g. in Refs. [14, 21, 22].

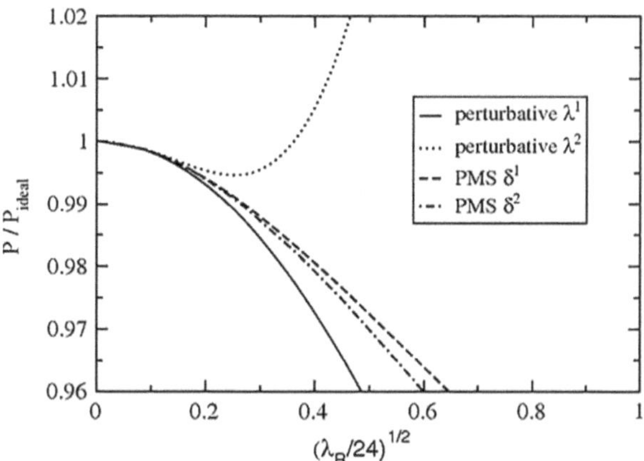

Fig. 7.3 The pressure normalized to the pressure of ideal gas with $\lambda = 0$ vs the renormalized coupling constant in the symmetric phase. Reprinted with permission from [14]. © 2008, American Physical Society. All rights reserved

strong. The method introduces an arbitrary trial parameter that prevents infrared-divergence problems. Nonperturbative results are generated when one optimizes the theory with respect to this variational parameter. Convergence of the theory is achieved by introducing control functions, by means of which the sequence of approximations is reorganized to become convergent for any given parameters. The main idea of Optimized Perturbation Theory is to reorganize a divergent sequence into a convergent sequence by introducing control functions. Such a reorganization of the approximants for the sought quantity is analogous to the renormalization of these approximants. This approximation has been shown to be a powerful nonperturbative method and sufficiently simple to use in very different applications, including the study of nonperturbative high-temperature effects, as shown very recently in the context of finite temperature quantum field theory.

In the next chapter, we shall apply the version of OPT-linear delta approximation to study BEC beyond bilinear approximation.

References

1. M. Bellac, *Thermal Field Theory*, Cambridge Monographs on Mathematical Physics (Cambridge University Press, 2000). https://books.google.co.uz/books?id=00_x6GR8GXoC
2. C. Itzykson, J. Zuber, *Quantum Field Theory*, Dover Books on Physics (Dover Publications, 2012). https://books.google.co.uz/books?id=CxYCMNrUnTEC
3. V.I. Yukalov, Theory of perturbations with a strong interaction. Mosc. Univ. Phys. Bull. **31**, 10–15 (1976). https://doi.org/10.48550/arXiv.1901.09598
4. V.I. Yukalov, Interplay between approximation theory and renormalization group. Phys. Part. Nucl. **50**(2), 141–209 (2019). https://doi.org/10.1134/S1063779619020047
5. A. Rakhimov, J.H. Yee, Optimized post-Gaussian approximation in the background field method. Int. J. Mod. Phys. A **19**(10), 1589–1607 (2004). https://doi.org/10.1142/S0217751X04017884
6. A. Rakhimov, M. Nishonov, L. Rani, B. Tanatar, Characteristic temperatures of a triplon system of dimerized quantum magnets. Int. J. Mod. Phys. B **35**(02), 2150018 (2021). https://doi.org/10.1142/S0217979221500181

7. P.M. Stevenson, Optimized perturbation theory. Phys. Rev. D **23**, 2916–2944 (1981). https://link.aps.org/doi/10.1103/PhysRevD.23.2916

8. S. Chiku, T. Hatsuda, Optimized perturbation theory at finite temperature. Phys. Rev. D **58**, 076001 (1998). https://link.aps.org/doi/10.1103/PhysRevD.58.076001

9. P.M. Stevenson, Gaussian effective potential. ii. $\lambda\varphi^4$ field theory. Phys. Rev. D **32**, 1389–1408 (1985). https://link.aps.org/doi/10.1103/PhysRevD.32.1389

10. I. Feranchuk, A. Ivanov, V.-H. Le, A. Ulyanenkov, *Non-perturbative Description of Quantum Systems*, Lecture Notes in Physics (Springer International Publishing, Springer Cham, Switzerland, 2015). https://doi.org/10.1007/978-3-319-13006-4

11. H. Kleinert, V. Schulte-Frohlinde, *Critical Properties of Φ^4-Theories*, G-Reference, Information and Interdisciplinary Subjects Series (World Scientific, 2001). https://books.google.co.uz/books?id=rBfWk4vVi-cC

12. F.F. de Souza Cruz, M.B. Pinto, R.O. Ramos, Transition temperature for weakly interacting homogeneous Bose gases. Phys. Rev. B **64**, 014515 (2001). https://link.aps.org/doi/10.1103/PhysRevB.64.014515

13. V. Yukalov, E. Yukalova, Self-similar perturbation theory. Ann. Phys. **277**(2), 219–254 (1999). https://www.sciencedirect.com/science/article/pii/S0003491699959535

14. R.L.S. Farias, G. Krein, R.O. Ramos, Applicability of the linear δ expansion for the $\lambda\phi^4$ field theory at finite temperature in the symmetric and broken phases. Phys. Rev. D **78**, 065046 (2008). https://link.aps.org/doi/10.1103/PhysRevD.78.065046

15. M.B. Pinto, R.O. Ramos, F.F. de Souza Cruz, Effective potential and thermodynamics for a coupled two-field Bose-gas model. Phys. Rev. A **74**, 033618 (2006). https://link.aps.org/doi/10.1103/PhysRevA.74.033618

16. S. Flügge, *Practical Quantum Mechanics*, Classics in Mathematics (Springer Berlin Heidelberg, 2012). https://books.google.co.uz/books?id=Md5sCQAAQBAJ

17. A.S. Davydov, *Quantum Mechanics: International Series in Natural Philosophy*, International series in natural philosophy (Elsevier Science, 2013). https://books.google.co.uz/books?id=rLE3BQAAQBAJ

18. F. Hioe, D. Macmillen, E. Montroll, Quantum theory of anharmonic oscillators: energy levels of a single and a pair of coupled oscillators with quartic coupling. Phys. Rep. **43**(7), 305–335 (1978). https://www.sciencedirect.com/science/article/pii/0370157378900972

19. A. Sharma, O.S.K.S. Sastri, Numerical simulation of quantum anharmonic oscillator, embedded within an infinite square well potential, by matrix methods using Gnumeric spread sheet. Eur. J. Phys. **41**(5), 055402 (2020). https://dx.doi.org/10.1088/1361-6404/ab988c

20. J. Killingbeck, Renormalised perturbation series. J. Phys. A: Math. Gen. **14**(5), 1005 (1981). https://dx.doi.org/10.1088/0305-4470/14/5/020

21. P. Stevenson, Exploring arbitrarily high orders of optimized perturbation theory in QCD with $n_f \to 16\frac{1}{2}$. Nucl. Phys. B **910**, 469–495 (2016). https://www.sciencedirect.com/science/article/pii/S0550321316302000

22. D.C. Duarte, R.L.S. Farias, P.H.A. Manso, R.O. Ramos, Optimized perturbation theory applied to the study of the thermodynamics and BEC-BCS crossover in the three-color Nambu-Jona-Lasinio model. Phys. Rev. D **96**, 056009 (2017). https://link.aps.org/doi/10.1103/PhysRevD.96.056009

BEC in Optimal Perturbation Theory

8

8.1 Introduction

From the previous chapter, it is understood that one can study properties of Bose gases as well as quantum liquids more accurately in OPT rather than in Bilinear approximation. In the present chapter, we show that such an accurate investigation of BEC reveals its some "hidden" peculiarities leading to better predictions. In the practical application of OPT opens the way to the solution of Hohenberg-Martin dilemma [1] and describes the experimental data much better than the simple Bogolyubov approximation [2,3]. In the next sections, we obtain the thermodynamic potential in OPT, outline the Hohenberg-Martin dilemma, and then show its possible solution. In the last sections, we shall calculate some physical observables, such as the condensate and superfluid densities, the static structure factor, etc.

8.2 Thermodynamic Potential

The Lagrangian density of the system is given by Eqs. (4.80) and (4.85). Following the previous chapter, a reader may expect that we shall do "manipulations" with this Lagrangian immediately. However, we shall use more subtle strategy. In fact, there is no rigorous restriction in OPT to introduce trial parameters. One may introduce them to the action or the Lagrangian, introduce them before or after the shift of fields. Thus, we make preference to start the LDE procedure from the action

$$S[\psi, \psi^+] = \int_0^\beta d\tau \int d\mathbf{r} \left\{ \psi^+(\tau, \mathbf{r})[\partial_\tau - \frac{\nabla^2}{2m} - \mu]\psi(\tau, \mathbf{r}) + \frac{g}{2}[\psi^+(\tau, \mathbf{r})\psi(\tau, \mathbf{r})]^2 \right\},$$

(8.1)

A. Rakhimov and S. Mardonov, *Theory of Quantum Bose Liquids Beyond Bogoliubov Approximation*, Lecture Notes in Physics 1045, https://doi.org/10.1007/978-3-032-05096-0_8

after performing the Bogoluybov shift

$$\psi(\tau, \mathbf{r}) = \sqrt{\rho_0} + \tilde{\psi}(\tau, \mathbf{r}), \qquad \psi^+(\tau, \mathbf{r}) = \sqrt{\rho_0} + \tilde{\psi}^\dagger(\tau, \mathbf{r}) \qquad (8.2)$$

where $\rho_0 = N_0/V = const$ is still the density of condensed particles, and $\tilde{\psi}(\tau, \mathbf{r})$ is the fluctuating field. So, after the insertion of (8.2) into the action (8.1) the latter can be divided into the following parts

$$S = S_0 + S_1 + S_2 + S_3 + S_4,$$

$$S_0 = \int_0^\beta d\tau \int d\mathbf{r}\{-\mu\rho_0 + \frac{g\rho_0^2}{2}\},$$

$$S_1 = \int_0^\beta d\tau \int d\mathbf{r}\left\{\left[g\rho_0^{3/2} - \mu\sqrt{\rho_0}\right]\tilde{\psi} + h.c\right\},$$

$$S_2 = \int_0^\beta d\tau \int d\mathbf{r}\left\{\tilde{\psi}^+\left[\partial_\tau - \frac{\nabla^2}{2m} + 2g\rho_0 - \mu\right]\tilde{\psi} + \frac{g\rho_0}{2}(\tilde{\psi}^+\tilde{\psi}^+ + \tilde{\psi}\tilde{\psi})\right\},$$

$$S_3 = g\sqrt{\rho_0}\int_0^\beta d\tau \int d\mathbf{r}\left\{\tilde{\psi}^+\tilde{\psi}\tilde{\psi} + \tilde{\psi}^+\tilde{\psi}^+\tilde{\psi}\right\},$$

$$S_4 = \frac{g}{2}\int_0^\beta d\tau \int d\mathbf{r}\tilde{\psi}^+\tilde{\psi}^+\tilde{\psi}\tilde{\psi}.$$

$$(8.3)$$

Now employing the linear delta expansion method, we add to the total action the term

$$S_\delta = (1 - \delta)\int_0^\beta d\tau \int d\mathbf{r}\left[S_n(\tilde{\psi}^+\tilde{\psi}) + \frac{S_{an}}{2}(\tilde{\psi}^+\tilde{\psi}^+ + \tilde{\psi}\tilde{\psi})\right], \qquad (8.4)$$

and make replacement $g \to \delta g$ everywhere but in S_0. It can be shown that, in the first order of δ, this strategy leads to the Hartree-Fock-Bogolyubov (HFB) approximation, developed by Yukalov [4]. Now we have

$$S = S_0 + S_1 + S_2 + S_3 + S_4 + S_\delta. \qquad (8.5)$$

Note that this time there are two variational parameters S_n and S_{an}. Clearly, when $\delta = 0$ we arrive to the action of an ideal gas ($g = 0$) while for $\delta = 1$ we are supposed to include the contact interaction as a whole. Thus, the action in Eq. (8.5) corresponds to the interpolation between the ideal and interacting gases with the action (8.1).

Now we pass to the real field representation, $\{\psi_1, \psi_2\}$ introduced by the Eq. (4.55) to separate the "free" and "interacting" parts of the action as follows:

$$S = S_0 + S_{\text{free}} + \delta S_{\text{int}},$$

$$S_{\text{free}} = \frac{1}{2} \int_0^\beta d\tau \int d\mathbf{r} \left\{ i\epsilon_{ab}\psi_a \partial_\tau \psi_b + \psi_1(-\frac{\nabla^2}{2m} + X_1)\psi_1 \right.$$

$$\left. +\psi_2(-\frac{\nabla^2}{2m} + X_2)\psi_2 \right\},$$

$$S_{\text{int}} = S_{\text{int}}^{(1)} + S_{\text{int}}^{(2)} + S_{\text{int}}^{(3)} + S_{\text{int}}^{(4)},$$

$$S_{\text{int}}^{(1)} = \int_0^\beta d\tau \int d\mathbf{r} \left\{ \psi_1 \sqrt{2}\sqrt{\rho_0}(-\mu + g\rho_0) \right\}, \tag{8.6}$$

$$S_{\text{int}}^{(2)} = \frac{1}{2} \int_0^\beta d\tau \int d\mathbf{r} \left\{ (-\mu - X_1 + 3g\rho_0)\psi_1^2 + (-\mu - X_2 + g\rho_0)\psi_2^2 \right\},$$

$$S_{\text{int}}^{(3)} = \frac{g\sqrt{2}\rho_0}{2} \int_0^\beta d\tau \int d\mathbf{r} \left\{ (\psi_1^2 + \psi_2^2)\psi_1 \right\},$$

$$S_{\text{int}}^{(4)} = \frac{g}{8} \int_0^\beta d\tau \int d\mathbf{r} \left\{ \psi_1^4 + 2\psi_1^2\psi_2^2 + \psi_2^4 \right\},$$

where new variational parameters X_1 and X_2 are the linear combinations of S_n and S_{an}:

$$X_1 = S_n + S_{an} - \mu, \qquad X_2 = S_n - S_{an} - \mu. \tag{8.7}$$

Following the strategy outlined in Chap. 4, one can immediately obtain from the S_{free} the Green function D in real field representation, which is given by the Eq. (4.82). The energy spectrum of bogolons is given by

$$E_k = \sqrt{(\varepsilon_k + X_1)(\varepsilon_k + X_2)}, \tag{8.8}$$

with $\varepsilon_k = \mathbf{k}^2/2m$. Knowing the Green functions, we can also evaluate normal

$$\rho_1 = \frac{1}{V} \int d\mathbf{r}\langle \tilde{\psi}^\dagger(\mathbf{r})\tilde{\psi}(\mathbf{r})\rangle, \tag{8.9}$$

and anomalous densities

$$\sigma = \frac{1}{2V} \int d\mathbf{r} \left[\langle \tilde{\psi}^\dagger(\mathbf{r})\tilde{\psi}^\dagger(\mathbf{r})\rangle + \langle \tilde{\psi}(\mathbf{r})\tilde{\psi}(\mathbf{r})\rangle \right]. \tag{8.10}$$

Clearly, they have formally the same expressions as in the Bilinear approximation, discussed in Chap. 4

$$\rho_1 = \frac{D_{11}(0) + D_{22}(0)}{2} = \frac{1}{V} \sum_k \frac{W_k(2\varepsilon_k + X_1 + X_2)}{E_k}, \tag{8.11}$$

and

$$\sigma = \frac{D_{11}(0) - D_{22}(0)}{2} = \frac{X_2 - X_1}{2V} \sum_k \frac{W_k}{E_k}, \tag{8.12}$$

(see Eqs. (4.98) and (6.96)). Here, X_1 and X_2 are supposed to be fixed by PMS conditions

$$\frac{d\Omega}{dX_1} = 0, \qquad \frac{d\Omega}{dX_2} = 0, \tag{8.13}$$

where the free energy Ω can be evaluated as

$$\Omega = -T \ln Z[J_1, J_2]|_{J_1=0, J_2=0}, \tag{8.14}$$

with the grand partition function defined as[1]

$$Z(J_1, J_2) = e^{-S_0} \int D\psi_1 D\psi_2 e^{-\frac{1}{2} \int dx \int dx' \psi_a(x) D_{ab}^{-1}(x,x') \psi_b(x')} e^{-\delta S_{\mathrm{int}}}$$
$$\times e^{\int dx [J_1(x)\psi_1(x) + J_2(x)\psi_2(x)]}, \tag{8.15}$$

in which we introduced $x = (\tau, \mathbf{r})$ and $\int dx \equiv \int_0^\beta d\tau \int d\mathbf{r}$. The "free" action gives rise to the "noninteracting" partition function Z_0 as

$$Z_0[J_1, J_2] = \int D\psi_1 D\psi_2 e^{-\frac{1}{2} \int dx dx' \psi_a(x) D_{ab}^{-1}(x,x') \psi_b(x')} e^{\int dx J_a(x)\psi_a(x)}$$
$$= (Det\ D^{-1})^{-1/2} \exp\left[\frac{1}{2} \int dx dx'\, J_a(x)\bar{D}_{ab}(x, x') J_b(x')\right], \tag{8.16}$$

where $\bar{D}_{ab}(x, y) = [D_{ab}(x, y) + D_{ba}(y, x)]/2$. Clearly, due to the presence of the factor $\exp[-\delta S_{\mathrm{int}}]$, the path integral in (8.15) can not be evaluated exactly, so we have to make an expansion similar to the Eq. (7.45):

$$e^{-\delta S_{\mathrm{int}}} = 1 - \delta S_{\mathrm{int}}^{(1)} - \delta S_{\mathrm{int}}^{(2)} - \delta S_{\mathrm{int}}^{(3)} - \delta S_{\mathrm{int}}^{(4)} + O(\delta^2), \tag{8.17}$$

Therefore, the first order of LDE gives

$$Z(J_1, J_2) = e^{-S_0} \left[Z_0(J_1, J_2) - \delta\langle S_{\mathrm{int}}^{(1)} \rangle - \delta\langle S_{\mathrm{int}}^{(2)} \rangle - \delta\langle S_{\mathrm{int}}^{(3)} \rangle - \delta\langle S_{\mathrm{int}}^{(4)} \rangle \right], \tag{8.18}$$

where the expectation values are defined in Sect. 6.6. Bearing in mind that $Z_0(J_1 = 0, J_2 = 0) = 1/\sqrt{Det\ D^{-1}}$, the expectation values in (8.18) can be easily calculated by using Eq. (6.94) as

---

[1] Below summation over repeating indices is assumed.

$$\langle S_{\text{int}}^{(1)} \rangle = 0, \qquad \langle S_{\text{int}}^{(3)} \rangle = 0,$$

$$\langle S_{\text{int}}^{(2)} \rangle = \frac{1}{2} \int dx [(-\mu - X_1 + 3g\rho_0) D_{11}(0) + (-\mu - X_2 + g\rho_0) D_{22}(0)]$$

$$= \frac{\beta}{2} [(-\mu - X_1 + 3g\rho_0) A_1 + (-\mu - X_2 + g\rho_0) A_2],$$

$$\langle S_{\text{int}}^{(4)} \rangle = \frac{g\beta V}{8} \left[3D_{11}^2(0) + 3D_{22}^2(0) + 2D_{11}(0) D_{22}(0) \right]$$

$$= \frac{g\beta}{8V} [3A_1^2 + 3A_2^2 + 2A_1 A_2].$$

$$(8.19)$$

where

$$A_1 = V D_{11}(0) = V(\rho_1 + \sigma), \qquad A_2 = V D_{22}(0) = V(\rho_1 - \sigma). \qquad (8.20)$$

Thus, from (8.14) and (8.18) we obtain

$$\Omega = -T \ln Z(j)|_{j=0} = -T \ln e^{-S_0} - T \ln Z_0 + T \langle S_{\text{int}}^{(2)} \rangle + T \langle S_{\text{int}}^{(4)} \rangle, \qquad (8.21)$$

where we used the formula $\ln(1 + x) \approx x$ and set $\delta = 1$. Finally, inserting here (8.19) and using (6.60) gives

$$\Omega = \Omega_0 + \Omega_{ln} + \Omega_2 + \Omega_4,$$

$$\Omega_0 = -\mu V \rho_0 + \frac{g V \rho_0^{\,2}}{2},$$

$$\Omega_{ln} = \frac{1}{2} \sum_k (E_k - \varepsilon_k) + T \sum_k \ln(1 - e^{-\beta E_k}), \qquad (8.22)$$

$$\Omega_2 = \frac{1}{2} [(-\mu - X_1 + 3g\rho_0) A_1 + (-\mu - X_2 + g\rho_0) A_2],$$

$$\Omega_4 = \frac{g}{8V} [3A_1^2 + 3A_2^2 + 2A_1 A_2].$$

Note that, the divergent integrals in (8.11), (8.12) and in Ω_{ln} can be easily regularized by using the rules outlined in Sect. 4.9. Comparing (8.22) with Ω given by Bilinear approximation, (6.99), we see that OPT gives two additional terms, denoted here as Ω_2 and Ω_4, which include still unknown parameters X_1 and X_2. In the next section we shall obtain certain equations for them, starting from PMS conditions (8.13).

8.3 Fixing Trial Parameters

Below, we show that the Eq. (8.13) can be presented in following compact form:

$$X_1 = -\mu + 3g\rho_0 + 2g\rho_1 + g\sigma, \qquad X_2 = -\mu + g\rho_0 + 2g\rho_1 - g\sigma. \qquad (8.23)$$

For this purpose we rewrite Ω in a more convenient form for differentiation:

$$\Omega(X_1, X_2, \rho_0) = \Omega_0 + \Omega_{ln} + \Omega_2 + \Omega_4,$$

$$\Omega_0 = -\mu V \rho_0 + \frac{g V \rho_0^2}{2},$$

$$\Omega_{ln} = \frac{1}{2} \sum_k (E_k(X_1, X_2) - \varepsilon_k) + T \sum_k \ln(1 - e^{-\beta E_k(X_1, X_2)}),$$

$$\Omega_2 = \frac{1}{2} \sum_{i=1,2} A_i(X_1, X_2)[\Lambda_i(\rho_0) - X_i],$$

$$\Omega_4 = \sum_{i,j=1,2} C_{ij} A_i(X_1, X_2) A_j(X_1, X_2). \tag{8.24}$$

where following notations are introduced

$$\Lambda_1(\rho_0) = -\mu + 3g\rho_0, \quad \Lambda_2(\rho_0) = -\mu + g\rho_0,$$

$$C_{11} = C_{22} = \frac{3g}{8V}, \quad C_{12} = C_{21} = \frac{g}{8V}. \tag{8.25}$$

Now taking derivatives we have ($m = 1, 2$)

$$\frac{\partial \Omega_0}{\partial X_m} = 0, \quad \frac{\partial \Omega_{ln}}{\partial X_m} = \frac{1}{2} \sum_{i=1,2} \delta_{mi} A_i,$$

$$\frac{\partial \Omega_2}{\partial X_m} = \frac{1}{2} \sum_{i=1,2} \left[\frac{\partial A_i}{\partial X_m} \Lambda_i - \delta_{mi} A_i - \frac{\partial A_i}{\partial X_m} X_i \right],$$

$$\frac{\partial \Omega_4}{\partial X_m} = 2 \sum_{i,j=1,2} C_{ij} \frac{\partial A_i}{\partial X_m} A_j. \tag{8.26}$$

All together, the PMS equations have the form

$$\frac{\partial \Omega(X_1, X_2, \rho_0)}{\partial X_m} = \frac{1}{2} \sum_i \left[\frac{\partial A_i}{\partial X_m} \Lambda_i - \frac{\partial A_i}{\partial X_m} X_i + 4 \sum_j C_{ij} \frac{\partial A_i}{\partial X_m} A_j \right],$$

$$= \frac{1}{2} \sum_i \frac{\partial A_i}{\partial X_m} \left[\Lambda_i - X_i + 4 \sum_j C_{ij} A_j \right] = 0. \tag{8.27}$$

The last equations have simple solutions as

$$\Lambda_1 - X_1 + 4[C_{11} A_1 + C_{12} A_2] = 0, \quad \Lambda_2 - X_2 + 4[C_{21} A_1 + C_{22} A_2] = 0. \tag{8.28}$$

Now inserting here the explicit expressions for Λ_i and C_{ij}, from Eqs (8.25) as well as for A_i from (8.20) one comes to the equation (8.23) which can be also rewritten as

$$X_1 = -\mu + g\rho_0 + 2g\rho + g\sigma, \qquad X_2 = -\mu - g\rho_0 + 2g\rho - g\sigma, \qquad (8.29)$$

where we used the normalization condition $\rho_0 + \rho_1 = \rho = N/V$. Note that, in the light of explicit expressions (8.11) and (8.12), the main equations (8.29) are, in fact, the system of two nonlinear algebraic equations with respect to c-numbers X_1 and X_2.

Remarkably, these equations hold both in condensed $T \leq T_c$ as well as in normal $T > T_c$ phases. Particularly, in the normal phase, when $\rho_0(T \geq T_c) = \sigma(T \geq T_c) = 0$ they are simplified as:

$$X_1(T \geq T_c) = X_2(T \geq T_c) = -\mu + 2g\rho. \qquad (8.30)$$

On the other side, in the BEC phase, the equations may also be simplified owing to Hugenholtz-Pines theorem. We shall discuss this point in the next section.

8.4   Hugenholtz-Pines Theorem and Hohenberg-Martin Dilemma

Firstly, it would be interesting to realize the physical meaning of the trial parameters. For this purpose, we come back to the Chap. 4 and concentrate on Eqs. (4.81) and (4.89). These equations affirm that if the inverse propagator has the general form

$$D^{-1}(\omega_n, \mathbf{k}) = \begin{pmatrix} \varepsilon_k + X_1 & \omega_n \\ -\omega_n & \varepsilon_k + X_2 \end{pmatrix}, \qquad (8.31)$$

then the self-energy in the real field representation will have the following diagonal matrix elements: $\Pi_{ii}(\omega_n = 0, \mathbf{k} = 0) = \mu + X_i$, $(i = 1, 2)$ and hence $X_2 = 0$ due to the Hugenholtz-Pines theorem. Moreover, from Eqs. (4.84) and (8.7) one may conclude that the trial parameters S_n and S_{an} are exactly equal to the normal and anomalous self energies, namely, $S_n = \Sigma_n(0, 0)$ and $S_{an} = \Sigma_{an}(0, 0)$. Therefore, using Eq (8.7) one can make sure that the Hugenholtz-Pines relations

$$\Sigma_n(0, 0) - \Sigma_{an}(0, 0) = S_n - S_{an} = \mu, \qquad (8.32)$$

really lead to the condition

$$X_2(T \leq T_c) = 0, \qquad (8.33)$$

which speaks in favor of the self-consistency of the theory.

But this is not the end of the story. As we have underlined in Chap. 4, Hugenholtz-Pines relations should be consistent with the stability condition of the equilibrium system

$$\frac{\partial \Omega}{\partial \rho_0} = 0. \tag{8.34}$$

In Chap. 4 we have seen that there is no such problem within the Bilinear approximation. What about our OPT? Are the Eqs. (8.33) and (8.34) compatible?

8.4.1   Hohenberg-Martin Dilemma

First, we exploit the Eqs. (8.23) and (8.33) to fix the chemical potential from the Hohenberg-Pines relations as

$$\mu = g\rho_0 + 2g\rho_1 - g\sigma \equiv \mu_{HP}. \tag{8.35}$$

This choice corresponds to gapless quasiparticle dispersion $E_k = \sqrt{\varepsilon_k(\varepsilon_k + X_1)} \approx ck + O(k^3)$. Second, we make an attempt to define μ from the condition (8.34) which will have the form

$$\frac{\partial \Omega}{\partial \rho_0} = V(-\mu + g\rho_0 + 2g\rho_1 + g\sigma) = 0. \tag{8.36}$$

This equation leads to

$$\mu = g\rho_0 + 2g\rho_1 + g\sigma \equiv \mu_{stab}. \tag{8.37}$$

It is seen that $\Delta\mu = \mu_{HP} - \mu_{stab} = -2g\sigma \neq 0$, and hence, the energy dispersion $E_k = \sqrt{(\varepsilon_k + X_1)(\varepsilon_k + X_2)} \approx ck^2 + O(k^4)$ is not gapless and moreover, violates the HP theorem also. That is, one cannot satisfy simultaneously the HP theorem and the stability condition. This is the essence of the Hohenberg-Martin (HM) dilemma. This principle problem, raised not only in the quantum mechanics of BEC, but also in relativistic field theory [5], has been still actual. So, it is worth to discuss it in a more detail.

HM dilemma is referred to in the literature also a dilemma of conserving versus gapless approximations. As we have seen in previous chapters it is absent in Bilinear (one loop) and hence in Bogolyubov approximations (see Eqs. (6.90) and (6.100)) which are good only in the limit of asymptotically weak interactions.

Above we illustrated the dilemma in the framework of the first order of OPT, However, actually it has a more general origin which could be explained as follows [6].

Bose-Einstein condensation implies the macroscopic occupation of the ground-state energy level, when the number of condensed particles N_0 is such that the condensate fraction N_0/N is nonzero in the thermodynamic limit. The total number of particles in the system, $N = N_0 + N_1$, becomes a sum of the condensed-particle number N_0 and the number N_1 of uncondensed particles. According to Bogolyubov [7], the number of condensed particles N_0 is such that it provides thermodynamic stability for the system, minimizing the thermodynamic potential Ω. At the same time, $N_0 = N - N_1$, where $N_1 = N_1(\mu, T, \rho)$ is a function of the chemical potential μ, temperature T, and density ρ. Hence, $N_0 = N_0(\mu, T, \rho)$ is also a function of the same variables. Conversely, the chemical potential $\mu = \mu(T, \rho)$ is a function of T and ρ. Thus, the condition of stability, realized as the minimization of the thermodynamic potential, defines the chemical potential μ. It is important to emphasize that this procedure of minimization does not depend on approximations, but is generally valid, being strictly proved by Ginibre [8].

On the other hand, the particle spectrum, under the broken gauge symmetry, has to be gapless. This is, actually, the condition for the existence of a stable Bose-Einstein condensate. Since, if there would be a gap in the spectrum, there could be no macroscopic occupation of a single ground-state level. In accordance with the HP theorem, the chemical potential should satisfy the relation $\mu = \Sigma_n(0, 0) - \Sigma_{an}(0, 0)$, which makes the spectrum gapless for any Bose system. As far as the self-energies are the functions of T and ρ, the Hugenholtz–Pines relation defines the chemical potential as a function $\mu = \mu(T, \rho)$. In this way, the condition of thermodynamic stability, in the frame of the Bogolyubov–Ginibre minimization procedure, defines a chemical potential that we denoted above as μ_{stab}. At the same time, the existence of a gapless spectrum, connected to the Hugenholtz–Pines relation, also defines a chemical potential, which we have denoted as μ_{HP}. These chemical potentials, found in two different ways do not necessarily coincide, and, generally, they are different. Therefore, if one accepts as the system chemical potential the Bogolyubov–Ginibre value $\mu = \mu_{stab}$, providing thermodynamic stability, then one acquires a nonphysical gap in the spectrum, proportional to the difference $\Delta\mu = \mu_{HP} - \mu_{stab}$. Conversely, accepting as the chemical potential the Hugenholtz–Pines form μ_{HP}, one gets a nonconserving theory with incorrect thermodynamics, since the stability condition does not hold. This is the origin of the Hohenberg–Martin dilemma of conserving versus gapless theories [1].

8.5   Solution of Hohenberg–Martin Dilemma: Yukalov Receipt

8.5.1   The Role of Anomalous Density

We have seen that the essence of the dilemma is related to the difference between the chemical potentials calculated in different ways, i.e. $\Delta\mu = \mu_{HP} - \mu_{stab} = -2g\sigma \neq 0$ in our LDE approach. Looking at this equation, the first thing that comes to mind is just neglecting the anomalous average by setting $\sigma = 0$. This version of the approach is referred in the literature as the Hartree-Fock-Popov approximation, although, as pointed out by Yukalov, in reality, Popov had nothing to do with it. However, neglecting σ, whose absolute value has the meaning of the density of pair-correlated particles, would be a very gross mistake. The reasons are following.

(1) The occurrence of the global gauge symmetry breaking is necessary and sufficient not merely for the condensate existence, but also for the appearance of the anomalous averages. Both the condensate itself and the anomalous averages either exist together or are together zero. The anomalous density enjoys the natural limiting property when $\rho_0 \to 0$, then $\sigma \to 0$. The physics of this property is evident. The existence of both the condensate density ρ_0 and the anomalous density σ is caused by the gauge symmetry breaking. Both of them are nonzero as soon as the symmetry is broken, while both become zero if the symmetry is restored. Any of these quantities could be treated as an order parameter for the broken symmetry phase. So, both of these quantities, ρ_0 and σ, have to nullify simultaneously, when one of them tends to zero.

(2) As it was shown, at zero temperature σ is larger than the uncondensed fraction (see Eq. (6.97)). Thus, omitting the larger terms, while keeping the smaller ones is mathematically wrong. Such a contradiction results in the system instability and in the incorrect description of the Bose-Einstein condensation transition

(3) As we shall see in Chap. 13, neglecting σ in the BEC of triplons leads to an anomalous jump in the magnetization, which is in contradiction to the experimental observations.

There have been several other attempts to cure the problem [9]. Below, we follow the method proposed by Yukalov [10].

This approach relies on introducing two different chemical potentials—one for the condensed phase and a second one for the normal phase. The two chemical potentials are distinct below the critical temperature, and allow us to simultaneously satisfy the selfconsistency condition for the mean field while having a gapless Goldstone mode. The two chemical potentials become equal at the critical temperature. Above the critical temperature, there is, of course, a single chemical potential associated with the conserved particle number. Like the other aforementioned approaches to the Hohenberg-Martin dilemma, the framework of Yukalov introduces an additional variable to account for the different constraints. One might argue, however, that the usage of two chemical potentials can be physically motivated, and can be naturally extended to a quantum field theory exhibiting spontaneous symmetry breaking.

The method entails that we have one chemical potential μ_0 for the condensate and a separate chemical potential μ_1 for the excitations. These encode the fact that the fraction of particles in the condensate and the excitations are fixed at a given T, V and N for (NVT) type of systems. Mathematically, let us minimize the free energy $F(N_0, N_1)$

$$\delta F = \frac{\partial F}{\partial N_0}\delta N_0 + \frac{\partial F}{\partial N_1}\delta N_1 = 0. \tag{8.38}$$

The derivative of the free energy with respect to the number of particles is the chemical potential: $\mu_0 = \partial F / \partial N_0$ and $\mu_1 = \partial F / \partial N_1$. Furthermore, since the total number of particles is conserved, $\delta N = 0$ and hence, $\delta N_0 = -\delta N_1$. Thus (8.38) becomes

$$(\mu_0 - \mu_1)\delta N_0 = 0. \tag{8.39}$$

If, in a phase transition at a particular critical temperature, the fraction of particles in either of the phases is not fixed, $\delta N_0 \neq 0$, then $\mu_0 = \mu_1$. Instead, in our case the fraction of particles in the condensate and in excited states are each fixed, $\delta N_0 = 0$, therefore the two chemical potentials are no longer required to be equal. We will see that they are indeed different for $T < T_c$ and become equal at $T = T_c$, when the condensate vanishes.

This prescription of two chemical potentials will allow us to circumvent the Hohenberg-Martin dilemma discussed previously, with each chemical potential satisfying a different constraint. The condensate chemical potential, μ_0, enforces the mean-field equation of motion, and the one associated with the excitations, μ_1, ensures that Goldstone's theorem is satisfied.

The corollary of the above statements is that we can set $\mu_{HP} = \mu_1$ and $\mu_{stab} = \mu_0$, such that, in general, $\mu_0 \neq \mu_1 \neq \mu$. In the next section, we apply this prescription to the BEC of atomic gases in the LDE approach.

8.6   The Effective Action and Thermodynamic Potential in Yukalov Prescription

Therefore, making replacement $\mu N \rightarrow \mu_0 N_0 + \mu_1 N_1$ we have following grand Hamiltonian [4]:

$$H_{gr} = H - \mu_0 N_0 - \mu_1 N_1. \tag{8.40}$$

In the theoretical field formalism, making $\mu \psi^\dagger \psi \rightarrow \mu_0 \rho_0 + \mu_1 \tilde{\psi}^\dagger \tilde{\psi}$ we rewrite the action (8.1) as follows

$$
\begin{aligned}
S[\psi, \psi^\dagger] = \int_0^\beta d\tau \int d\mathbf{r} \Bigg\{ &\psi^\dagger(\tau, \mathbf{r}) \left[\partial_\tau - \frac{\nabla^2}{2m} \right] \\
&-\mu_0 \rho_0 - \mu_1 \tilde{\psi}^\dagger(\tau, \mathbf{r})\tilde{\psi}(\tau, \mathbf{r}) + \frac{g}{2}[\psi^\dagger(\tau, \mathbf{r})\psi(\tau, \mathbf{r})]^2 \Bigg\},
\end{aligned}
\tag{8.41}
$$

where ρ_0 and $\tilde{\psi}$ come from (8.2). Next, introducing S_δ (8.4) and passing to the Cartesian i.e. real field representation (4.55) this can be represented as

$$S = S_0 + S_{\text{free}} + \delta S_{\text{int}},$$

$$S_{\text{free}} = \frac{1}{2} \int_0^\beta d\tau \int d\mathbf{r} \left\{ i\epsilon_{ab} \psi_a \partial_\tau \psi_b + \psi_1 \left(-\frac{\nabla^2}{2m} + X_1 \right) \psi_1 \right.$$

$$\left. + \psi_2 \left(-\frac{\nabla^2}{2m} + X_2 \right) \psi_2 \right\},$$

$$S_{\text{int}} = S_{\text{int}}^{(1)} + S_{\text{int}}^{(2)} + S_{\text{int}}^{(3)} + S_{\text{int}}^{(4)},$$

$$S_{\text{int}}^{(1)} = g \int_0^\beta d\tau \int d\mathbf{r} \rho_0^{3/2} \psi_1,$$

$$S_{\text{int}}^{(2)} = \frac{1}{2} \int_0^\beta d\tau \int d\mathbf{r} \left\{ (-\mu_1 - X_1 + 3g\rho_0)\psi_1^2 + (-\mu_1 - X_2 + g\rho_0)\psi_2^2 \right\},$$

$$S_{\text{int}}^{(3)} = \frac{g}{\sqrt{2}} \int_0^\beta d\tau \int d\mathbf{r} \sqrt{\rho_0} \psi_1 \left(\psi_1^2 + \psi_2^2 \right),$$

$$S_{\text{int}}^{(4)} = \frac{g}{8} \int_0^\beta d\tau \int d\mathbf{r} \left(\psi_1^4 + 2\psi_1^2 \psi_2^2 + \psi_2^4 \right),$$

$$(8.42)$$

where $X_{1,2}$ are the linear combinations of the trial parameters via

$$S_n = \mu_1 + \frac{X_1 + X_2}{2}, \qquad S_{an} = \frac{X_1 - X_2}{2}. \tag{8.43}$$

The inverse propagator has the same form as it has been predicted in (8.31):

$$D^{-1}(\omega_n, \mathbf{k}) = \begin{pmatrix} \varepsilon_k + X_1 & \omega_n \\ -\omega_n & \varepsilon_k + X_2 \end{pmatrix}, \tag{8.44}$$

so that

$$D(\omega_n, \mathbf{k}) = \frac{1}{(E_k^2 + \omega_n^2)} \begin{pmatrix} \varepsilon_k + X_2 & -\omega_n \\ \omega_n & \varepsilon_k + X_1 \end{pmatrix}, \tag{8.45}$$

with the dispersion:

$$E_k = \sqrt{(\varepsilon_k + X_1)(\varepsilon_k + X_2)}. \tag{8.46}$$

Thus, the normal and anomalous densities are also formally the same as in previous sections:

$$\rho_1 = \frac{1}{V} \sum_k \left[\frac{W_k(2\varepsilon_k + X_1 + X_2)}{E_k} - \frac{1}{2} \right], \tag{8.47}$$

and

$$\sigma = \frac{X_2 - X_1}{2V} \sum_k \left[\frac{W_k}{E_k} - \frac{1}{\varepsilon_k} \right], \tag{8.48}$$

where the proper counter terms have been included. Similarly to the previous case with unique μ, now one obtains the thermodynamic potential $\Omega(\mu_0, \mu_1)$ given by

$$
\begin{aligned}
\Omega &= \Omega_0 + \Omega_{ln} + \Omega_2 + \Omega_4, \\
\Omega_0 &= -V\mu_0\rho_0 + \frac{Vg\rho_0{}^2}{2}, \\
\Omega_{ln} &= \frac{1}{2}\sum_k (E_k - \epsilon_k) + T\sum_k \ln(1 - e^{-\beta E_k}), \\
\Omega_2 &= \frac{1}{2}[(-\mu_1 - X_1 + 3g\rho_0)A_1 + (-\mu_1 - X_2 + g\rho_0)A_2], \\
\Omega_4 &= \frac{g}{8V}[3A_1^2 + 3A_2^2 + 2A_1 A_2],
\end{aligned}
\tag{8.49}
$$

where $A_{1,2}$ are given by (8.20).

The solution of PMS Eq. (8.13) may be found by using the same strategy as in the previous section, yielding the result

$$
\begin{aligned}
X_1 &= -\mu_1 + 3g\rho_0 + 2g\rho_1 + g\sigma = -\mu_1 + g\rho_0 + 2g\rho + g\sigma, \\
X_2 &= -\mu_1 + g\rho_0 + 2g\rho_1 - g\sigma = -\mu_1 - g\rho_0 + 2g\rho - g\sigma.
\end{aligned}
\tag{8.50}
$$

As to the stability condition $(\partial\Omega/\partial\rho_0) = 0$, it leads to the following equation with respect to the "BEC" part of the chemical potential μ_0

$$
\mu_0 = g\rho_0 + 2g\rho_1 + g\sigma = g(\rho + \rho_1 + \sigma).
\tag{8.51}
$$

8.6.1   Critical Temperature and Normal Phase

Let's define the critical temperature T_c as $\rho_0(T \geq T_c) = 0$ and $\sigma(T \geq T_c) = 0$. Then for the temperatures higher than T_c, i.e. for the normal phase from Eq. (8.50) we have

$$
X_1(T \geq T_c) = X_2(T \geq T_c) = -\mu_1 + 2g\rho = -\mu + 2g\rho
\tag{8.52}
$$

where the relations $\mu\rho = \mu_0\rho_0 + \mu_1\rho_1$, $\rho_1(T \geq T_c) = \rho(T \geq T_c) - \rho_0(T \geq T_c) = \rho(T \geq T_c)$ and

$$
\mu_0(T \geq T_c) = \mu_1(T \geq T_c) = \mu(T \geq T_c)
\tag{8.53}
$$

have been used. Since the BEC phase transition is expected as a second-order one, the self energies $\Sigma_{n,an}$ and hence $X_{1,2}$ should be continuous at the critical point:

$$
X_1(T = T_c - 0) = X_1(T = T_c + 0), \qquad X_2(T = T_c - 0) = X_2(T = T_c + 0).
\tag{8.54}
$$

On the other hand in the condensed phase, due to the HP theorem, X_2 should vanish: $X_2(T = T_c - 0) = 0$. Therefore from Eq. (8.52)–(8.54) one obtains

$$X_1(T = T_c) = X_2(T = T_c) = 0,$$
$$\mu_0(T = T_c) = \mu_1(T = T_c) = \mu(T = T_c) = 2g\rho. \tag{8.55}$$

This leads to simple dispersion.

$$E_k(T = T_c) = \left(\sqrt{(\varepsilon_k + X_1)(\varepsilon_k + X_2)}\right)_{T=T_c} = \varepsilon_k = \frac{k^2}{2\,m}. \tag{8.56}$$

For this reason, the critical temperature given by the equation

$$\rho = \frac{1}{(2\pi)^3} \int \frac{d^3k}{e^{E_k/T_c} - 1} = \frac{1}{2\pi^2} \int_0^\infty \frac{k^2 dk}{e^{\varepsilon_k/T_c} - 1} \tag{8.57}$$

will coincide with that of the ideal Bose gas:

$$T_c = \frac{2\rho^{2/3}\pi(\hbar c)^2}{mc^2 k_B(\zeta(3/2))^{2/3}} \approx \frac{3.31\rho^{2/3}(\hbar c)^2}{mc^2 k_B} \equiv T_c^0. \tag{8.58}$$

Therefore, we have to accept that, unfortunately, OPT, or more precisely, the first order of LDE, is not able to predict the shift of the critical temperature due to the contact interaction.

8.7   Condensed Phase, $T \leq T_c$

Here, as we underlined above, the HP theorem requires

$$\Sigma_n(0, 0) - \Sigma_{an}(0, 0) = \mu_1, \quad i.e. \quad X_2(T \leq T_c) = 0. \tag{8.59}$$

Note that, in the Yukalov framework HP theorem includes μ_1 instead of μ. This requirement together with the Eq. (8.50) defines μ_1 as

$$\mu_1(T \leq T_c) = 2g\rho_1 + g\rho_0 - g\sigma = g(\rho + \rho_1 - \sigma). \tag{8.60}$$

Comparing this with Eq. (8.51) we again see that $\mu_1(T \leq T_c) \neq \mu_0(T \leq T_c)$, which is natural in Yukalov's formalism, due to the fact that the anomalous density has been accurately taken into account by means of OPT. Next, inserting (8.60) into the first of Eq. (8.50) we obtain following compact expression for X_1:

$$X_1(T \leq T_c) = 2g\rho_0 + 2g\sigma = 2g(\rho_0 + \sigma), \tag{8.61}$$

which is consistent with $X_1(T = T_c) = 0$. It is seen that, in comparison with the Bilinear approximation, where $X_1(T \leq T_c) = 2g\rho_0$, the first order of LDE gives an additional term proportional to σ. For convenience, below we work in terms of another variable, Δ, defined as

$$\Delta \equiv \frac{X_1}{2} = g(\rho_0 + \sigma) = g\rho + g(\sigma - \rho_1), \tag{8.62}$$

where the densities $\rho_1(\Delta)$ and $\sigma(\Delta)$ are given by following explicit expressions:

$$\rho_1 = \frac{1}{V} \sum_k \left[\frac{W_k(\varepsilon_k + \Delta)}{E_k} - \frac{1}{2} \right] = \rho_1(T = 0)$$

$$+ \frac{1}{2\pi^2} \int_0^\infty dk\, k^2 \frac{f_B(E_k)(\varepsilon_k + \Delta)}{E_k},$$

$$\sigma = -\frac{\Delta}{V} \sum_k \left[\frac{W_k}{E_k} - \frac{1}{2\varepsilon_k} \right] = \sigma(T = 0) - \frac{\Delta}{2\pi^2} \int_0^\infty dk\, k^2 \frac{f_B(E_k)}{E_k},$$

$$\rho_1(T = 0) = \frac{(m\Delta)^{3/2}}{3\pi^2}, \quad \sigma(T = 0) = 3\rho_1(T = 0) \approx \frac{\Delta m^{3/2}}{\pi^2} \sqrt{g\rho_0}, \tag{8.63}$$

with the boson distribution $f_B(x)$ defined in (4.23).

Actually, the Eq. (8.62) is a nonlinear algebraic equation with respect to Δ with given input parameters $\gamma = \rho a_s^3$, T and m. For each set of these parameters, one may find the physical solution ($\Delta \geq 0$) and evaluate the sound velocity as

$$s = \sqrt{\frac{\Delta}{m}}. \tag{8.64}$$

In some cases, the main Eq. (8.62) may be solved approximately or even analytically at $T = 0$ as it was shown in Sect. 4.9 leading to the power series in γ similar to (4.118). Particularly, at low temperatures one may realize an expansion in powers of (T/Δ) to get [4]

$$\rho_1 \approx \rho_1(T = 0) + \frac{(ms)^3}{12} \frac{T^2}{\Delta^2},$$

$$\sigma \approx \sigma(T = 0) - \frac{(ms)^3}{12} \frac{T^2}{\Delta^2}, \tag{8.65}$$

$$n_0 = \frac{\rho - \rho_1}{\rho} \approx 1 - \frac{(m\Delta)^{3/2}}{3\pi^2 \rho} - \frac{(ms)^3}{12\rho} \frac{T^2}{\Delta^2}.$$

8.7.1 The Main Equation in Dimensionless Form

Below we show that, for a homogeneous system, all observables, particularly the main equation with respect to Δ, (8.62):

$$\Delta = g(\rho_0 + \sigma), \tag{8.66}$$

depend mainly on two parameters of the system: $\gamma = \rho a_s^3$ and $t = T/T_c$, where T_c is given by (8.58).

Introducing dimensionless variable, $Z = \Delta/g\rho$, we may present (8.66) as follows:

$$Z + n_1(Z) - m_1(Z) - 1 = 0. \tag{8.67}$$

Here the reduced densities, $n_1 = \rho_1/\rho = 1 - \rho_0/\rho, m_1 = \sigma/\rho$ are given by

$$n_1 = n_1(T = 0) + n_1(T), \quad m_1 = m_1(T = 0) + m_1(T). \tag{8.68}$$

Zero temperature values of densities can be evaluated analytically as:

$$n_1(T = 0) = \frac{1}{2\rho V} \sum_k \left[\frac{\varepsilon_k + \Delta}{E_k} - 1 \right] = \frac{8Z^{3/2}\sqrt{\gamma}}{3\sqrt{\pi}},$$

$$m_1(T = 0) = -\frac{\Delta}{2\rho V} \sum_k \left[\frac{1}{E_k} - \frac{1}{\varepsilon_k} \right] = \frac{8Z^{3/2}\sqrt{\gamma}}{\sqrt{\pi}}. \tag{8.69}$$

To evaluate $n_1(T)$ and $m_1(T)$:

$$n_1(T) = \frac{1}{\rho V} \sum_k \frac{\varepsilon_k + \Delta}{E_k} f_B(E_k),$$

$$m_1(T) = -\frac{\Delta}{\rho V} \sum_k \frac{f_B(E_k)}{E_k}, \tag{8.70}$$

it is convenient to introduce the dimensionless variable $x = \varepsilon_k/T_c$ in the integral

$$\sum_k f(k^2) = \frac{4\pi V}{(2\pi)^3} \int_0^\infty f(k^2)k^2 dk. \tag{8.71}$$

As a result $n_1(T)$ and $m_1(T)$ can be rewritten as

$$n_1(T) = \frac{\sqrt{2}\tilde{c}}{2\pi^2\gamma^{2/3}} \int_0^\infty dx \frac{f_B(E_x)f_1(x)}{f_2(x)},$$

$$m_1(T) = -\frac{2\sqrt{2}Z\tilde{c}\gamma^{1/3}}{\pi} \int_0^\infty dx \frac{f_B(E_x)}{f_2(x)},$$

$$f_1(x) = x\tilde{c}\gamma^{2/3} + 4Z\pi\gamma, \quad f_2 = \sqrt{x\tilde{c} + 8Z\pi\gamma^{1/3}}, \tag{8.72}$$

$$f_B(E_x) = \frac{1}{e^{E_x/t} - 1}, \quad E_x = \frac{\sqrt{x}}{\sqrt{\tilde{c}}} f_2(x).$$

where $\tilde{c} = 3.3125$. Clearly, in practice, it is convenient dealing with dimensionless Eqs. (8.67)–(8.72), where dimensionless input parameters γ and t are supposed to be known and fixed.

8.7.2 Thermodynamic Quantities

Using Eqs. (6.62) (8.20) and (8.49), one may present the thermodynamic potential F as

$$
F = \Omega + \mu N = \Omega + \mu_0 N_0 + \mu_1 N_1 = \frac{V g \rho^2}{2} \left(1 + n_1^2 - 2n_1 \tilde{\sigma} - \tilde{\sigma}^2\right)
$$
$$
+ E_{LHY} + T \sum_k \ln\left(1 - e^{-\beta E_k}\right),
\tag{8.73}
$$

where

$$
E_k = \sqrt{\varepsilon_k(\varepsilon_k + 2\Delta)},
\tag{8.74}
$$

and $\tilde{\sigma} = \sigma/\rho$, $n_1 = \rho_1/\rho$. Here by E_{LHY} we denoted the Lee-Huang-Yang term:

$$
E_{LHY} = \frac{8V m^{3/2} \Delta^{5/2}}{15\pi^2}.
\tag{8.75}
$$

Then the entropy is given by

$$
S = -\left(\frac{\partial F}{\partial T}\right)_V = -\sum_k \ln\left(1 - e^{-\beta E_k}\right) + \beta \sum_k E_k f_B(E_k),
\tag{8.76}
$$

where $f_B(x) = 1/(\exp(\beta x) - 1)$ is the Boson distribution. The total energy of the system can be determined from the above equations as

$$
E = F + TS = \frac{V g \rho^2}{2}\left(1 + n_1^2 - 2n_1 \tilde{\sigma} - \tilde{\sigma}^2\right) + E_{LHY} + \sum_k E_k f_B(E_k).
\tag{8.77}
$$

Task 8.1 Derive Eqs. (8.73)–(8.77) in detail.

As to the heat capacity, it can be found from the Maxwell relation $C_v = T(\partial S/\partial T)$ and will be presented explicitly later.

Meanwhile, below we discuss particular cases of our first order of OPT, which is usually referred in the literature as Hartree-Fock-Bogolyubov (HFB) approximation [11]. In fact, HFB approximation is so general that it covers by itself almost all mean-field approaches, describing Bose condensation [9]. Actually, neglecting anomalous density σ, one obtains the so-called Hartree-Fock-Popov (HFP) approximation, and neglecting further quadratic fluctuations e.g. $\rho_1^2 \sim (\tilde{\psi}^\dagger \tilde{\psi})^2$, one comes to the Bilinear approximation. And finally, neglecting all fluctuations, one ends with the Bogolyubov approximation, where all particles are supposed to be condensed. In Table 8.1 we illustrate these particular cases with the example of energy.

Table 8.1 The energy of the Bose gas in condensed in units of $E_{\text{classic}} = Vg\rho^2/2$. In all three cases, the densities are given by Eq. (8.63) with the universal dispersion (8.74).

Approximation	E/E_{classic}	Δ	ρ_0
HFB(OPT)	$1 + n_1^2 - 2n_1\tilde{\sigma} - \tilde{\sigma}^2 + E_{LHY} + \sum_k E_k f_B(E_k)$	$g(\rho_0 + \sigma)$	$\rho_0 = \rho - \rho_1$
HFP	$1 + n_1^2 + E_{LHY} + \sum_k E_k f_B(E_k)$	$g\rho_0$	$\rho_0 = \rho - \rho_1$
Bilinear	$1 + E_{LHY} + \sum_k E_k f_B(E_k)$	$g\rho_0$	$\rho_0 = \rho - \rho_1$
Bogolyubov	1	$g\rho$	$\rho_0 \approx \rho$

8.7.3 Superfluid Density

It has been generally accepted to describe superfluidity in the Landau two-fluid model, where a superfluid consists of two components with two types of densities: normal ρ_n and superfluid ρ_s, such that $\rho_s + \rho_n = \rho$. Clearly, there is no superfluidity when $\rho_s = 0$.

Then the total mass current of the liquid, i.e. the momentum per unit volume, $\mathbf{j}$ takes the form

$$\mathbf{j} = \rho_s \mathbf{v}_s + \rho_n \mathbf{v}_n, \tag{8.78}$$

where $\mathbf{v}_s$ and $\mathbf{v}_n$ are the velocities of corresponding components. How to calculate ρ_s at a microscopical level starting from the Hamiltonian (4.18)? At first glance, it seems that the superfluid density should be proportional to the condensate density, $\rho_s \sim \rho_0$. However, unfortunately, there is no apparent relation between these two quantities. To substantiate this statement, we follow the review article [12].

So far, we have considered a resting system with zero total velocity. Now, we let the system move being boosted by the velocity $\mathbf{v}$. In the frame of moving with the velocity $\mathbf{v}$, the Hamiltonian is a functional $H[\psi]$ of the field operators ψ. In the laboratory frame, the Hamiltonian $H_v = H[\psi_v]$ is the functional of the field operators ψ_v given by the Galilean transformation

$$\psi_v(\mathbf{r}, t) = \psi(\mathbf{r}, t) \exp\left[i\left(m(\mathbf{rv}) - \frac{m\mathbf{v}^2 t}{2}\right)\right]. \tag{8.79}$$

The super-fluid component is interpreted as that part of the system, which nontrivially responds to the velocity boost with the superfluid density

$$\rho_s = \frac{\rho}{3mN} \lim_{v \to 0} \sum_{\alpha=1}^{3} \frac{\partial}{\partial v_\alpha} \langle \mathbf{P} \rangle, \tag{8.80}$$

where $\mathbf{P}$ is the total momentum

$$\mathbf{P} = \int \psi_v^+(\mathbf{r})(-i\nabla)\psi_v(\mathbf{r}) d\mathbf{r} \tag{8.81}$$

Performing the limit $v \to 0$ in (8.80), one has

$$\rho_n = \frac{2\rho Q}{3T},$$ (8.82)

so that the superfluid fraction will be given by

$$n_s = \frac{\rho_s}{\rho} = 1 - \frac{2Q}{3T}.$$ (8.83)

It is understood that the existence of the normal component leads to the friction and hence to the dispersed heat, which is denoted in Eqs. (8.82)–(8.83) as Q. For this quantity We have the following expression:

$$Q = \frac{1}{2mN} \langle \hat{\mathbf{P}}^2 \rangle = \frac{1}{2mN} \int \lim_{\mathbf{r}_3 \to \mathbf{r}_1} \lim_{\mathbf{r}_4 \to \mathbf{r}_2} \nabla_3 \nabla_4 \left\{ \langle \psi^+(\mathbf{r}_3) \psi(\mathbf{r}_4) \rangle \delta(\mathbf{r}_1 - \mathbf{r}_2) \right.$$
$$\left. - \langle \psi^+(\mathbf{r}_1) \psi^+(\mathbf{r}_2) \psi(\mathbf{r}_3) \psi(\mathbf{r}_4) \rangle \right\} d\mathbf{r}_1 d\mathbf{r}_2,$$ (8.84)

where the Eqs. (8.79) and (8.81) are used. The expectation values in (8.84) can be found by using Eqs. (4.55), (6.82), (6.94), (6.95) and (8.2). Thus, having performed tedious calculations, one may obtain

$$Q = \frac{1}{\rho} \int_0^\infty \frac{k^2 dk}{(2\pi)^3} \left[n_k + n_k^2 - \sigma_k^2 \right],$$ (8.85)

where n_k and σ_k are introduced by

$$\rho_1 \equiv \frac{1}{V} \sum_k n_k, \qquad \sigma \equiv \frac{1}{V} \sum_k \sigma_k.$$ (8.86)

Now using explicit expressions for the densities (8.63) we finally obtain [6]:

$$n_s = 1 - \frac{2\beta}{3(4\pi)^2 m\rho} \int_0^\infty \frac{k^4 dk}{\sinh^2(\beta E_k/2)}.$$ (8.87)

It is seen that, in spite of our expectation, the superfluid density is not just proportional to the condensate fraction. Note that the latter is about 10% in superfluid helium four, while $n_s \approx 1$ near absolute zero. Further, at low temperatures, where we may use expansion in powers of (T/Δ), the equation (8.87) yields

$$n_s \approx 1 - \frac{\beta m^4 s^5}{360\rho} \left(\frac{T}{\Delta} \right)^5.$$ (8.88)

Here it should be underlined that our final expression (8.87) for the superfluid density is valid only for the dilute Bose gases, whose interatomic potential has a contact form (4.1). So, for liquid helium one has to use a more realistic potential, such as the Aziz potential [13]. However, in this case, it would be rather difficult to calculate e.g. the expectation values presented in Eq. (8.84). This is why the problem of superfluid helium has not been completely solved yet.

8.8 Structure Factor and Correlation Function

In Chap. 5, we have introduced the static structure factor $S(\mathbf{q})$ and correlation function $g(\mathbf{r})$, which are served as a good tool to verify the appropriateness of any theoretical approximation in the view of experimental observations. In the framework of the field theoretical approach, they are formally written as [14]

$$
\begin{aligned}
g(\mathbf{r}_1, \mathbf{r}_2) &= \frac{\langle \psi^\dagger(\mathbf{r}_1)\psi^\dagger(\mathbf{r}_2)\psi(\mathbf{r}_2)\psi(\mathbf{r}_1)\rangle}{\rho(\mathbf{r}_1)\rho(\mathbf{r}_2)}, \\
S(\mathbf{q}) &= 1 + \frac{1}{N}\int d\mathbf{r}_1 d\mathbf{r}_2 e^{-i\mathbf{q}(\mathbf{r}_1 - \mathbf{r}_2)}\left[\langle\hat{n}(\mathbf{r}_1)\hat{n}(\mathbf{r}_2)\rangle - \langle\hat{n}(\mathbf{r}_1)\rangle\langle\hat{n}(\mathbf{r}_2)\rangle\right],
\end{aligned}
\tag{8.89}
$$

where $\hat{n}(\mathbf{r}) = \psi^\dagger(\mathbf{r})\psi(\mathbf{r})$. For a uniform system, these can be presented as

$$
\begin{aligned}
g(\mathbf{r}) &= \frac{\langle\psi^\dagger(\mathbf{r})\psi(\mathbf{r})\psi^\dagger(0)\psi(0)\rangle}{\rho^2} = \frac{\langle\hat{n}(\mathbf{r})\hat{n}(0)\rangle}{\rho^2} \\
&= 1 + \frac{1}{\rho(2\pi)^3}\int d\mathbf{q}\,e^{-i\mathbf{q}\mathbf{r}}\left[S(\mathbf{q}) - 1\right], \\
S(\mathbf{q}) &= 1 + \frac{1}{\rho}\int d\mathbf{r}\,e^{-i\mathbf{q}\mathbf{r}}\left[\rho^2 g(\mathbf{r}) - \langle\hat{n}(\mathbf{r})\rangle\langle\hat{n}(0)\rangle\right] \\
&= 1 + \rho\int d\mathbf{r}\,e^{-i\mathbf{q}\mathbf{r}}\left[g(\mathbf{r}) - 1\right].
\end{aligned}
\tag{8.90}
$$

Note that, by definition, $S(\mathbf{q})$ is normalized to unity as $\lim_{q\to\infty} S(\mathbf{q}) = 1$.

Calculation of the expectation values in the HFB approximation is straightforward. Having made Bogolyubov shift (8.2) in (8.90) one can obtain

$$
\rho^2 g(\mathbf{r}) = \rho_0^2 + 2\rho_0(2\rho_1 + \sigma) + \langle\tilde{\psi}^\dagger(\mathbf{r})\tilde{\psi}(\mathbf{r})\tilde{\psi}^\dagger(0)\tilde{\psi}(0)\rangle
\tag{8.91}
$$

where the following relations, which hold for a uniform system, are used:

$$
\langle\tilde{\psi}^\dagger(\mathbf{r})\tilde{\psi}(\mathbf{r})\rangle = \langle\tilde{\psi}^\dagger(0)\tilde{\psi}(0)\rangle = \rho_1, \qquad \langle\tilde{\psi}^\dagger(\mathbf{r})\tilde{\psi}^\dagger(\mathbf{r})\rangle + \langle\tilde{\psi}(\mathbf{r})\tilde{\psi}(\mathbf{r})\rangle = 2\sigma.
\tag{8.92}
$$

The last term in (8.91) can be easily evaluated by using Eqs. (4.55), (6.82), (6.93) and (6.94) leading to following result:

$$
\langle\tilde{\psi}^\dagger(\mathbf{r})\tilde{\psi}(\mathbf{r})\tilde{\psi}^\dagger(0)\tilde{\psi}(0)\rangle = \rho_1^2 + \frac{D_{11}^2(\mathbf{r}) + D_{22}^2(\mathbf{r})}{2},
\tag{8.93}
$$

where in accordance with Eqs. (4.95) and (6.93)

$$D_{11}(\mathbf{r}) \equiv D_{11}(\tau_1, \mathbf{r}_1, \tau_2, \mathbf{r}_2)|_{(\tau_1=\tau_2, \mathbf{r}_1-\mathbf{r}_2=\mathbf{r})} = \frac{1}{V} \sum_k e^{i\mathbf{kr}} \frac{W_k(\varepsilon_k + X_2)}{E_k}$$

$$\equiv \frac{1}{V} \sum_k e^{i\mathbf{kr}} D_{11}(\mathbf{k})$$

$$D_{22}(\mathbf{r}) \equiv D_{22}(\tau_1, \mathbf{r}_1, \tau_2, \mathbf{r}_2)|_{(\tau_1=\tau_2, \mathbf{r}_1-\mathbf{r}_2=\mathbf{r})} = \frac{1}{V} \sum_k e^{i\mathbf{kr}} \frac{W_k(\varepsilon_k + X_1)}{E_k}$$

$$\equiv \frac{1}{V} \sum_k e^{i\mathbf{kr}} D_{22}(\mathbf{k}). \tag{8.94}$$

In the BEC phase $X_2 = 0$ and $E_k = \sqrt{\varepsilon_k(\varepsilon_k + X_1)}$.

It seems to be so far so good. Alas, this is not the case. In fact, inserting Eqs. (8.93) and (8.94) into the Eq. (8.90), one has to evaluate the integral similar to

$$\int d\mathbf{r} d\mathbf{k} d\mathbf{k}' \exp[-i\mathbf{qr} + i\mathbf{kr} + i\mathbf{k}'\mathbf{r}] D_{11}(\mathbf{k}) D_{11}(\mathbf{k}') \sim \int d\mathbf{k} D_{11}(\mathbf{k}) D_{11}(\mathbf{k} - \mathbf{q}), \tag{8.95}$$

which turns out to be divergent! Can we use a regularization scheme here? Unfortunately, no. At first glance, we may improve the situation by introducing adequate counterterms. However, this is not the case, since these counter terms will be dependent on the external parameter-momentum $\mathbf{q}$, which contradicts to the fundamental ideology of any renormalization procedure. Thus, we have to accept that this divergence is inevitable.

Actually, the reason for such failure is hidden in the character of our interaction potential. If we hadn't used the contact interaction potential (4.1), then we wouldn't have faced such a problem. So, hopefully, a finite range interaction would make the above integrals finite. In the meantime, following Yukalov's proposal, [15], we limit ourselves to omit $D_{ii}^2(\mathbf{r})$ terms in (8.93) to present $S(\mathbf{q})$ in the following final form

$$S(\mathbf{q}) = 1 + 2(n_q + \sigma_q) = \frac{2\varepsilon_q W_q}{E_q} = \frac{\varepsilon_q}{E_q} \coth\left(\frac{E_q}{2T}\right), \tag{8.96}$$

where we used explicit expressions for ρ_1 and σ from (8.63):

$$n_q = \frac{W_q(\varepsilon_q + \Delta)}{E_q} - \frac{1}{2}, \qquad \sigma_q = -\frac{W_q \Delta}{E_q}. \tag{8.97}$$

At low momentum, $E_q \sim sq$, this equation leads to the relation (5.12) brought in Chap. 5.

Unfortunately, the expression (8.96) fails to describe experimental data on $S(\mathbf{q})$, and hence on $g(r)$. In our knowledge, the best description has been obtained only within numerical Monte-Carlo type calculations (see Figs. 5.3 and 5.5).

There is one more physical quantity which is worth considering. It is C-Tan's contact, introduced in Chap. 5. We shall evaluate it in next chapter.

8.9   Conclusion

In the present chapter, we have applied optimal perturbation theory to describe the BEC of dilute Bose gases with the contact interaction. We have derived explicit expressions for thermodynamic quantities for a uniform system in $D = 3$. It has been proven that the present version of OPT is self-consistent: The theory satisfies Hugenholtz-Pines theorem and overcomes the problem caused by the Hohenberg-Martin dilemma.

References

1. P. Hohenberg, P. Martin, Microscopic theory of superfluid helium. Ann. Phys. **34**(2), 291–359 (1965). https://www.sciencedirect.com/science/article/pii/0003491665902800
2. A. Rakhimov, Tan's contact as an indicator of completeness and self-consistency of a theory. Phys. Rev. A **102**, 063306 (2020). https://link.aps.org/doi/10.1103/PhysRevA.102.063306
3. A. Rakhimov, T. Abdurakhmonov, Z. Narzikulov, V.I. Yukalov, Self-consistent theory of a homogeneous binary Bose mixture with strong repulsive interspecies interaction. Phys. Rev. A **106**, 033301 (2022). https://link.aps.org/doi/10.1103/PhysRevA.106.033301
4. V. Yukalov, Representative statistical ensembles for Bose systems with broken gauge symmetry. Ann. Phys. **323**(2), 461–499 (2008). https://www.sciencedirect.com/science/article/pii/S0003491607000723
5. A. Sharma, G. Kartvelishvili, J. Khoury, Finite temperature description of an interacting Bose gas. Phys. Rev. D **106**, 045025 (2022). https://link.aps.org/doi/10.1103/PhysRevD.106.045025
6. V. Yukalov, Representative statistical ensembles for Bose systems with broken gauge symmetry. Ann. Phys. **323**(2), 461–499 (2008). https://www.sciencedirect.com/science/article/pii/S0003491607000723
7. N. Bogoliubov, *Lectures on Quantum Statistics*, Lectures on Quantum Statistics (Gordon and Breach Science Publishers, 1967). https://books.google.co.uz/books?id=sZsOAAAAQAAJ
8. J. Ginibre, On the asymptotic exactness of the Bogoliubov approximation for many boson systems. Commun. Math. Phys. **8**(1), 26–51 (1968). https://doi.org/10.1007/BF01646422
9. J.O. Andersen, Theory of the weakly interacting Bose gas. Rev. Mod. Phys. **76**, 599–639 (2004). https://link.aps.org/doi/10.1103/RevModPhys.76.599
10. V.I. Yukalov, Principal problems in Bose-Einstein condensation of dilute gases. Laser Phys. Lett. **1**(9), 435–461 (2004). https://onlinelibrary.wiley.com/doi/abs/10.1002/lapl.200410097
11. V.I. Yukalov, Interplay between approximation theory and renormalization group. Phys. Part. Nucl. **50**(2), 141–209 (2019). https://doi.org/10.1134/S1063779619020047
12. P.W. Courteille, V.S. Bagnato, V. Yukalov, Bose-Einstein condensation of trapped atomic gases. Laser Phys. **11**(6), 659–800 (2001)
13. R.A. Aziz, V.P.S. Nain, J.S. Carley, W.L. Taylor, G.T. McConville, An accurate intermolecular potential for helium. J. Chem. Phys. **70**(9), 4330–4342 (1979). https://doi.org/10.1063/1.438007
14. L. Pitaevskii, S. Stringari, *Bose-Einstein Condensation*, International Series of Monographs on Physics (Clarendon Press, 2003). https://books.google.co.uz/books?id=rIobbOxC4j4C
15. V. Yukalov, Structure factor of Bose-condensed systems. J. Phys. Stud. **11**, 55–62 (2007)

Tan's Contact in Mean-Field Theories

9

9.1 Introduction

Apart from the theories considered in the present lectures, various theories to describe a system of particles with mutual contact interaction have been developed recently [1,2]. Naturally, to stand the test of time, every physical theory or approach should be at least reliable and self-consistent. Their reliability can be verified by a simple comparison of their predictions with existing (or forthcoming) His Majesty's experiments. However, the verification of their self-consistency is not a simple business. In the present chapter, we shall discuss the self-consistency of mean field approaches by exploiting the idea about the equivalence of Tan's contact, calculated in three different ways outlined in Chap. 5.

So, let's rewrite the Eqs. (5.17), (5.19) and (5.20) introducing following notations

$$C_n = \lim_{k \to \infty} k^4 n_k, \tag{9.1}$$

$$C_E = \frac{8\pi m a_s^2}{V} \left(\frac{\partial F}{\partial a_s} \right), \tag{9.2}$$

$$C_\psi = \frac{m^2 g^2}{V} \int d\mathbf{r} \langle \psi^\dagger(\mathbf{r}) \psi^\dagger(\mathbf{r}) \psi(\mathbf{r}) \psi(\mathbf{r}) \rangle. \tag{9.3}$$

Obviously, regardless of the way of evaluation (or measurement) of Tan's contact by using any of the above three Eqs. (9.1), (9.2) or (9.3), one is supposed to obtain the same unique result, i.e.

$$C = C_n = C_E = C_\psi. \tag{9.4}$$

© The Author(s), under exclusive license to Springer Nature Switzerland AG 2026   141
A. Rakhimov and S. Mardonov, *Theory of Quantum Bose Liquids Beyond Bogoliubov Approximation*, Lecture Notes in Physics 1045,
https://doi.org/10.1007/978-3-032-05096-0_9

This trivial statement gives us an opportunity to check the self-consistency of an applied approach. Below, as examples, we shall consider Gaussian (Bilinear), Optimized Gaussian (OPT), and Bogolyubov approaches.[1]

9.2 Bilinear Approximation

(1) $C_n(Bil)$. The condensate depletion ρ_1 at zero temperature is given by (4.98) i.e.

$$\rho_1(T = 0) = \frac{1}{2V} \sum_k \left[\frac{(\varepsilon_k + g\rho_0)}{E_k} - 1 \right] \equiv \frac{1}{V} \sum_k n_k, \tag{9.5}$$

with $E_k = \sqrt{\varepsilon_k(\varepsilon_k + 2g\rho_0)}$, $\varepsilon_k = \mathbf{k}^2/2m$, which immediately leads to

$$C_n(Bil) = \lim_{k \to \infty} k^4 n_k = (gm\rho_0)^2 = (4\pi a_s\rho)^2 n_0^2 \equiv n_0^2 C_{\text{class}}, \tag{9.6}$$

where $n_0 = \rho_0/\rho$ and $C_{\text{class}} = 16\pi^2\gamma^2/a_s^4$ is Tan's contact, corresponding to the case when all fluctuations are neglected. Note that, in the Bilinear approximation at zero temperature, and particularly in (9.6), ρ_0 satisfies following algebraic equation:

$$\rho_0 = \rho - \rho_1 = \rho - \frac{(mg\rho_0)^{3/2}}{3\pi^2}, \tag{9.7}$$

which follows from the Eq. (4.113) with $\Delta = g\rho_0$.

(2) $C_{E(Bil)}$. At zero temperature the free energy F equals to the total energy of the system which is given by

$$E = \frac{Vg\rho^2}{2} - \frac{Vg\rho_1^2}{2} + E_{LHY} = \frac{Vg}{2}\left[\rho^2 - (\rho - \rho_0)^2 + \frac{8Vm^{3/2}(g\rho_0)^{5/2}}{15\pi^2} \right], \tag{9.8}$$

where we used (8.75) with $\Delta = g\rho_0$. The differentiation of E with respect to a_s requires an explicit expression for $d\rho_0/da_s$, which could be performed by differentiation of both sides of the Eq. (9.7) and solving the resulting equation with respect to $d\rho_0/da_s$. This gives

$$\frac{d\rho_0}{da_s} = -\frac{\rho_0}{a_s}\left[1 + \frac{\sqrt{\pi}}{4\sqrt{\gamma n_0}} \right]^{-1} = -\frac{n_0\gamma}{a_s^4}\left[1 + \frac{\sqrt{\pi}}{4\sqrt{\gamma n_0}} \right]^{-1}. \tag{9.9}$$

Thus, by using Eqs. (9.2), (9.8) and (9.9) one obtains

$$C_E(Bil) = C_{\text{class}}\left[1 + \frac{64n_0^{5/2}\sqrt{\gamma}}{3(\sqrt{\pi} + 4\sqrt{\gamma n_0})} \right], \tag{9.10}$$

---

[1] Actually, the terms "Gaussian" and "HFB" stand for "Bilinear" and "OPT", respectively.

which is firstly derived by Shakel [3].

(3) $C_\psi(Bil)$. Actually, in Chap. 8, we have obtained explicit expressions for the expectation value $\langle \psi^\dagger(\mathbf{r})\psi^\dagger(\mathbf{r})\psi(\mathbf{r})\psi(\mathbf{r})\rangle$. So, using Eqs. (8.89)–(8.97) in (9.3) gives

$$C_\psi(Bil) = C_{\text{class}}\left[1 + \frac{2(\rho_1 + \sigma)}{\rho}\right], \tag{9.11}$$

where $\rho_1(T = 0)$ and $\sigma(T = 0)$ are given by Eqs. (8.63).

Task 9.1. Prove following limit

$$\lim_{x \to \infty} x^4\left[\frac{x^2 + \alpha}{x\sqrt{x^2 + 2\alpha}} - 1\right] = \frac{\alpha^2}{2}. \tag{9.12}$$

$(\alpha > 0)$.

9.3   OPT

(1) $C_n(OPT)$. This time n_k at $T = 0$ is given by Eqs. (8.63) as

$$n_k = \frac{1}{2V}\left[\frac{\varepsilon_k + \Delta}{E_k} - 1\right], \tag{9.13}$$

where

$$E_k = \sqrt{\varepsilon_k(\varepsilon_k + 2\Delta)}, \qquad \varepsilon_k = \mathbf{k}^2/2m, \tag{9.14}$$

and

$$\Delta = g(\rho_0 + \sigma). \tag{9.15}$$

Therefore,

$$C_n(OPT) = \lim_{k \to \infty} k^4 n_k = (\Delta m)^2 = (sm)^4. \tag{9.16}$$

This result shows that, at zero temperature, Tan's contact is directly related to the sound velocity and hence could be measured also by sound velocity experiments [4,5].

(2) $C_E(OPT)$. To evaluate dE/da_s, where the energy at zero temperature is given by (8.77):

$$E = \frac{Vg\rho^2}{2}\left(1 + n_1^2 - 2n_1\tilde{\sigma} - \tilde{\sigma}^2\right) + \frac{8Vm^{3/2}\Delta^{5/2}}{15\pi^2}, \tag{9.17}$$

evaluation of $\Delta'_a \equiv d\Delta/da_s$ is needed. The latter can be found by differentiating both sides of Eq. (9.15) with respect to a_s leading to

$$\Delta'_a = \frac{\Delta}{a_s}\left[1 + \frac{6\pi a_s(\rho_1 - \sigma)}{\Delta m}\right]^{-1}. \tag{9.18}$$

Now using the last two equations in (9.2), we finally obtain

$$\mathcal{C}_E(OPT) = \mathcal{C}_{\text{class}}(1 + W_E), \qquad W_E = n_\sigma + \Delta'_a\left[\frac{2mn_1}{\pi\rho} + \frac{3a_s n_\sigma}{\Delta}\right], \tag{9.19}$$

$$n_\sigma \equiv n_1^2 - \tilde{\sigma}^2 - 2n_1\tilde{\sigma}.$$

Note that, $\tilde{\sigma} = \sigma/\rho$, $n_1 = \rho_1/\rho$, where σ and ρ_1 are defined in Eq. (8.63).

(3) $\mathcal{C}_\psi(OPT)$. The expectation value $\langle \psi^\dagger(\mathbf{r})\psi^\dagger(\mathbf{r})\psi(\mathbf{r})\psi(\mathbf{r})\rangle$ with $\psi(\mathbf{r}) = \sqrt{\rho_0} + \tilde{\psi}(\mathbf{r})$ can be easily evaluated in a similar way that we have followed in the derivation of (8.93). As a result we obtain [6]:

$$\mathcal{C}_\psi(OPT) = \mathcal{C}_{\text{class}}(1 + W_\psi), \qquad W_\psi = 2(n_1 + \tilde{\sigma} - 2n_1\tilde{\sigma}) - n_\sigma. \tag{9.20}$$

Task 9.2. Derive the Eq. (9.20)

9.4   Bogolyubov Approximation

(1) & (2). $\mathcal{C}_n(Bog)$ and $\mathcal{C}_\psi(Bog)$. If one defines Bogolyubov approximation as the case when all fluctuations are neglected, then we come to the following conclusion:

$$\mathcal{C}_n(Bog) = \mathcal{C}_\psi(Bog) = \mathcal{C}_{\text{class}} = (4\pi\gamma/a_s^2)^2. \tag{9.21}$$

(3) $\mathcal{C}_n(Bog)$. On the other hand, Bogolyubov approximation can also be defined as the particular case of OPT with $\Delta \approx g\rho$ and $\sigma = 0$. This leads to the following total energy in (9.17):

$$E = \frac{Vg\rho^2}{2} + \frac{1}{2}\sum\left(E_k - \varepsilon_k - g\rho + \frac{g^2\rho^2}{2\varepsilon_k}\right) = \frac{2V\pi\gamma^2}{ma_s^5}\left(1 + \frac{128\sqrt{\gamma}}{15\sqrt{\pi}}\right), \tag{9.22}$$

and hence [6]

$$\mathcal{C}_E(Bog) = \mathcal{C}_{\text{class}}\left(1 + \frac{64\sqrt{\gamma}}{3\sqrt{\pi}}\right). \tag{9.23}$$

Remarkably, the expression for the total energy in (9.22) coincides with that one obtained long years ago by Lee, Huang and Yang (LHY) [7] in the hard core boson model, and the Eq. (9.23) for $\mathcal{C}_E$ does with the result by Schakel [3] derived in a similar way.

The question arises, what is the difference between Bilinear (Gaussian) and Bogolyubov approximations? The main difference is that in the Gaussian approximation, one has to preliminarily solve the Eq. (9.7) with respect to ρ_0 for a given γ, while in the Bogolyubov one, there is no need to solve any equation. This fact makes the Bogolyubov approximation attractive and the most practical one in order to make a fast estimation of a physical quantity in the BEC regime. Moreover, the Bogolyubov approximation takes into account the gas parameter up to the first order in the expansion by $\sqrt{\gamma}$ in the evaluation of the condensed fraction.

9.5   Results and Discussions

Now we are in the position of studying three versions of MFT for self-consistency in the spirit of the requirement in Eq. (9.4). In Fig. 9.1 we present Tan's contact obtained in Bogolyubov (Fig. 9.1a), Gaussian (Fig. 9.1b) and HFB (Fig. 9.1c) approximations. Here dashed, solid and dotted curves correspond to $\mathcal{C}_n$, $\mathcal{C}_E$ and $\mathcal{C}_\psi$ defined by Eqs. (9.1), (9.2) and (9.3) respectively. From Fig. 9.1a it is seen that Bogolyubov approximation satisfies the first equality $\mathcal{C}_n = \mathcal{C}_\psi$, but does not the second one, i.e. $\mathcal{C}_\psi \neq \mathcal{C}_E$. As to the Gaussian (one loop) approximation, the difference between these three quantities is rather notable (see Fig. 9.1b). In this sense, the Bogolyubov approximation seems more reliable than the Gaussian one. This fact can explain the popularity of Bogolyubov approximation, including LHY terms [7] in the literature [8]. From Fig. 9.1c it is seen that the discrepancy between $\mathcal{C}_n$, $\mathcal{C}_E$ and $\mathcal{C}_\psi$ is rather small for the Optimized Perturbation Theory, i.e. Variational Gaussian approximation. Hence, one may conclude that this approximation can be regarded as the most complete and self-consistent one among other existing MFT-based approaches. Nevertheless, strongly speaking, HFB is also needed corrections, especially for $\gamma > 0.002$, arising from the high-order quantum fluctuations. The intensity of such fluctuations is almost proportional to the fraction of uncondensed particles n_1. As it is seen from Fig. 9.1d, even at $\gamma \sim 0.005$ the depletion is about 15%. For comparison, we note that, in superfluid helium ^{4}He $n_1 \approx 90\%$.

On the other hand, one may judge about an appropriateness of any theory just by comparing its predictions with experimental measurements. In Fig. 9.2a we compare our predictions for Tan's contact given by the HFB approach with the experimental data on ^{85}Rb atomic condensate at fixed density $\rho = 5.8\mu m^{-3}$. It is seen that HFB approximation is able to describe $\mathcal{C}$ rather satisfactory up to $a/a_B < 1200$, which corresponds to $\gamma \approx 0.0015$. Moreover, HFB predictions for Tan's contact are in good agreement with path-integral ground-state (PIGS) Monte Carlo calculations performed by Rossi and Salasnich [9].

Unfortunately, presently Tan's contact for a Bose gas has been measured at very small values of the gas parameter, $\gamma \leq 0.002$. To predict its behavior at larger γ we calculated Tan's contact in the region $0 \leq \gamma \leq 0.2$ and presented the results in Fig. 9.2b. It is seen that PIGS Monte Carlo method predicts a smooth increase of $\mathcal{C}$, while the latter remains practically unchanged in the HFB approximation for $\gamma > 0.05$. Note that, for superfluid helium $\gamma \approx 0.6$.

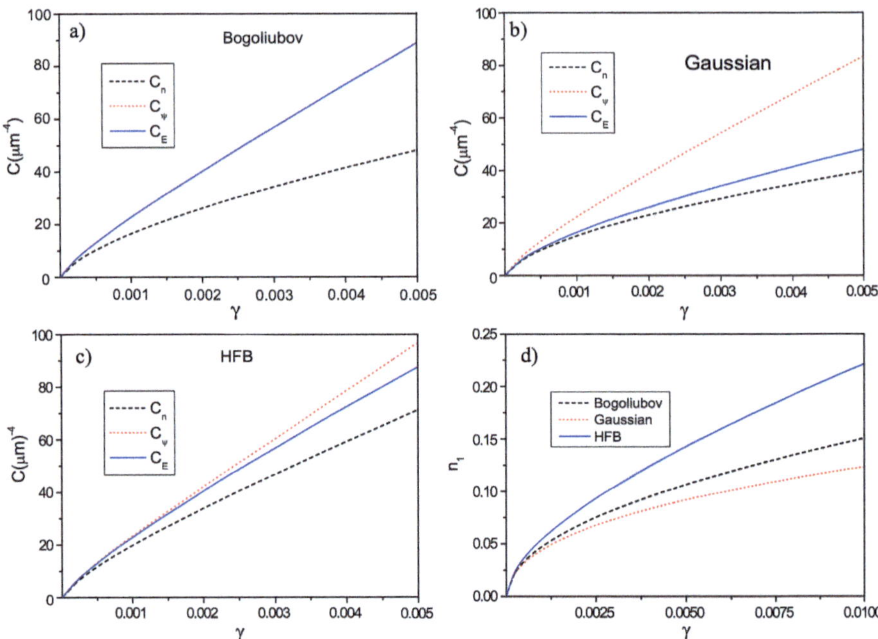

Fig. 9.1 Tan's contact as a function of the gas parameter $\gamma = \rho a^3$ in Bogolyubov (**a**), Gaussian i.e. Bilinear (**b**), and OPT i.e. HFB approximations (**c**). Dashed, solid and dotted lines are obtained with Eqs. (9.1), (9.2) and (9.3), respectively. The corresponding condensate depletions, $n_1 = N_1/N$, are presented in Fig. 9.1d

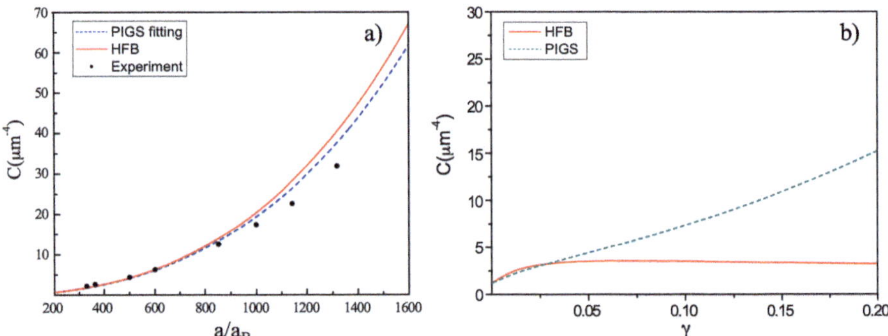

Fig. 9.2 a The contact in natural units $(\mu m)^{-4}$ vs scattering length $a = a_s$ computed in PIGS Monte Carlo method [9] (dashed curve) and HFB approximation (solid curve). Filled circles are experimental data of Wild et al. [8] obtained for ^{85}Rb atomic condensate (see Chap. 5). The scattering length a_s is given in units of the Bohr radius $a_B = 5.3 \cdot 10^{-5} \mu m$; **b** Tan's contact at large values of the gas parameter $\gamma = \rho a^3$ in HFB (solid line) and PIGS Monte Carlo (dashed line) approaches [9]. In both figures, the density is fixed in its typical value as $\rho = 5.8 \mu m^{-3}$

9.6 Conclusion

In the present chapter, we have derived explicit expressions for Tan's contact for bosons at zero temperature within various approximations based on mean field theory. Numerical analysis, made with these equations, gave us an opportunity to study such approximations for completeness and self-consistency. We have shown that in this concept, the Hartree-Fock-Bogolyubov approximation, derived within optimized perturbation theory, satisfies the requirement $C_n = C_E = C_\psi$ better than one-loop or Bogolyubov approximations. Moreover, HFB predictions are in good agreement with existing experimental data as well as with Monte Carlo calculations for small values of the gas parameter.

At the end of this section, we answer the question "What is the validation of present theory, HFB, that is the first order of LDE, with respect of the gas parameter $\gamma = \rho a_s^3$?". Actually, comparison of HFB calculations with Monte-Carlo ones has revealed that the present theory can accurately describe, say, $n_0(\gamma, T = 0)$ for $\gamma \leq 0.1$ [9]. Therefore, for large values of γ present version of OPT needs serious corrections. These could be performed by extension of the present approach in the spirit of Post Gaussian Perturbative approximation, which includes the second order of δ-expansion [10] (see Chap. 17). It is expected that such an extension would give rise to a desired logarithmic term, which is used in the literature [3,9,11].

On the other hand, there is one more requirement for a self-consistent theory. In accordance with general rules of quantum mechanics, predicted physical observables should not depend on the phase of the wave function. We shall discuss this point for OPT in the next chapter.

References

1. J.O. Andersen, Theory of the weakly interacting Bose gas. Rev. Mod. Phys. **76**, 599–639 (2004). https://doi.org/10.1103/RevModPhys.76.599
2. B. Svistunov, E. Babaev, N. Prokof'ev, *Superfluid States of Matter* (1st ed.) (CRC Press, Boca Raton, 2015). https://doi.org/10.1201/b18346
3. A.M.J. Schakel, Tan relations in dilute Bose gasses (2010). arXiv:1007.3452
4. A.R. Fritsch, P.E.S. Tavares, F.A.J. Vivanco, G.D. Telles, V.S. Bagnato, E.A.L. Henn, Thermodynamic measurement of the sound velocity of a Bose gas across the transition to Bose–Einstein condensation. J. Stat. Mech.: Theory Exp. **2018**(5), 053108 (2018). https://doi.org/10.1088/1742-5468/aabbcf
5. M.R. Andrews, D.M. Kurn, H.-J. Miesner, D.S. Durfee, C.G. Townsend, S. Inouye, W. Ketterle, Propagation of sound in a Bose-Einstein condensate. Phys. Rev. Lett. **79**, 553–556 (1997). https://doi.org/10.1103/PhysRevLett.79.553
6. A. Rakhimov, Tan's contact as an indicator of completeness and self-consistency of a theory. Phys. Rev. A **102**, 063306 (2020). https://doi.org/10.1103/PhysRevA.102.063306
7. T.D. Lee, K. Huang, C.N. Yang, Eigenvalues and eigenfunctions of a Bose system of hard spheres and its low-temperature properties. Phys. Rev. **106**, 1135–1145 (1957). https://doi.org/10.1103/PhysRev.106.1135

8. R.J. Wild, P. Makotyn, J.M. Pino, E.A. Cornell, D.S. Jin, Measurements of Tan's contact in an atomic Bose-Einstein condensate. Phys. Rev. Lett. **108**, 145305 (2012). https://doi.org/10.1103/PhysRevLett.108.145305

9. M. Rossi, L. Salasnich, Path-integral ground state and superfluid hydrodynamics of a bosonic gas of hard spheres. Phys. Rev. A **88**, 053617 (2013). https://doi.org/10.1103/PhysRevA.88.053617

10. I. Stancu, P.M. Stevenson, Second-order corrections to the Gaussian effective potential of $\lambda\varphi^4$ theory. Phys. Rev. D **42**, 2710–2725 (1990). https://doi.org/10.1103/PhysRevD.42.2710

11. E. Braaten, A. Nieto, Quantum corrections to the energy density of a homogeneous Bose gas. Eur. Phys. J. B-Condens. Matter Complex Syst. **11**(1), 143–159 (1999). https://doi.org/10.1007/s100510050925

The Condensate Phase and Its Relation to the Anomalies Density

10

10.1 Introduction

In previous chapters, we exploited SSB for a homogeneous BEC in a standard way by introducing the following shift

$$\psi(\mathbf{r}, t) = \sqrt{\rho_0} + \tilde{\psi}(\mathbf{r}, t), \tag{10.1}$$

where ρ_0 is the condensate density and $\tilde{\psi}(\mathbf{r}, t)$ is the fluctuating complex field, which will be integrated out. In this case, the Hugenholtz-Pines relation is given by

$$\Sigma_n(0) - \Sigma_{an}(0) = \mu_1, \tag{10.2}$$

where $\Sigma_n(\mathbf{k})$ and $\Sigma_{an}(\mathbf{k})$ are normal and anomalous self energies and μ_1 has been involved due to Yukalov's prescription, outlined in Sect. 8.5. On the other hand, strictly speaking, there is no physical requirement that the first term of (10.1) should be real. So, in general, this representation must be replaced by

$$\psi(\mathbf{r}, t) = e^{i\theta}\sqrt{\rho_0} + \tilde{\psi}(\mathbf{r}, t) \equiv \psi(\mathbf{r}, t) = \eta\sqrt{\rho_0} + \tilde{\psi}(\mathbf{r}, t), \tag{10.3}$$

where the condensate phase η is related to its real phase angle θ as $\eta = \exp(i\theta)$. Will the relation (10.2) need any modification for this case?

In their original work [1], Hugenholtz and Pines proved their theorem (10.2) using the real representation (10.1) by default. Later on, the general case with (10.3) has been considered by Watabe [2], who derived the following extension of the Hugenholtz-Pines theorem:

$$\Sigma_n(0) - e^{-2i\theta}\Sigma_{an}(0) = \mu_1. \tag{10.4}$$

© The Author(s), under exclusive license to Springer Nature Switzerland AG 2026
A. Rakhimov and S. Mardonov, *Theory of Quantum Bose Liquids Beyond Bogoliubov Approximation*, Lecture Notes in Physics 1045,
https://doi.org/10.1007/978-3-032-05096-0_10

Below, we rewrite this relation in real field representation

$$\tilde{\psi} = \frac{1}{\sqrt{2}}(\psi_1 + i\psi_2), \qquad \tilde{\psi}^{\dagger} = \frac{1}{\sqrt{2}}(\psi_1 - i\psi_2), \tag{10.5}$$

where the inverse Green function has the general form:

$$D^{-1}(\omega_n, \mathbf{k}) = \begin{pmatrix} \varepsilon_k + X_1 & \omega_n \\ -\omega_n & \varepsilon_k + X_2 \end{pmatrix}, \tag{10.6}$$

leading to the energy dispersion

$$E_k = \sqrt{\varepsilon_k + X_1}\sqrt{\varepsilon_k + X_2}, \qquad \varepsilon_k = \mathbf{k}^2/2\,m. \tag{10.7}$$

In these equations the self energies $X_{1,2}$ are related to $\Sigma_{n,an}$ as:

$$\Sigma_n = \mu_1 + \frac{X_1 + X_2}{2}, \qquad \Sigma_{an} = \frac{X_1 - X_2}{2}, \tag{10.8}$$

Thus, the Eqs. (10.4) and (10.8) lead to

$$X_1(1 - \eta^2) + X_2(1 + \eta^2) = 0. \tag{10.9}$$

Therefore, if the phase angel is chosen e.g. as $\theta = 2\pi n$, (n- integer) i.e. $\eta^2 = 1$, the Hugenholtz-Pines-Watabe (HPW) theorem requires $X_2 = 0$, while for $\theta = \pi/2 + 2\pi n$, i.e. $\eta^2 = i$ the condition will be $X_1 = 0$.

Can physical quantities depend on the phase of the condensate? Well, in accordance with the general rules of quantum mechanics, the wave function of a particle is determined with precision to the phase, since only its modulu has a physical sense. Below, we prove that the thermodynamic characteristics of a single Bose condensed system are not sensitive to θ. Note that only the difference of the phases of two condensates can be observed, especially in the interference experiments outlined in Chap. 5.

10.2   Variational Perturbation Theory with Arbitrary Phase

Using (10.3) in the action (8.1) and introducing variational parameters $X_{1,2}$ in the spirit of linear delta expansion one may present the action as follows

$$S = S_0 + S_{\text{free}} + \delta S_{\text{int}},$$

$$S_0 = \int_0^\beta d\tau \int d\mathbf{r}[-\mu_0\rho_0 + \frac{g\rho_0^2}{2}],$$

$$S_{\text{free}} = \frac{1}{2}\int_0^\beta d\tau \int d\mathbf{r}\left\{i\epsilon_{ab}\psi_a\partial_\tau\psi_b + \psi_1\left(-\frac{\nabla^2}{2m} + X_1\right)\psi_1\right.$$

$$\left.+\psi_2\left(-\frac{\nabla^2}{2m} + X_2\right)\psi_2\right\}, \tag{10.10}$$

$$S_{\text{int}} = S_{\text{int}}^{(1)} + S_{\text{int}}^{(2)} + S_{\text{int}}^{(3)} + S_{\text{int}}^{(4)},$$

$$S_{\text{int}}^{(2)} = \frac{1}{2}\int_0^\beta d\tau \int d\mathbf{r}\left\{\beta_1\psi_1^2 + \beta_2\psi_2^2 + 2\beta_{12}\psi_1\psi_2\right\},$$

$$S_{\text{int}}^{(4)} = \frac{g}{8}\int_0^\beta d\tau \int d\mathbf{r}\left\{\psi_1^2 + \psi_2^2\right\}^2,$$

where we omit explicit expressions for $S_{\text{int}}^{(1)}$ and $S_{\text{int}}^{(3)}$, since their averages are equal to zero: $< S_{\text{int}}^{(1)} >=< S_{\text{int}}^{(3)} >= 0$. Moreover, following notations are introduced

$$\beta_1 = -\mu_1 - X_1 + \frac{g\rho_0}{2}(\eta^2 + \bar\eta^2 + 4),$$

$$\beta_2 = -\mu_1 - X_2 - \frac{g\rho_0}{2}(\bar\eta^2 + \eta^2 - 4), \tag{10.11}$$

$$\beta_{12} = \frac{ig\rho_0(\bar\eta^2 - \eta^2)}{2}, \quad \bar\eta = \eta^+.$$

The term S_{free} corresponds to the inverse Green function given by (10.6), and hence the expressions for the densities remain formally unchanged:

$$\rho_1 = \frac{1}{V}\sum_k\left[\frac{W_k(\varepsilon_k + X_1/2 + X_2/2)}{E_k} - \frac{1}{2}\right], \tag{10.12}$$

$$\sigma = \frac{(X_2 - X_1)}{2V}\tilde S, \quad \tilde S \equiv \sum_k\left[\frac{W_k}{E_k} - \frac{1}{2\varepsilon_k}\right], \tag{10.13}$$

where E_k is given by (10.7) and the counter term in σ is included. The total density is clearly the sum of condensed and uncondensed fractions:

$$\rho = \frac{N}{V} = \rho_0 + \rho_1. \tag{10.14}$$

The thermodynamic potential derived in the standard way has the form

$$\Omega = \Omega_0 + \Omega_{ln} + \Omega_2 + \Omega_4,$$

$$\Omega_0 = -\mu\rho_0 + \frac{g\rho_0^2}{2},$$

$$\Omega_{ln} = \frac{1}{2}\sum_k (E_k - \varepsilon_k) + T\sum_k \ln(1 - e^{-\beta E_k}), \tag{10.15}$$

$$\Omega_2 = \frac{1}{2}(\beta_1 B + \beta_2 A), \qquad \Omega_4 = \frac{g}{8V}(3A^2 + 3B^2 + 2AB),$$

where $A = V(\rho_1 - \sigma)$ and $B = V(\rho_1 + \sigma)$. Particularly, for $\eta^2 = 1$ the Eqs. (10.15) coincide with (8.49). Variational parameters $X_{1,2}$ corresponding to the minimum of Ω may be calculated as the positive solutions of the following algebraic equations

$$X_1 = 2g\rho + g\sigma - \mu_1 + \frac{g\rho_0(\eta^2 + \bar{\eta}^2)}{2},$$
$$X_2 = 2g\rho - g\sigma - \mu_1 - \frac{g\rho_0(\eta^2 + \bar{\eta}^2)}{2}. \tag{10.16}$$

Moreover, we have two additional variational parameters: θ and ρ_0. So, the corresponding minimization procedure results in

$$\frac{\partial \Omega}{\partial \rho_0} = g\sigma \cos 2\theta + g(\rho_0 + 2\rho_1) - \mu_1 = 0,$$
$$\frac{\partial \Omega}{\partial \theta} = 2\rho_0 g\sigma \sin \theta = 0, \quad -2g\sigma\rho_0 \cos 2\theta \geq 0. \tag{10.17}$$

The first equation is convenient to determine ρ_0, while the other two are used to fix the phase angle. For the latter, we have two solutions: $\theta = \pi n$ and $\theta = \pi/2 + 2\pi n$, i.e. $\eta^2 = 1$ and $\eta^2 = i$. Consequently, since $\eta^2 = \bar{\eta}^2$ the Eq. (10.16) are simplified as

$$X_1 = 2g\rho + g\sigma - \mu_1 + g\rho_0\eta^2,$$
$$X_2 = 2g\rho - g\sigma - \mu_1 - g\rho_0\eta^2. \tag{10.18}$$

From these equations, we obtain

$$X_2 + X_1 = 4g\rho - 2\mu_1, \qquad X_2 - X_1 = -2g\sigma - 2g\rho_0\eta^2. \tag{10.19}$$

Now putting these into Eqs. (10.12) and (10.13) we have

$$\rho_1 = \frac{1}{V}\sum_k \left[\frac{W_k(\varepsilon_k + 2g\rho - \mu_1)}{E_k} - \frac{1}{2}\right], \tag{10.20}$$

and

$$\sigma = -g\tilde{S}(\sigma + \rho_0\eta^2). \tag{10.21}$$

The last equation can be formally solved with respect to σ:

$$\sigma = -\frac{\tilde{S}g\rho_0\eta^2}{1+\tilde{S}g} \equiv \bar{\sigma}\eta^2. \tag{10.22}$$

This equation demonstrates the fact that anomalous density vanishes in the normal phase together with the condensed fraction, i.e. $\sigma(T \geq T_c) = \rho_0(T \geq T_c) = 0$. Now using the last equation in Eq. (10.18) we can present them in the following compact form:

$$\begin{aligned}
X_1 &= 2g\rho - \mu_1 + g\bar{\sigma}\eta^2 + g\rho_0\eta^2 \equiv a + b\eta^2, \\
X_2 &= 2g\rho - \mu_1 - g\bar{\sigma}\eta^2 - g\rho_0\eta^2 \equiv a - b\eta^2,
\end{aligned} \tag{10.23}$$

where a and b do not depend on η. Inserting these equations to (10.7) we see that the dispersion simplifies as

$$E_k^2 = \varepsilon_k^2 + \varepsilon_k(X_1 + X_2) + X_1 X_2 = \varepsilon_k^2 + 2a\varepsilon_k + a^2 - b^2. \tag{10.24}$$

Thus, we have proven that the energy dispersion as well as the normal density ρ_1 in the Eq. (10.20) are not sensitive to the condensate phase angle. An alternative explanation of this statement is the following. The rotation of the phase angle, say, to $90°$ i.e. $\theta \to \theta + \pi/2$ is equivalent to the transformation $\eta^2 = 1 \to \eta^2 = i$. Such transformation, as it is seen From Eqs. (10.23) leads to interchange of parameters $X_{1,2}$ as

$$\begin{aligned}
X_1(\theta + \pi/2) &= X_1(\eta^2 = i) = X_2(\eta^2 = 1), \\
X_2(\theta + \pi/2) &= X_2(\eta^2 = i) = X_1(\eta^2 = 1).
\end{aligned} \tag{10.25}$$

Since any physical quantity $\mathcal{F}$ in the present OPT of BEC is symmetric in $X_{1,2}$, i.e. $\mathcal{F}(X_1, X_2) = \mathcal{F}(X_2, X_1)$, one may conclude that a single BEC is a phase invariant object, as it should be. This provides an additional argument in favour of the self-consistency of the present theory. Particularly, one may make sure that the phase invariance is not violated in the bilinear approximation, also. This is why in practical calculations the choice $\psi(\mathbf{r}, t) = \sqrt{\rho_0} + \tilde{\psi}(\mathbf{r}, t)$ is always accepted by default.

Task 10.1 Prove phase invariance of Ω given by (10.15).

10.3   How to Measure $|\sigma|$?

Thus, in actual calculations, it does not matter which value of $\eta^2 = \pm 1$ will be chosen. Particularly, if one chooses the case $\eta^2 = +1$, then HPW relations (10.9) lead to $X_2 = 0$, while the case with $\eta^2 = -1$, imposes $X_1 = 0$. In each case, one is left, actually, with the same equations for physical quantities.

Let's, for concreteness, fix the choice $\eta^2 = 1$. Then the main Eq. (10.18) have the form

$$X_1 = -\mu_1 + 2g\rho + g(\rho - \rho_1 + \bar{\sigma}), \tag{10.26}$$
$$X_2 = -\mu_1 + 2g\rho - g(\rho - \rho_1 + \bar{\sigma}), \tag{10.27}$$

with $\bar{\sigma}$ defined in (10.22). In accordance with (10.9) we set here $X_2 = 0$ to obtain

$$\mu_1 = 2g\rho - g(\rho - \rho_1 + \bar{\sigma}). \tag{10.28}$$

Inserting this to (10.26) leads to the equation

$$\Delta \equiv \frac{X_1}{2} = g(\rho - \rho_1 + \bar{\sigma}) = g\rho(n_0 + \bar{m}_1), \tag{10.29}$$

where

$$n_0 = \frac{N_0}{N} = \frac{\rho_0}{\rho} = 1 - n_1, \quad \bar{m}_1 = \bar{\sigma}/\rho,$$
$$n_1 = \frac{1}{V\rho} \sum_k \left[\frac{W_k(\varepsilon_k + \Delta)}{E_k} - \frac{1}{2} \right], \tag{10.30}$$

with the dispersion

$$E_k = \sqrt{\varepsilon_k}\sqrt{\varepsilon_k + 2\Delta}. \tag{10.31}$$

In practice the set of nonlinear algebraic Eqs. (10.29)–(10.31) can be easily solved numerically with respect to Δ, being expressed in terms of only two input parameters, $\gamma = \rho a_s^3$ and $t = T/T_c$, as it has been proven in Chap. 8.

Considering the above, we propose the following strategy to estimate experimental values of $|\sigma|$ using the data on the condensed fraction n_0 and the sound velocity s. Thus, for each set of γ and t

1. Evaluate $\Delta(\gamma, t)$ from the experimental data on the sound velocity s, by using the equation $\Delta = ms^2(\gamma, t)$;
2. Find $|\sigma|$ from the equation

$$m_1 \equiv \frac{|\sigma|}{\rho} = n_0 - \frac{\Delta}{g\rho} = n_0 - \frac{ms^2}{g\rho}, \tag{10.32}$$

where $\Delta(\gamma, t)$ has been already fixed, and $n_0(\gamma, t)$ is given by experiments.

Unfortunately, at the present time, this strategy has not been realized for the following reasons:

1. Strictly speaking, the present theory is valid only for homogeneous dilute Bose gases with constant density, $\rho = const$;
2. Although inhomogeneous BECs in magnetic traps have been well studied inside and out, the data on homogeneous condensates are quite scarce.

We failed to find a report on experimental measurements of the condensate fraction n_0 and the sound velocity s on the same material in a uniform box. Although the first homogeneous BEC in an optical box trap was created twenty years ago [3], only the work by Gaunt et al. [4], provides with experimental data on the condensate fraction, but not on the sound velocity. In the next section, we make an attempt to describe this data within the present theory and bring our predictions for $|\sigma|$.

10.4 Numerical Results and Discussions

In 2013, experimenters in Cavendish Laboratory [4] observed the BEC of atomic gas of ^{87}Rb in the state $|F, m_F\rangle = |2, 2\rangle$ with constant density $\rho \approx 0.11 \times 10^{14}/cm^3$. Bearing in mind the experimental value of the s- wave scattering length $a_s = 5\,nm$, obtained by Brazilian physicists [5], we have $\gamma = \rho a_s^3 \approx 1.4 \times 10^{-6}$. It is understood that, for such a small value of the gas parameter, the critical temperature T_c can be evaluated from the simple formula (8.58):

$$T_c = \frac{3.3125\rho^{2/3}(\hbar c)^2}{k_B mc^2} \tag{10.33}$$

with $k_B = 8.617 \times 10^{-5}ev/K$, $(\hbar c) \approx 197.33\,nm\,ev$ and $mc^2 \approx 87 \times 939$ Mev. Actually, for $\rho \approx 0.11 \times 10^{-7}/nm^3$ we get $T_c = 92.17\,K$, which coincides with the experimental one with great accuracy.

Task 10.1 Evaluate N_1/N for sodium atoms ^{23}Na. The input parameters may be found e.g. in Ref. [6].

Using above parameters, we have calculated n_0, Δ, s and $|\sigma|$ within present theory. As it is seen from Fig. 10.1a, the theory describes existing data on $n_1 = 1 - n_0$ nicely. Note that there was no need to optimize input parameters. In Fig. 10.1b we present our predictions for the sound velocity and the absolute value of the anomalous density vs temperature for homogeneous BEC condensate in ^{87}Rb dilute gas. It is seen that, in contrast to the sound velocity, $|\sigma|$ does not decrease monotonically vs temperature. Actually, it first increases, reaches its maximum near $t \sim 0.7$ and then

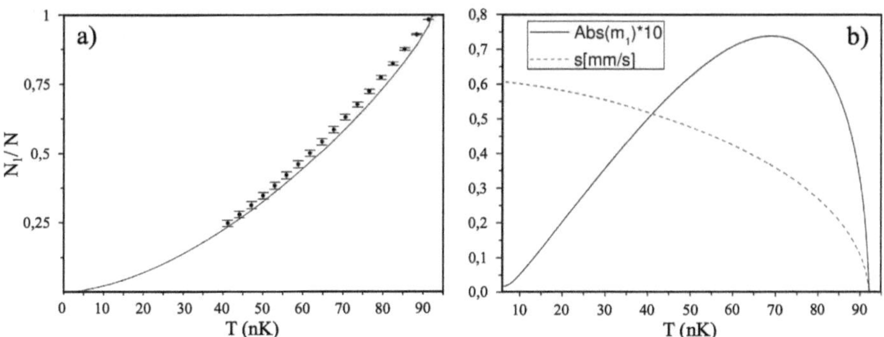

Fig. 10.1 **a** Fraction of uncondensed particles, $n_1 = 1 - n_0$ vs temperature. Experimental points are taken from Ref. [4]. **b** Our prediction for the absolute value of the anomalous density, $|m_1| = |\sigma|/\rho$ (solid line) and sound velocity (dashed line) for ^{87}Rb condensate in a uniform box potential

rapidly decreases to reach zero at $t = 1$. Hopefully, such behavior of the anomalous density will also be revealed in future experiments. Here it should be underlined that, as it has been outlined in the previous chapter, optimal perturbation theory is able to accurately describe, say, $n_0(\gamma, T = 0)$ for $\gamma \leq 0.1$. Nevertheless, obviously, our conclusion, e.g. on the sign of σ remains true [7].

10.5   Conclusion

- We have shown that the unique wave function of condensed particles has no fixed phase.
- In the framework of variational perturbation theory of Bose-Einstein condensation all physical observables are phase-independent.
- The phase of the condensate wave function defines the sign of the anomalous density (see Eq. (10.22)). Therefore, σ, and, especially, its sign can not be considered as a physical observable parameter.
- We have shown that the experimental data on the fraction of the condensate in a uniform box can be nicely described within the present theory-the first order of delta expansion. We also calculated the sound velocity and the absolute value of the anomalous density for finite temperatures. These numerical results serve as a prediction for future experimental studies.
- Present theory establishes a mathematical relation between the sound velocity and the anomalous density. However, a physical mechanism by which anomalous density affects the speed of sound remains still not transparent.

References

1. N.M. Hugenholtz, D. Pines, Ground-state energy and excitation spectrum of a system of interacting bosons. Phys. Rev. **116**, 489–506 (1959). https://link.aps.org/doi/10.1103/PhysRev.116.489
2. S. Watabe, Identities and many-body approaches in Bose-Einstein condensates. Acta Phys. Pol. A **135**(6), 1222–1230 (2019), publisher Copyright: 2019 Polish Academy of Sciences. All rights reserved
3. T.P. Meyrath, F. Schreck, J.L. Hanssen, C.-S. Chuu, M.G. Raizen, Bose-Einstein condensate in a box. Phys. Rev. A **71**, 041604 (2005). https://link.aps.org/doi/10.1103/PhysRevA.71.041604
4. A.L. Gaunt, T.F. Schmidutz, I. Gotlibovych, R.P. Smith, Z. Hadzibabic, Bose-Einstein condensation of atoms in a uniform potential. Phys. Rev. Lett. **110**, 200406 (2013). https://link.aps.org/doi/10.1103/PhysRevLett.110.200406
5. E. Henn, J. Seman, G. Seco, E. Olimpio, P. Castilho, G. Roati, D.V. Magalhães, K.M.F. Magalhães, V.S. Bagnato, Bose-Einstein condensation in ^{87}Rb: characterization of the Brazilian experiment. Braz. J. Phys. **38**, 279–286 (2008). https://www.redalyc.org/pdf/464/46413553012.pdf
6. J.H. Kim, D. Hong, Y. Shin, Observation of two sound modes in a binary superfluid gas. Phys. Rev. A **101**, 061601 (2020). https://link.aps.org/doi/10.1103/PhysRevA.101.061601
7. A. Rakhimov, M. Nishonov, On the anomalous density of a dilute homogeneous Bose gas. Phys. Lett. A **531**, 130164 (2025). https://www.sciencedirect.com/science/article/pii/S0375960124008582

Two-Component Bose Mixture 11

11.1 Introduction

In Chap. 2, we have outlined that mixtures of atomic gases are much more flexible due to the large variety of atomic species, characterized by different states, the possibility of generating coherently coupled configurations, and tuning the interaction between the different components of the mixture. To study such exotic gases at ultra-low temperatures, Bose-Einstein condensed mixtures were first realized experimentally by the JILA group in 1997 [1].

Owing to the possibility of tuning the intercomponent scattering lengths by using Feshbach resonances, two-component quantum liquids exhibit rich physics that is not accessible in a single-component fluid. Therefore, theoretical [2,3] and experimental studies [4–7] have revealed that, the nature of this physics dramatically depends on the sign of the intercomponent coupling constant g_{12}. Actually, for $g_{12} < 0$ ($g_{11} > 0$, $g_{22} > 0$) quantum liquid droplets may arise [2], while for $g_{12} > 0$, a phase transition between miscible and immiscible states may occur [3] (see Fig. 11.1). In some sense, the situation is similar to two-body physics: when the interparticle interaction is negative, one is mainly interested in the properties of bound states, otherwise one studies scattering angles and cross sections. It is also interesting that, regardless of the sign of g_{12}, there exist two distinct sound waves, propagating with different speeds as it is illustrated in Fig. 11.2. These sounds, i.e. the propagation of a perturbation in the two-component mixture, correspond to in-phase S_n and out-of-phase S_s oscillations. The former characterizes an ordinary pressure wave, whose propagation speed is determined by the compressibility of the system, similarly to Eq. (5.13). The latter is a wave of the density difference between the two components, which is referred to in the literature a *spin* sound, regarding the two components as two opposite spin states $|\uparrow\rangle$ and $|\downarrow\rangle$. Remarkably, both types of sound waves have been detected e.g. in the mix of ^{23}Na atoms with two spin states, $|\uparrow\rangle = |F = 1, m_F = 1\rangle$ and $|\downarrow\rangle = |F = 1, m_F = -1\rangle$ [6].

A. Rakhimov and S. Mardonov, *Theory of Quantum Bose Liquids Beyond Bogoliubov Approximation*, Lecture Notes in Physics 1045, https://doi.org/10.1007/978-3-032-05096-0_11

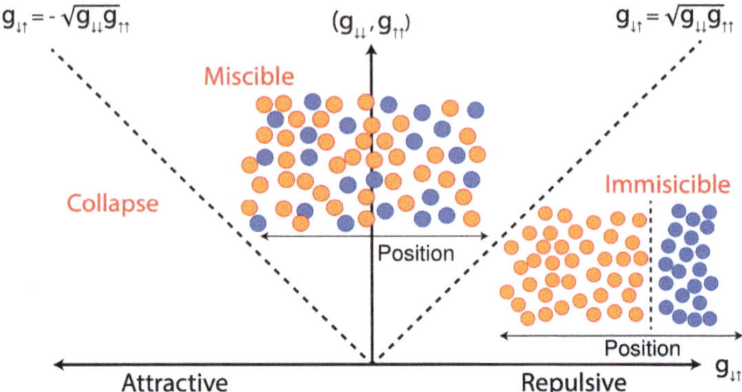

Fig. 11.1 Phase diagram of a Bose-Bose mixture. The horizontal axis represents the inter-species scattering length, and the vertical axis represents the intra-species interactions on each condensate. Three different regimes are observed: miscibility, immiscibility and collapse. The miscible phase is bounded from the right and the left by the condition $g_{12} = \sqrt{g_{11}g_{22}}$. Reprinted with permission from [8]. © 2018, C. Cabrera. All rights reserved

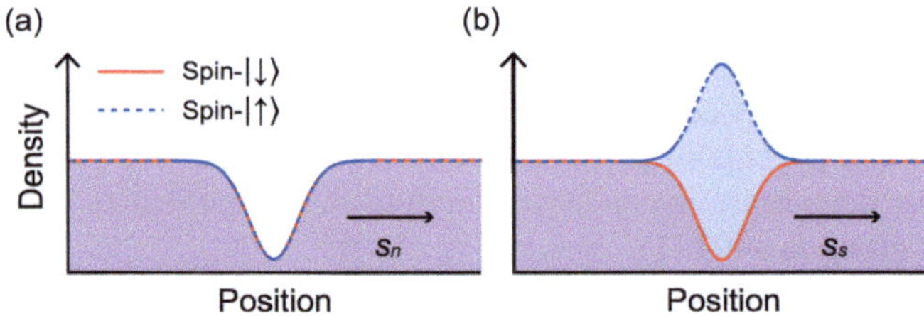

Fig. 11.2 Sound waves in a symmetric binary superfluid. **a** Ordinary density sound where the two superfluid components oscillate in phase, and **b** spin sound wave where they oscillate out of phase. The red solid and blue dashed lines indicate the spatial density profiles of the two components. Here S_n and S_s denote the propagation speeds of the density and spin sounds, respectively. Reprinted with permission from [6]. © 2020, American Physical Society. All rights reserved

In general, the existence of these two sound modes may lead to an additional instability of the two-component system even for the case when all coupling constants are positive. In fact, for some values of coupling constants and temperature, the speed of a sound, say s_s, may become imaginary $s_s^2 < 0$, so that the collective excitation of the mode will grow dramatically, being a signal of an instability. Particularly, at zero temperature, this may happen when the intercomponent coupling constant g_{12} exceeds a critical value, $g_{12} > g_c \equiv \sqrt{g_{11}g_{22}}$. As it is seen from Fig. 11.1, this point corresponds to the miscible-immiscible transition. Nowadays, this criterion is so widely accepted that some authors consider it as a definition of miscible ($g_{12} < g_c$) or immiscible ($g_{12} > g_c$) states [9, 10] in spite of the fact that it was obtained in the rather crude Bogolyubov approximation, which is valid only for very dilute gases, with the gas parameter $\gamma = \rho a^3 \sim 10^{-5}$ [11]. On the other hand, as we know from previous chapters, the Bogolyubov approximation is rather rude, since it does not take into account neither quantum fluctuations nor anomalous densities.

In the present chapter, we develop a self-consistent mean-field-based approach in the framework of optimized perturbation theory. Firstly, we shall derive general explicit expressions beyond Bogolyubov approximation for thermodynamic quantities, and then consider $g_{12} > 0$ and $g_{12} < 0$ cases separately.[1]

11.2 OPT for Two-Component Bose Mixtures

The Lagrangian density for two-species complex scalar fields ψ and ϕ, with contact self-couplings g_a and g_b and interspecies coupling g_{ab} is given as

$$\mathcal{L} = \psi^\dagger (i\partial_t + \frac{\nabla^2}{2m_a} + \mu_a)\psi - \frac{g_a}{2}(\psi^\dagger\psi)^2 + \phi^\dagger (i\partial_t + \frac{\nabla^2}{2m_b} + \mu_b)\phi$$
$$- \frac{g_b}{2}(\phi^\dagger\phi)^2 - g_{ab}(\psi^\dagger\psi)(\phi^\dagger\phi), \tag{11.1}$$

where the associated chemical potentials are represented by $\mu_{a,b}$ while $m_{a,b}$ represent the masses. In terms of the corresponding s-wave scattering lengths a_s, the coupling constants can be written as $g_{a,b} = 4\pi a_{a,b}/m_{a,b}$, while the cross coupling is $g_{ab} = 2\pi a_{ab}/m_{ab}$, where $m_{ab} = m_a m_b/(m_a + m_b)$ represents reduced mass.

The grand canonical thermodynamic potential Ω can be calculated in the path integral formalism as

$$\Omega = -T \ln Z, \tag{11.2}$$

$$Z = \int \mathcal{D}\psi^\dagger \mathcal{D}\psi \mathcal{D}\phi^\dagger \mathcal{D}\phi \, e^{-S[\psi^\dagger,\psi,\phi^\dagger,\phi]} \tag{11.3}$$

where the equivalent finite-temperature Euclidean ($\tau = it$) space time action to (11.1) is given by

$$S = \int_0^\beta d\tau \int d\mathbf{r} \left\{ \psi^\dagger \hat{K}_a \psi + \phi^\dagger \hat{K}_b \phi + \frac{g_a}{2}(\psi^\dagger\psi)^2 \right.$$
$$\left. + \frac{g_b}{2}(\phi^\dagger\phi)^2 + g_{ab}(\psi^\dagger\psi)(\phi^\dagger\phi) \right\},$$
$$\hat{K}_{a,b} = \frac{\partial}{\partial\tau} - \hat{O}_{a,b}, \tag{11.4}$$
$$\hat{O}_{a,b} = \frac{\nabla^2}{2m_{a,b}} + \mu_{a,b}.$$

In Eq. (11.4) the fields $\psi(\mathbf{r}, \tau)$ and $\phi(\mathbf{r}, \tau)$ are periodic in τ with period $\beta = 1/T$. We know that, the path integral in (11.3) cannot be exactly evaluated due to the terms of fourth order in fields, so we have to apply optimized perturbation theory, outlined in Chaps. 7 and 8. For the two-component Bose system, this method involves following steps:

---

[1] The latter case will be considered in the next chapter.

- **Step 1.** Introduce fluctuating fields $\tilde{\psi}$ and $\tilde{\phi}$ by the Bogolyubov shift:

$$\psi(\mathbf{r}, \tau) = \sqrt{\rho_{0a}} + \tilde{\psi}(\mathbf{r}, \tau) \quad \phi(\mathbf{r}, \tau) = \sqrt{\rho_{0b}} + \tilde{\phi}(\mathbf{r}, \tau) \tag{11.5}$$

where the order parameters ρ_{0a} and ρ_{0b} correspond to the condensate fractions of the components a and b, respectively. As in previous chapters, for a uniform system at equilibrium ρ_{0a} and ρ_{0b} are real variational constants, which are fixed by the minimum of the free energy Ω as $\partial\Omega/\partial\rho_{0,a,b} = 0$, $\partial^2\Omega/\partial^2\rho_{0,a,b} \geq 0$. As to the numbers of uncondensed particles N_{1a} and N_{1b}, they are related to the fields $\tilde{\psi}$ and $\tilde{\phi}$:

$$N_{1a} = V\rho_{1a} = \int d\mathbf{r}\langle\tilde{\psi}^\dagger(r)\tilde{\psi}(r)\rangle, \quad N_{1b} = V\rho_{1b} = \int d\mathbf{r}\langle\tilde{\phi}^\dagger(r)\tilde{\phi}(r)\rangle \tag{11.6}$$

so that

$$N_a = \int d\mathbf{r}\langle\psi^\dagger(r)\psi(r)\rangle, \quad N_b = \int d\mathbf{r}\langle\phi^\dagger(r)\phi(r)\rangle, \tag{11.7}$$

with the normalization conditions $N = N_a + N_b$, $N_a = V\rho_a = V(\rho_{0a} + \rho_{1a})$, $N_b = V\rho_b = V(\rho_{0b} + \rho_{1b})$, where $N_{a(b)}$ is the number of particles in the component a, (b) and N is the particle number in the whole two-component system and V is the total volume of the system. Since we are considering a homogeneous system, the densities ρ_a and ρ_b are uniform.

- **Step 2.** Make the following replacement in the action: $g_a \to \tilde{\delta}g_a$, $g_b \to \tilde{\delta}g_b$, $g_{ab} \to \tilde{\delta}g_{ab}$
- **Step 3.** Add to the action the term:

$$S_\Sigma = (1 - \tilde{\delta}) \int d\tau d\mathbf{r}\{\Sigma_n^{(a)}(\tilde{\psi}^\dagger\tilde{\psi}) + \Sigma_n^{(b)}(\tilde{\phi}^\dagger\tilde{\phi}) + \frac{1}{2}\Sigma_{an}^{(a)}(\tilde{\psi}\tilde{\psi} + \tilde{\psi}^\dagger\tilde{\psi}^\dagger)$$

$$+ \frac{1}{2}\Sigma_{an}^{(b)}(\tilde{\phi}\tilde{\phi} + \tilde{\phi}^\dagger\tilde{\phi}^\dagger) + \Sigma_n^{(ab)}(\tilde{\psi}^\dagger\tilde{\phi} + \tilde{\phi}^\dagger\tilde{\psi}) + \Sigma_{an}^{(ab)}(\tilde{\psi}\tilde{\phi} + \tilde{\phi}^\dagger\tilde{\psi}^\dagger)\}$$

$$\tag{11.8}$$

where the variational parameters Σ_n and Σ_{an} can be naturally interpreted as normal and anomalous self-energies, respectively.

- **Step 4.** Now in the Cartesian representation

$$\tilde{\psi} = \frac{1}{\sqrt{2}}(\psi_1 + i\psi_2), \quad \tilde{\phi} = \frac{1}{\sqrt{2}}(\psi_3 + i\psi_4) \tag{11.9}$$

such that

$$\int \mathcal{D}\tilde{\psi}^\dagger\mathcal{D}\tilde{\psi}\mathcal{D}\tilde{\phi}^\dagger\mathcal{D}\tilde{\phi} \to \int \prod_{i=1}^{4}\mathcal{D}\psi_i \tag{11.10}$$

the action (11.4) can be written as

$$S = S_0 + S_{\text{free}} + S_{\text{int}},$$

$$S_{\text{int}} = S_{\text{int}}^{(1)} + S_{\text{int}}^{(2)} + S_{\text{int}}^{(3)} + S_{\text{int}}^{(4)},$$

$$S_0 = \int d\tau d\mathbf{r} \{ -\mu_{0a}\rho_{0a} - \mu_{0b}\rho_{0b} + \frac{g_a \rho_{0a}^2}{2} + \frac{g_b \rho_{0b}^2}{2} + g_{ab}\rho_{0a}\rho_{0b} \},$$

$$S_{\text{free}} = \frac{1}{2} \int d\tau d\mathbf{r} \{ \psi_1 [X_1 + \hat{K}_a] \psi_1 + \psi_2 [X_2 + \hat{K}_a] \psi_2 + \psi_3 [X_3 + \hat{K}_b] \psi_3$$

$$+ \psi_4 [X_4 + \hat{K}_b] \psi_4 + X_5 [\psi_1 \psi_3 + \psi_3 \psi_1] + X_6 [\psi_2 \psi_4 + \psi_4 \psi_2] \},$$

$$S_{\text{int}}^{(2)} = \frac{1}{2} \int d\tau d\mathbf{r} \left\{ \sum_{i=1}^{4} \Lambda_i \psi_i^2 + \Lambda_5 (\psi_1 \psi_3 + \psi_3 \psi_1) + \Lambda_6 (\psi_2 \psi_4 + \psi_4 \psi_2) \right\},$$

$$S_{\text{int}}^{(4)} = \frac{1}{8} \int d\tau d\mathbf{r} \left\{ g_a (\psi_1^2 + \psi_2^2)^2 + g_b (\psi_3^2 + \psi_4^2)^2 \right.$$

$$\left. + 2g_{ab}(\psi_1^2 + \psi_2^2)(\psi_3^2 + \psi_4^2) \right\},$$

$$(11.11)$$

where we have introduced the following notations:

$$X_1 = \Sigma_n^{(a)} + \Sigma_{an}^{(a)} - \mu_{1a}, \qquad X_3 = \Sigma_n^{(b)} + \Sigma_{an}^{(b)} - \mu_{1b},$$

$$X_2 = \Sigma_n^{(a)} - \Sigma_{an}^{(a)} - \mu_{1a}, \qquad X_4 = \Sigma_n^{(b)} - \Sigma_{an}^{(b)} - \mu_{1b},$$

$$X_5 = \Sigma_n^{(ab)} + \Sigma_{an}^{(ab)}, \qquad\qquad X_6 = \Sigma_n^{(ab)} - \Sigma_{an}^{(ab)},$$

$$\Lambda_1 = -\mu_{1a} - X_1 + 3g_a \rho_{0a} + g_{ab}\rho_{0b},$$

$$\Lambda_2 = -\mu_{1a} - X_2 + g_a \rho_{0a} + g_{ab}\rho_{0b},$$

$$\Lambda_3 = -\mu_{1b} - X_3 + 3g_b \rho_{0b} + g_{ab}\rho_{0a}, \qquad (11.12)$$

$$\Lambda_4 = -\mu_{1b} - X_4 + g_b \rho_{0b} + g_{ab}\rho_{0a},$$

$$\Lambda_5 = -X_5 + 2g_{ab}\sqrt{\rho_{0a}\rho_{0b}}, \qquad \Lambda_6 = -X_6.$$

Above we have omitted explicit expressions for $S^{(1)}$ and $S^{(3)}$ since their mean values vanish. Moreover, Yukalov's prescription with two chemical potentials for each component is used. It is understood that, the six variational parameters $X_1...X_6$ should be fixed by the minimization of Ω, $\partial\Omega/(\partial X_i) = 0$, $(i = 1 \div 6)$.

- **Step 5.** Now, passing to the momentum space,

$$\psi_i(\mathbf{r}, \tau) = \frac{1}{\sqrt{V\beta}} \sum_{n=-\infty}^{\infty} \sum_{k} \psi_i(\omega_n, \mathbf{k}) e^{i\omega_n \tau + i\mathbf{k}\mathbf{r}} \qquad (11.13)$$

one may present S_{free} as

$$S_{\text{free}} = \frac{(2\pi)^4}{2V\beta} \sum_{\mathbf{k},\mathbf{p},m,n} \sum_{i,j=1}^{4} \psi_i(\omega_n, \mathbf{k}) \mathcal{D}_{ij}^{-1}(\omega_n, \mathbf{k}; \omega_m, \mathbf{p})$$
$$\times \psi_j(\omega_m, \mathbf{p}) \delta(\mathbf{k}+\mathbf{p}) \delta(\omega_n + \omega_m) \tag{11.14}$$

which leads to following inverse propagator

$$\mathcal{D}^{-1}(\omega_n, \mathbf{k}) = \begin{pmatrix} \varepsilon_a(\mathbf{k}) + X_1 & \omega_n & X_5 & 0 \\ -\omega_n & \varepsilon_a(\mathbf{k}) + X_2 & 0 & X_6 \\ X_5 & 0 & \varepsilon_b(\mathbf{k}) + X_3 & \omega_n \\ 0 & X_6 & -\omega_n & \varepsilon_b(\mathbf{k}) + X_4 \end{pmatrix} \tag{11.15}$$

where $\varepsilon_{a,b}(\mathbf{k}) = \mathbf{k}^2/2m_{a,b}$. The condition $Det(\hat{\mathcal{D}}^{-1}) = 0$ gives two branches of collective excitations:

$$\omega_{1,2}(\mathbf{k}) = \sqrt{\frac{E_a^2 + E_b^2}{2} + X_5 X_6 \pm \frac{\sqrt{D}}{2}}$$
$$D = (E_a^2 - E_b^2)^2 + 4E_{13}^2 X_6^2 + 4E_{24}^2 X_5^2 + 4X_5 X_6(E_a^2 + E_b^2), \tag{11.16}$$

where

$$E_a^2 = (\varepsilon_a(\mathbf{k}) + X_1)(\varepsilon_a(\mathbf{k}) + X_2), \qquad E_b^2 = (\varepsilon_b(\mathbf{k}) + X_3)(\varepsilon_b(\mathbf{k}) + X_4),$$
$$E_{13}^2 = (\varepsilon_a(\mathbf{k}) + X_1)(\varepsilon_b(\mathbf{k}) + X_3), \qquad E_{24}^2 = (\varepsilon_a(\mathbf{k}) + X_2)(\varepsilon_b(\mathbf{k}) + X_4). \tag{11.17}$$

- **Step 6.** The perturbation scheme is considered as an expansion in powers of $\tilde{\delta}$ by using the propagators

$$\mathcal{D}_{ij}(\mathbf{r}, \tau; \mathbf{r}', \tau') = \frac{1}{V\beta} \sum_{n,\mathbf{k}} e^{i\mathbf{k}(\mathbf{r}-\mathbf{r}')} e^{i\omega_n(\tau-\tau')} \mathcal{D}_{ij}(\omega_n, \mathbf{k}). \tag{11.18}$$

Following the ideology of OPT, the expansion parameter $\tilde{\delta}$ will be set to $\tilde{\delta} = 1$ at the end of calculations.

Task 11.1 Prove the Eqs. (11.15)–(11.17) and derive explicit expressions for $\mathcal{D}_{ij}(\omega_n, \mathbf{k})$.

- **Step 7.** The detailed calculation of the generating functional and hence Ω in the first order of $\tilde{\delta}$ can be performed in a similar way as it has been done in Chap. 8 for the one-component system. Therefore, one obtains

$$\Omega = \Omega_0 + \Omega_{ln} + \Omega_2 + \Omega_4,$$

$$\Omega_0 = V\left[-\mu_{0a}\rho_{0a} - \mu_{0b}\rho_{0b} + \frac{g_a\rho_{0a}^2}{2} + \frac{g_b\rho_{0b}^2}{2} + g_{ab}\rho_{0a}\rho_{0b}\right.$$

$$\Omega_{ln} = \frac{T}{2}\sum_{\mathbf{k},\omega_n}\ln[(\omega_n^2 + \omega_1^2)(\omega_n^2 + \omega_2^2)] = \frac{1}{2}\sum_{\mathbf{k}}(\omega_1(\mathbf{k}) + \omega_2(\mathbf{k}))$$

$$+ T\sum_{\mathbf{k}}\ln(1 - e^{-\beta\omega_1(\mathbf{k})}) + T\sum_{\mathbf{k}}\ln(1 - e^{-\beta\omega_2(\mathbf{k})}),$$

(11.19)

$$\Omega_2 = \frac{1}{2}\sum_{i=1}^{6} A_i\Lambda_i,$$

$$\Omega_4 = \frac{1}{8V}\left\{g_a[3A_1^2 + 3A_2^2 + 2A_1A_2] + g_b[3A_3^2 + 3A_4^2 + 2A_3A_4]\right.$$

$$\left. + 2g_{ab}\left[(A_1 + A_2)(A_3 + A_4) + \frac{A_5^2 + A_6^2}{2}\right]\right\},$$

where $A_i = V\mathcal{D}_{ii}(x,x)$ $(i = 1 \div 4)$, $A_5 = 2V\mathcal{D}_{13}(x,x)$, $A_6 = 2V\mathcal{D}_{24}(x,x)$, Λ_i are given by Eq. (11.12) and the matrix elements of the propagator $\mathcal{D}_{ij}(x,x)$ are presented explicitly in the Appendix D. Feynman diagrams contributing to Ω by means of $\tilde{\delta}$ expansion are illustrated in Refs. [11,12]. The variational parameters are determined by the minimization of the thermodynamic potential $\Omega(X_1...X_6)$ as $\partial\Omega(X_1...X_6)/\partial X_i = 0$ $(i = 1 \div 6)$. By using the method, explained in Sect. 8.3, these equations can be rewritten in the following compact form:

$$X_1 = g_a[3\rho_{0a} + 2\rho_{1a} + \sigma_a] + g_{ab}\rho_b - \mu_{1a},$$
$$X_2 = g_a[\rho_{0a} + 2\rho_{1a} - \sigma_a] + g_{ab}\rho_b - \mu_{1a},$$
$$X_3 = g_b[3\rho_{0b} + 2\rho_{1b} + \sigma_b] + g_{ab}\rho_a - \mu_{1b},$$
$$X_4 = g_b[\rho_{0b} + 2\rho_{1b} - \sigma_b] + g_{ab}\rho_a - \mu_{1b},$$
$$X_5 = 2g_{ab}\sqrt{\rho_{0a}\rho_{0b}} + g_{ab}\frac{\rho_{ab} + \sigma_{ab}}{2},$$
$$X_6 = \frac{g_{ab}}{2}(\rho_{ab} - \sigma_{ab}),$$

(11.20)

where the densities ρ_1 and σ will be presented in the next subsection explicitly. Note that, in the derivation of Eq. (11.20), we used the relation $\partial\Omega_{ln}/\partial X_i = A_i/2$, $(i = 1 \div 6)$, which can be checked by using Mathematica or Maple. In general, the system of Eq. (11.20) with the given set of input parameters, such as coupling parameters, the total densities of atoms, is the system of nonlinear algebraic equations with respect to unknown variational parameters $(X_1...X_6)$. As it is seen from their definition in Eq. (11.12), the latter can be considered as self energies in the Cartesian representation (11.9).

11.2.1 Normal and Anomalous Densities

Fluctuating fields $\tilde{\psi}(\mathbf{r}))$ and $\tilde{\phi}(\mathbf{r})$ define the density of uncondensed particles, which may be calculated as

$$
\begin{aligned}
\rho_{1a} &= \frac{1}{V} \int d\mathbf{r} \langle \tilde{\psi}^{\dagger}(\mathbf{r})\tilde{\psi}(\mathbf{r})\rangle = \frac{1}{2V} \int d\mathbf{r}[D_{11}(\mathbf{r},\mathbf{r}) + D_{22}(\mathbf{r},\mathbf{r})] \\
&= \frac{1}{2V}(A_1 + A_2), \\
\rho_{1b} &= \frac{1}{V} \int d\mathbf{r} \langle \tilde{\phi}^{\dagger}(\mathbf{r})\tilde{\phi}(\mathbf{r})\rangle \\
&= \frac{1}{2V} \int d\mathbf{r}[D_{33}(\mathbf{r},\mathbf{r}) + D_{44}(\mathbf{r},\mathbf{r})] \\
&= \frac{1}{2V}(A_3 + A_4).
\end{aligned}
\tag{11.21}
$$

In general, one may introduce the anomalous

$$
\begin{aligned}
\sigma_a &= \frac{1}{2V} \int d\mathbf{r}[\langle \tilde{\psi}^{\dagger}(\mathbf{r})\tilde{\psi}^{\dagger}(\mathbf{r})\rangle + \langle \tilde{\psi}(\mathbf{r})\tilde{\psi}(\mathbf{r})\rangle] = \frac{1}{2V}(A_1 - A_2), \\
\sigma_b &= \frac{1}{2V} \int d\mathbf{r}[\langle \tilde{\phi}^{\dagger}(\mathbf{r})\tilde{\phi}^{\dagger}(\mathbf{r})\rangle + \langle \tilde{\phi}(\mathbf{r})\tilde{\phi}(\mathbf{r})\rangle] = \frac{1}{2V}(A_3 - A_4)
\end{aligned}
\tag{11.22}
$$

and "mixed" densities:

$$
\begin{aligned}
\rho_{ab} &= \frac{1}{V} \int d\mathbf{r}[\langle \tilde{\psi}^{\dagger}(\mathbf{r})\tilde{\phi}(\mathbf{r})\rangle + \langle \tilde{\phi}^{\dagger}(\mathbf{r})\tilde{\psi}(\mathbf{r})\rangle] \\
&= \frac{1}{V} \int d\mathbf{r}[D_{13}(\mathbf{r},\mathbf{r}) + D_{24}(\mathbf{r},\mathbf{r})] = \frac{1}{2V}(A_5 + A_6), \\
\sigma_{ab} &= \frac{1}{V} \int d\mathbf{r}[\langle \tilde{\psi}(\mathbf{r})\tilde{\phi}(\mathbf{r})\rangle + \langle \tilde{\phi}^{\dagger}(\mathbf{r})\tilde{\psi}^{\dagger}(\mathbf{r})\rangle] = \frac{1}{2V}(A_5 - A_6).
\end{aligned}
\tag{11.23}
$$

These densities, which are explicitly given in the Appendix D, do not depend on the coordinate variables, i.e. they are constants for a uniform system. Physically, the pair densities ρ_{ab} and σ_{ab} describe the processes where, due to the presence of the reservoir, particles are exchanged or pairing correlations emerge between the two components.

11.2.2 Particular Cases of Present OPT

Above-presented HFB-type theory for a two-component Bose mixture is rather general, so some well-known approximations to this general theory can be easily derived as particular cases.

- Sometimes, one uses the trick (first suggested by Shohno [13]) by omitting anomalous averages, which corresponds to the case when in Eqs. (11.19)–(11.20) σ_a, σ_b, σ_{ab} are omitted by setting $\mu_{0a,b} = \mu_{1a,b} = \mu_{a,b}$. However, as it has been underlined in Chap. 8, this trick results in a non-self-consistent approach containing paradoxes.
- The quadratic (Bilinear) approximation corresponds to the case when, after the shift (11.5), only the quadratic terms of fluctuating fields are kept in the action (11.4): $S \approx S_0 + S_{\text{free}} + S_{\text{int}}^{(2)}$, with $\tilde{\delta} = 1$. The formal difference is that in this approximation one is left with

$$\Omega^{Bil} = \Omega_0 + \Omega_{ln} \tag{11.24}$$

where Ω_0 and Ω_{ln} have the same expressions as in (11.19) but with the self energies given by

$$
\begin{aligned}
X_1 &\approx X_1^{Bil} = 3g_a\rho_{0a} + g_{ab}\rho_{0b} - \mu_a, \\
X_2 &\approx X_2^{Bil} = g_a\rho_{0a} + g_{ab}\rho_{0b} - \mu_a, \\
X_3 &\approx X_3^{Bil} = 3g_b\rho_{0b} + g_{ab}\rho_{0a} - \mu_b, \\
X_4 &\approx X_4^{Bil} = g_b\rho_{0b} + g_{ab}\rho_{0a} - \mu_b, \\
X_5 &\approx X_5^{Bil} = 2g_{ab}\sqrt{\rho_{0a}\rho_{0b}}, \\
X_6 &\approx X_6^{Bil} = 0.
\end{aligned}
\tag{11.25}
$$

- In the Bogolyubov approximation, the thermodynamic potential is formally given by (11.24), while the self energies, by Eq. (11.25), with setting there $\rho_{0a,b} \approx \rho_{a,b}$, i.e. $X_i^{Bog} = X_i^{Bil}(\rho_{0a} = \rho_a, \rho_{0b} = \rho_b)$. In this case, the equations are uncoupled and the solutions are simple. Both these variants enjoy the same level of accuracy and are valid only for small gas parameters $\gamma \leq 10^{-5}$.

Note that, for all above cases, the expressions for the energy dispersions as well as for the densities are formally the same as given by Eqs. (11.16) and (11.21)–(11.23), respectively.

Which of the above approximations is optimal in practical calculations? This will depend on the intercomponent interaction, since the behavior of the two-component system strongly depends on its sign. In the present chapter we consider the case when all coupling constants are positive, while the case with $g_{12} < 0$ will be discussed in next chapter.

11.3 Miscible-Immiscible Transition

When $g_{12} > 0$, the main problem under study is related to the miscible-immiscible transition. For this purpose, it is convenient to introduce the overlap parameter η,

$$\eta = \frac{1}{2\sqrt{N_a N_b}} \int d\mathbf{r}\{\langle\psi^\dagger(\mathbf{r})\phi(\mathbf{r})\rangle + \langle\phi^\dagger(\mathbf{r})\psi(\mathbf{r})\rangle\} \tag{11.26}$$

where $\psi(\mathbf{r})$ and $\phi(\mathbf{r})$ are the field operators of the components a and b, respectively.
Using (11.5), (11.9), and (11.20)–(11.26), one may present η as follows:

$$\eta = \frac{1}{\sqrt{N_a N_b}} \int d\mathbf{r}\sqrt{\rho_{0a}\rho_{0b}} + \frac{1}{2\sqrt{N_a N_b}} \int d\mathbf{r}\{\langle \tilde{\psi}^\dagger(\mathbf{r})\tilde{\phi}(\mathbf{r})\rangle$$
$$+ \langle \tilde{\phi}^\dagger(\mathbf{r})\tilde{\psi}(\mathbf{r})\rangle\} \tag{11.27}$$
$$= \sqrt{n_{0a}n_{0b}} + \frac{\rho_{ab}}{2\sqrt{\rho_a\rho_b}} = \frac{X_5 + X_6}{2g_{ab}\sqrt{\rho_a\rho_b}},$$

where n_{0a} and n_{0b} are the normalized condensed fractions, $n_{0a} = \rho_{0a}/\rho_a$, $n_{0b} = \rho_{0b}/\rho_b$. Note that, when the fluctuations are neglected, i.e. $\tilde{\psi} = \tilde{\phi} = 0$, η in (11.27)
coincides with the miscibility parameter of Refs. [14, 15], introduced for nonuniform
coupled systems. Particularly, when at least one of the components is in the normal
phase, the parameter η is completely defined by the normal pair density, $\eta(T >
T_c) = \rho_{ab}/2\sqrt{\rho_a\rho_b} = X_5/g_{ab}\sqrt{\rho_a\rho_b}$.

Stability and miscibility properties can be studied by analyzing the spectrum of
collective excitations given by Eq. (11.16). At first glance, this procedure, especially
solving the system of six nonlinear algebraic equations, seems rather cumbersome.
However, in reality, the number of unknown variational parameters $[X_1...X_6]$ may
be reduced depending on the considered state (BEC or normal phase) and on the
existing symmetries in the system. In the next sections, we discuss these cases in
detail.

11.3.1 Condensed and Normal Phases

11.3.1.1 A. Condensed Phase

In this phase, the number of variational parameters is reduced due to the Hugenholtz-
Pines theorem, which has been extended for multicomponent Bose-Einstein condensates in Refs. [16, 17]. For a two-component Bose system, in our notation, it reads

$$\Sigma_n^{(a)} - \Sigma_{an}^{(a)} = \mu_{1a}, \qquad \Sigma_n^{(b)} - \Sigma_{an}^{(b)} = \mu_{1b}, \qquad \Sigma_n^{(ab)} = \Sigma_{an}^{(ab)}, \tag{11.28}$$

and hence

$$X_2 = 0, \qquad X_4 = 0, \qquad X_6 = 0, \qquad \sigma_{ab} = \rho_{ab}. \tag{11.29}$$

Therefore, in the BEC phase, instead of six equations, we are left with a system of
three equations:

$$\Delta_a \equiv X_1/2 = g_a(\rho_{0a} + \sigma_a) = g_a[\rho_a - \rho_{1a} + \sigma_a],$$
$$\Delta_b \equiv X_3/2 = g_b(\rho_{0b} + \sigma_b) = g_b[\rho_b - \rho_{1b} + \sigma_b], \tag{11.30}$$
$$\Delta_{ab} \equiv X_5/2 = g_{ab}\sqrt{\rho_{0a}\rho_{0b}} + \frac{g_{ab}}{2}\rho_{ab}.$$

Since the densities are fixed, one may determine the chemical potentials from Eqs.
(11.20) and (11.28) as

$$\mu_{1a} = g_a[\rho_a + \rho_{1a} - \sigma_a] + g_{ab}\rho_b$$
$$\mu_{1b} = g_b[\rho_b + \rho_{1b} - \sigma_b] + g_{ab}\rho_a. \tag{11.31}$$

The total chemical potentials defined as $\mu_a = (\partial F/\partial N_a)$ and $\mu_b = (\partial F/\partial N_b)$ (where F is the total free energy of the system) can be calculated as

$$\mu_a \rho_a = \mu_{1a}\rho_{1a} + \mu_{0a}\rho_{0a}, \qquad \mu_b \rho_b = \mu_{1b}\rho_{1b} + \mu_{0b}\rho_{0b}, \tag{11.32}$$

where

$$\mu_{0a} = g_a[\rho_a + \rho_{1a} + \sigma_a] + g_{ab}\left[\rho_b + \frac{\rho_{0b}\sigma_{ab}}{\sqrt{\rho_{0a}\rho_{0b}}}\right],$$
$$\mu_{0b} = g_b[\rho_b + \rho_{1b} + \sigma_b] + g_{ab}\left[\rho_a + \frac{\rho_{0a}\sigma_{ab}}{\sqrt{\rho_{0a}\rho_{0b}}}\right]. \tag{11.33}$$

The last two equations are derived from $\partial\Omega/\partial\rho_{0a} = 0$ and $\partial\Omega/\partial\rho_{0b} = 0$, where Ω is given by Eq. (11.19). As is expected, when one neglects anomalous densities by setting $\sigma_a = \sigma_b = \sigma_{ab} = 0$, then $\mu_{0a} = \mu_{1a} = \mu_a$, $\mu_{0b} = \mu_{1b} = \mu_b$.

With the constraints (11.29), the dispersions in Eq. (11.16) can be rewritten as

$$\omega_{1,2} = \sqrt{\frac{\varepsilon(\mathbf{k})^2}{2}(\nu_1^2 + \nu_2^2) + 2\varepsilon(\mathbf{k})\Lambda_{1,2}},$$
$$\Lambda_{1,2} = \frac{1}{2}(\Delta_a\nu_1 + \Delta_b\nu_2) \pm \frac{\sqrt{D_s}}{4}, \tag{11.34}$$
$$D_s = 16\nu_1\nu_2\Delta_{ab}^2 + [\nu_1^2\varepsilon(\mathbf{k}) - \nu_2^2\varepsilon(\mathbf{k}) + 2\Delta_a\nu_1 - 2\Delta_b\nu_2]^2$$
$$\omega_1^2 - \omega_2^2 = \varepsilon(\mathbf{k})\sqrt{D_s},$$

where we introduce the reduced mass $m_R = m_{ab} = m_a m_b/(m_a + m_b)$, and $\nu_1 = m_b/(m_a + m_b)$, $\nu_2 = m_a/(m_a + m_b)$, $\varepsilon(\mathbf{k}) = \mathbf{k}^2/2m_R$. Decomposing ω_i in powers of momenta gives the sound velocities through the equations

$$\omega_1 = s_1|\mathbf{k}| + O(k^3) \quad , \quad \omega_2 = s_2|\mathbf{k}| + O(k^3), \tag{11.35}$$

$$s_1^2 = \frac{\Delta_a m_b + \Delta_b m_a + \sqrt{4m_a m_b \Delta_{ab}^2 + (\Delta_b m_a - \Delta_a m_b)^2}}{2m_a m_b},$$
$$s_2^2 = \frac{\Delta_a m_b + \Delta_b m_a - \sqrt{4m_a m_b \Delta_{ab}^2 + (\Delta_b m_a - \Delta_a m_b)^2}}{2m_a m_b}. \tag{11.36}$$

The velocities s_1 (s_2) are referred to in the literature as density and pseudospin sound velocities, respectively. In a binary superfluid, the density sound corresponds to the oscillation of two superfluid components in phase, while the pseudospin sound corresponds to the out-of-phase oscillations as it has been illustrated in Fig. 11.2.

From the last equation, it is seen that when $\Delta_a\Delta_b < \Delta_{ab}^2$, c_2^2 becomes negative signalling the instability of the system. In particular, applying the Bogolyubov approximation, i.e. setting $\rho_{0a} \approx \rho_a$, $\rho_{0b} \approx \rho_b$, $\sigma_a = \sigma_b = \rho_{ab} \approx 0$, in the Eq. (11.30) one obtains

$$\Delta_a \approx g_a \rho_a, \qquad \Delta_b \approx g_b \rho_b, \qquad \Delta_{ab} \approx g_{ab}\sqrt{\rho_a \rho_b}, \qquad (11.37)$$

thus arriving at the well known stability condition $g_a g_b / g_{ab}^2 \geq 1$.

Explicit expressions for the densities may be obtained from Eqs. (11.21–11.23) by setting there $X_2 = X_4 = X_6 = 0$. Particularly, as it is expected, when the inter-component coupling constant goes to zero, we arrive at the well-known formulas of the single-component case:

$$\rho_{1a}(g_{ab} \to 0) = \frac{1}{V} \sum_k \left[\frac{\Delta_a + \varepsilon_a(\mathbf{k})}{\omega_a(\mathbf{k})} W_1(\mathbf{k}) - \frac{1}{2} \right],$$

$$\rho_{1b}(g_{ab} \to 0) = \frac{1}{V} \sum_k \left[\frac{\Delta_b + \varepsilon_b(\mathbf{k})}{\omega_b(\mathbf{k})} W_2(\mathbf{k}) - \frac{1}{2} \right],$$

$$\sigma_a(g_{ab} \to 0) = -\frac{\Delta_a}{V} \sum_k \frac{W_1(\mathbf{k})}{\omega_a(\mathbf{k})}, \qquad (11.38)$$

$$\sigma_b(g_{ab} \to 0) = -\frac{\Delta_b}{V} \sum_k \frac{W_2(\mathbf{k})}{\omega_b(\mathbf{k})},$$

$$\rho_{ab}(g_{ab} \to 0) = \sigma_{ab}(g_{ab} \to 0) = 0,$$

$$s_1^2 = \Delta_b / m_b, \qquad s_2^2 = \Delta_a / m_a,$$

where $W_{a,b}(\mathbf{k}) = 1/2 + 1/(e^{\omega_{a,b}(\mathbf{k})\beta} - 1)$, $\omega_{a,b} = \sqrt{\varepsilon_{a,b}(\mathbf{k})(\varepsilon_{a,b}(\mathbf{k}) + 2\Delta_{a,b})}$.

In the above discussion, we have assumed that both components are in the BEC state. In the next subsection, we consider the case where the whole system is in the normal phase, where there is no way for the HP theorem.

11.3.1.2 B. Normal Phase

The general criterion of miscibility or immiscibility is prescribed by the behavior of the spectrum of collective excitations. To be miscible, a binary Bose mixture has to possess all real branches of the collective spectrum positive (non-negative). In the case of a Bose-condensed system, the global gauge symmetry is broken and the spectra of single-particle and collective excitations coincide [18]. However, for a normal (uncondensed) system, these spectra are different. The positiveness of the single-particle spectrum, defined by the poles of the single-particle Green function, tells us that, on the level of single-particle properties, the system is stable, but this tells us nothing about whether it is mixed or separated. The spectrum of collective excitations of a normal system is defined by the poles of the Green function or the poles of the dynamic susceptibility (response function). The mixture starts to separate when the lowest branch of the collective spectrum crosses zero.

First, let us prove that the binary normal Bose mixture at $T > T_c = T_c^a = T_c^b$ enjoys a stable single-particle spectrum, that is positive at any temperature and the values of local interaction parameters, independently of whether the system is mixed or separated.

By definition, in the normal phase $\rho_{0a} = \rho_{0b} = \sigma_a = \sigma_b = \sigma_{ab} = 0$, and hence, $\rho_{1a} = \rho_a$, $\rho_{1b} = \rho_b$, $\mu_{0a} = \mu_{1a} = \mu_a$, $\mu_{0b} = \mu_{1b} = \mu_b$, $X_6 = X_5$, $X_2 = X_1$, $X_4 = X_3$. Thus, the main Eq. (11.20) are simplified as

$$X_1 = 2\rho_a g_a + \rho_b g_{ab} - \mu_a \equiv -\mu_{\text{eff}}^{(a)},$$
$$X_3 = 2\rho_b g_b + \rho_a g_{ab} - \mu_b \equiv -\mu_{\text{eff}}^{(b)}, \qquad (11.39)$$
$$X_5 = \frac{1}{2}\rho_{ab}g_{ab}$$

where the densities are given by the equations,

$$\rho_{1a} = \rho_a = \frac{1}{V}\sum_k \left\{ \frac{(X_5^2 E_b + E_a\omega_1^2 - E_a E_b^2)f(\omega_1)}{\sqrt{D}\omega_1} \right.$$
$$\left. + \frac{(X_5^2 E_b + E_a\omega_2^2 - E_a E_b^2)f(\omega_2)}{\sqrt{D}\omega_2} \right\},$$

$$\rho_{1b} = \rho_b = \frac{1}{V}\sum_k \left\{ \frac{(X_5^2 E_a + E_b\omega_1^2 - E_a^2 E_b)f(\omega_1)}{\sqrt{D}\omega_1} \right.$$
$$\left. + \frac{(X_5^2 E_a + E_b\omega_2^2 - E_a^2 E_b)f(\omega_2)}{\sqrt{D}\omega_2} \right\}, \qquad (11.40)$$

$$\rho_{ab} = \frac{2X_5}{V}\sum_k \left\{ \frac{(-X_5^2 + E_a E_b + \omega_1^2)f(\omega_1)}{\sqrt{D}\omega_1} \right.$$
$$\left. - \frac{(-X_5^2 + E_a E_b + \omega_2^2)f(\omega_2)}{\sqrt{D}\omega_2} \right\},$$

$$\eta = \frac{\rho_{ab}}{2\sqrt{\rho_a\rho_b}}, \qquad f(x) = 1/(e^{\beta x} - 1).$$

Then the dispersion relations (11.16) reduce to the form,

$$\omega_{1,2} = \sqrt{\frac{E_a^2 + E_b^2}{2} + X_5^2 \pm \frac{\sqrt{D}}{2}}, \quad D = (E_a + E_b)^2[4X_5^2 + (E_a - E_b)^2] \qquad (11.41)$$

with

$$E_a = \varepsilon_a(k) - \mu_{\text{eff}}^{(a)}, \qquad E_b = \varepsilon_b(k) - \mu_{\text{eff}}^{(b)}. \qquad (11.42)$$

This spectrum is real and positive, provided the expression under the square root does not become negative. It is convenient to test the positiveness of the expression $\omega_1^2 \cdot \omega_2^2$. Then from (11.41) one easily obtains the condition

$$\omega_1^2 \cdot \omega_2^2 = (E_a E_b - X_5^2)^2 \geq 0. \tag{11.43}$$

Since ω_1 is positive, both ω_1^2 and ω_2^2 are positive simultaneously for any g_{ab} and temperature $T \geq T_c$, hence the spectra are real and positive.

Note that, when $g_{ab} = 0$, Eq. (11.40) turn into the well known expressions

$$\rho_{1a} = \rho_a = \frac{1}{V} \sum_k \frac{1}{e^{\beta[\varepsilon_a(\mathbf{k}) - \mu_{\text{eff}}^{(a)}]} - 1},$$

$$\rho_{1b} = \rho_b = \frac{1}{V} \sum_k \frac{1}{e^{\beta[\varepsilon_b(\mathbf{k}) - \mu_{\text{eff}}^{(b)}]} - 1}, \tag{11.44}$$

where $\mu_{\text{eff}}^{(a)} = \mu_a - 2\rho_a g_a$.

Task 11.2 Starting from Eqs. (11.21)–(11.23), prove (11.44).

The spectrum of collective excitations of a binary mixture of normal components has also been studied in the random-phase approximation in Ref. [19]. In that approximation, for the case of contact interactions, the dynamic condition for mixture stability is found to coincide with the inequality $g_{ab}^2 < g_a g_b$, being practically independent of temperature.

11.3.1.3 C. Near Critical Temperature

Let T_c^a and T_c^b be the BEC transition temperatures for the corresponding components. For concreteness, we assume that $T_c^a \leq T_c^b$. In fact, T_c^a corresponds to the point where the effective chemical potential vanishes, $\mu_{\text{eff}}^{(a)} = -X_1 = 0$. From the stability condition $\Delta_a \Delta_b > \Delta_{ab}^2$, that is, $X_5^2 \leq X_1 X_3$, it is understood that the system can be stable only for $X_5 = 0$, with the sound velocities:

$$s_1^2 = \frac{\Delta_b}{m_b} \geq 0, \qquad s_2^2 = 0. \tag{11.45}$$

This temperature can be evaluated from (11.44) as

$$\rho_{1a} = \rho_a = \frac{1}{V} \sum_k \frac{1}{e^{\varepsilon_a(k)/T_c^a} - 1}. \tag{11.46}$$

From this equation, it is seen that in the present approximation, like in many versions of mean field approximations [11], there is no shift of critical temperature due to intercomponent coupling constant g_{ab}, i.e. $T_c(g_{ab}) = T_c(g_{ab} = 0)$.

11.3.2 Intermediate Summary

For completeness, at the end of this section, we present explicit expressions for the free energy $F = \Omega + \mu N = \Omega + \mu_a N_a + \mu_b N_b$, which has the form

$$F(T < T_c) = F_0 + F_{ZM} + F_T,$$

$$F_0 = \frac{V\rho_a^2 g_a}{2}(1 + n_{1a}^2 - \tilde{m}_a^2 - 2n_{1a}\tilde{m}_a) + \frac{V\rho_b^2 g_b}{2}(1 + n_{1b}^2 - \tilde{m}_b^2 - 2n_{1b}\tilde{m}_b)$$

$$+ \frac{V g_{ab}}{2}\rho_a\rho_b(2 - \tilde{m}_{ab}^2),$$

$$F_{ZM} = \frac{1}{2}\sum_k \left[\omega_1 + \omega_2 - \varepsilon_k - \Delta_a - \Delta_b + \frac{\nu_2^2\Delta_a^2 + \nu_1^2\Delta_b^2 + 4\Delta_{ab}^2\nu_1\nu_2}{2\nu_1\nu_2\varepsilon(k)} \right],$$

$$F_T = T\sum_k \left[\ln(1 - e^{-\omega_1\beta}) + \ln(1 - e^{-\omega_2\beta}) \right],$$

$$F(T > T_c) = V g_a\rho_a^2 + V g_b\rho_b^2 + V g_{ab}\rho_a\rho_b + F_T$$

$$(11.47)$$

where $n_{1a} = \rho_{1a}/\rho_a, \tilde{m}_a = \sigma_a/\rho_a, \tilde{m}_{ab} = \sigma_{ab}/\rho_a$, and the dispersions ω_1 and ω_2 for the BEC and normal phases are given in Eqs. (11.34) and (11.41), respectively. Note that, the present approach includes by itself not only the Lee-Huang-Yang term, but also the corrections beyond this approximation due to taking account of anomalous densities. Particularly, the LHY term can be obtained by expanding F_{ZM} in powers of the coupling parameters.

Concluding the present section, let us summarize the conditions of stability for a two component uniform Bose system with $g_{ab} > 0$:

(i) At temperatures below the critical one, when the system is in the condensed phase, the mixture is stable, provided that the general condition

$$\frac{\Delta_a(\gamma, T)\Delta_b(\gamma, T)}{\Delta_{ab}^2(\gamma, T)} \geq 1 \tag{11.48}$$

holds. Here the self-energies $\Delta_a(\gamma, T)$, $\Delta_b(\gamma, T)$ and $\Delta_{ab}(\gamma, T)$ are the solutions to the Eq. (11.30).

(ii) The inequality (11.48) may be replaced by

$$\frac{g_a g_b}{g_{ab}^2} \geq 1, \tag{11.49}$$

for very dilute gases, where the Bogolyubov approximation is valid.
In the next section, it will be shown that, for a balanced symmetric Bose mixture, the condition (11.48) may be represented as an expansion in powers of γ.

11.4   Balanced Symmetric Bose Mixtures

The case of a binary superfluid gas with two symmetric components consisting of ^{23}Na in an equal mixture of two hyperfine ground states was realized experimentally by Kim et al. in [6]. So we assume that $g_a = g_b = g$, $g_{ab} = \overline{g}_{ab}g_a$, $m_a = m_b = m$, $\varepsilon_a(\mathbf{k}) = \varepsilon_b(\mathbf{k}) = \varepsilon(\mathbf{k}) = k^2/2m$, $\rho_a = \rho_b = \rho/2$, where $N = \rho V$ is the total number of atoms in the mixture. Note that, treating the anomalous averages, we resort to the standard way of regularization by employing the method of counterterms outlined in Chap. 4.

11.4.1 A. Zero Temperature

At zero temperature, the densities are simplified to

$$
\begin{aligned}
n_{1a} &= \frac{\rho_{1a}(T=0)}{\rho_a} = \frac{\rho_{1b}(T=0)}{\rho_a} = \frac{1}{2V\rho_a}\sum_k \left\{ \frac{\Delta_a + \varepsilon(k) + \Delta_{ab}}{2\omega_1} \right. \\
&\left. + \frac{\Delta_a + \varepsilon(\mathbf{k}) - \Delta_{ab}}{2\omega_2} - 1 \right\} = \frac{m^3(s_1^3 + s_2^3)}{6\pi^2\rho_a} = n_{1b}, \\
\tilde{m}_b = \tilde{m}_a &= \frac{\sigma_a(T=0)}{\rho_a} = \frac{\sigma_b(T=0)}{\rho_b} = \\
&- \frac{1}{2V\rho_a}\sum_k \left\{ \frac{\Delta_a + \Delta_{ab}}{2\omega_1} + \frac{\Delta_a - \Delta_{ab}}{2\omega_2} - \frac{\Delta_a}{\varepsilon(\mathbf{k})} \right\} = 3n_{1a}(T=0), \\
n_{ab} &= \frac{\rho_{ab}(T=0)}{\sqrt{\rho_a\rho_b}} = \frac{1}{2V\rho_a}\sum_k \left\{ \frac{\varepsilon(\mathbf{k}) + \Delta_a + \Delta_{ab}}{\omega_1} - \frac{\varepsilon(\mathbf{k}) + \Delta_a - \Delta_{ab}}{\omega_2} \right\} \\
&= \frac{m^3(s_1^3 - s_2^3)}{3\pi^2\rho_a}, \\
\tilde{m}_{ab} &= \frac{\sigma_{ab}(T=0)}{\sqrt{\rho_a\rho_b}} = n_{ab},
\end{aligned}
\tag{11.50}
$$

where we have introduced the sound velocities

$$
s_{1,2}^2 = (\Delta_a \pm \Delta_{ab})/m,
\tag{11.51}
$$

which satisfy the following equations, derived from (11.30) and (11.36):

$$
\begin{aligned}
s_1^3 - s_2^3(\overline{g}_{ab} - 1) - \frac{3\pi^2 s_1^2}{gm^2} + \frac{3\pi^2\rho_a(\overline{g}_{ab} + 1)}{m^3} &= 0, \\
s_1^3 + s_2^3(\overline{g}_{ab} + 1) - \frac{3\pi^2 s_2^2}{gm^2} - \frac{3\pi^2\rho_a(\overline{g}_{ab} - 1)}{m^3} &= 0.
\end{aligned}
\tag{11.52}
$$

These equations can be rewritten in the dimensionless form as

$$\tilde{s}_1^3 + (1 - \overline{g}_{ab})\tilde{s}_2^3 - \frac{3\pi\tilde{s}_1^2}{4} + \frac{3\pi^2\gamma(\overline{g}_{ab} + 1)}{2} = 0,$$

$$\tilde{s}_1^3 + (1 + \overline{g}_{ab})\tilde{s}_2^3 - \frac{3\pi\tilde{s}_2^2}{4} - \frac{3\pi^2\gamma(\overline{g}_{ab} - 1)}{2} = 0,$$

$$(11.53)$$

where $\gamma = \rho a_s^3$, $\rho = 2\rho_a$ is the total density of the whole binary system, $a_s = mg/4\pi$, and $\tilde{s}_{1,2} = s_{1,2} \cdot ma_s$. From the Bogolyubov approximation, it is known that the system becomes unstable for $\overline{g}_{ab} > 1$. What are the possible corrections to this criterion due to quantum fluctuations?

To find an answer to this question one has to consider the dispersions:

$$\omega_1^2 = \varepsilon(\mathbf{k})[\varepsilon(\mathbf{k}) + 2(\Delta_a + \Delta_{ab})], \quad \omega_2^2 = \varepsilon(\mathbf{k})[\varepsilon(\mathbf{k}) + 2(\Delta_a - \Delta_{ab})]. \quad (11.54)$$

It is seen that the boundary of stability is defined by the condition $\Delta_a = \Delta_{ab}$, since for $\Delta_a < \Delta_{ab}$ the sound velocity s_2 becomes negative. This is also seen from the general condition $\Delta_a \Delta_b \leq \Delta_{ab}^2$, with $\Delta_a = \Delta_b$. In other words, the boundary of stability in the phase diagram $(\overline{g}_{ab}, \gamma)$ lies on the line $s_2 = 0$, i.e. $\tilde{s}_2(\overline{g}_{ab}^*, \gamma^*) \equiv 0$. From Eq. (11.53), one obtains

$$(s_1^*)^3 - \frac{3\pi}{8}(s_1^*)^2 + \frac{3\pi^2\gamma^*}{2} = 0, \qquad \overline{g}_{ab}^* = \frac{(s_1^*)^2}{4\pi\gamma^*}, \qquad (11.55)$$

where the "star" indicates the threshold values of the parameters, corresponding to the boundary of stability for the symmetric binary mixture. In Fig. 11.3a, we present the phase diagram on the $(\overline{g}_{ab}, \gamma)$ plane (solid line). It is seen that, due to quantum fluctuations, the system at $T = 0$ remains stable, even, for example, at $\overline{g}_{ab}^*(\gamma \approx 0.013) \approx 1.9$. This is one of the main results of the present chapter.

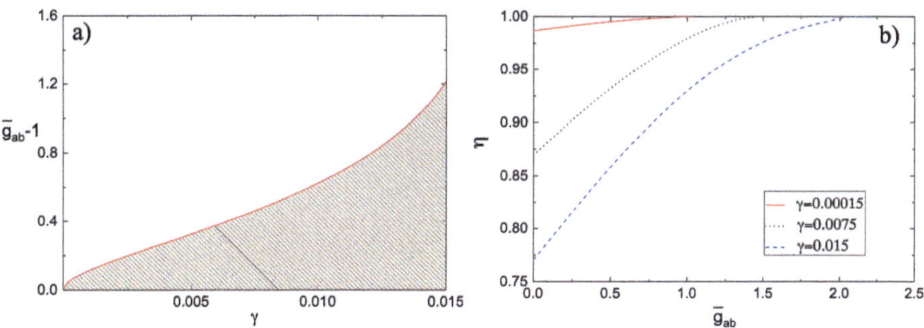

Fig. 11.3 a The phase diagram of a symmetric binary Bose system with repulsive interactions at zero temperature. The shaded region corresponds to the stable, miscible phase; **b** Overlap parameter η versus $\overline{g}_{ab} = g_{ab}/g$ for three different values of the gas parameter: $\gamma = 0.15 \cdot 10^{-3}$ (solid line), $\gamma = 0.75 \cdot 10^{-2}$ (dotted line), and $\gamma = 0.15 \cdot 10^{-1}$ (dashed line)

For small γ, one may use the following expansion for $\overline{g}^*_{ab}$:

$$\overline{g}^*_{ab} = 1 + \frac{16\sqrt{\gamma}}{3\sqrt{\pi}} + \frac{128\gamma}{3\pi} + \frac{3584\gamma^{3/2}}{9\pi^{3/2}} + O(\gamma^{5/2}), \tag{11.56}$$

to obtain the stability condition in the form

$$g_{ab} \leq g_a \left[1 + \frac{16\sqrt{\gamma}}{3\sqrt{\pi}} + O(\gamma) \right], \tag{11.57}$$

which is valid for $\gamma \leq 0.005$.

The overlap parameter for the symmetric case at $T = 0$ has the form:

$$\eta = n_{0a} + \frac{n_{ab}}{2} = 1 - n_{1a} + \frac{n_{ab}}{2} = 1 - \frac{2s_2^3}{3\pi^2\gamma}, \tag{11.58}$$

where we used Eqs. (11.50) and (11.51). On the boundary of stability, $s_2 = s_2^* = 0$, and η reaches its maximum value $\eta = 1$ (see Fig. 11.3b). Close to the phase transition point, the condensed fraction can be presented as

$$n_0^*(T = 0) = 1 - \frac{\rho_{1a}}{\rho_a} \bigg|_{\overline{g}_{ab} \to \overline{g}^*_{ab}} = 1 - \frac{8\sqrt{\gamma}}{3\sqrt{\pi}} - \frac{64\gamma}{3\pi} + O(\gamma^{3/2}). \tag{11.59}$$

In Fig. 11.4a, we present the condensed fraction versus $\overline{g}_{ab}$. It is seen that intercomponent repulsion g_{ab} tends to destroy BEC, repelling the condensed particles.

For completeness, we compare these results with the experiments performed by Kim et al. [6]. The authors studied the mixture of atoms with two hyperfine ground states of ^{23}Na and measured the sound velocities $s_1 = 3.23$ mm/s and $s_2 = 0.70$ mm/s by fixing the relative coupling constant $\overline{g}_{ab} = 0.93$ and the gas parameter $\gamma \approx 1.4 \cdot 10^{-6}$. For this set of parameters from Eq. (11.52), we get the following values for the sound velocities: $s_1 = 3.91$ mm/s, $s_2 = 0.75$ mm/s, which

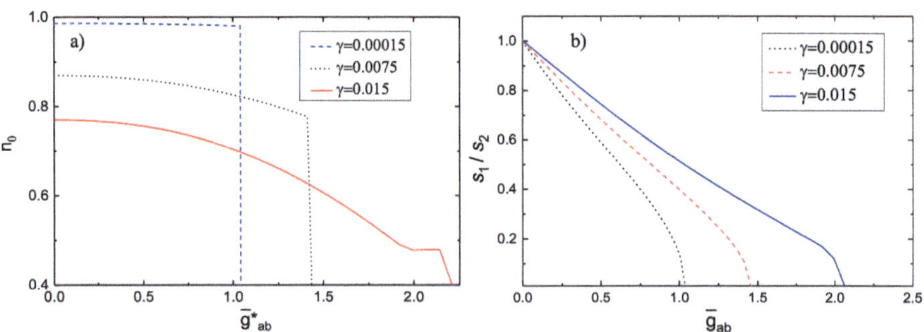

Fig. 11.4 a The condensed fraction, $n_0 = \rho_0/(\rho/2)$, at zero temperature versus $\overline{g}_{ab} = g_{ab}/g$ in the interval $0 < \overline{g}_{ab} < \overline{g}^*_{ab}(\gamma)$ for different values of γ; **b** The relative sound velocity s_2/s_1 versus $\overline{g}_{ab}$ for different values of γ at $T = 0$

are rather close to the experimental data. To make further predictions, we have calculated the relative sound velocity s_2/s_1 versus $\overline{g}_{ab}$ for three different values of γ. The results are presented in Fig. 11.4b. It is seen that s_2/s_1 reduces with increasing $\overline{g}_{ab}$ and vanishes at $\overline{g}_{ab} = \overline{g}_{ab}^*$, where the phase separation occurs.

11.5 Summary

In the present chapter, we have developed a self-consistent mean-field theory for a binary homogeneous mixture of two-component Bose systems. This theory, being conserving and gapless, imposes no restriction on the gas parameter γ, and hence, it is valid for arbitrarily strong interactions g_{ij}. The theory satisfies generalized HP theorem and takes into account anomalous densities σ_a, σ_b and σ_{ab}.

For numerical analysis, we have considered the balanced symmetric configuration of a two-component mixture of Bose gases. We have obtained the phase diagram for this system at zero as well as at finite temperatures for arbitrary gas parameters. The phase diagram at zero temperature on (g_{ab}, γ) plane shows that the system may remain stable and miscible even at $g_{ab}/g_{aa} > 1$, provided the anomalous densities are properly taken into account. Comparing this phase diagram with that at finite temperature (g_{ab}, γ, T), we see that the finite temperature can transform the phase-separated two-component BECs at $T = 0$ to a miscible state. This conclusion is in good agreement with the works by Roy et al. [20], Ota et al. [21], and Shi et al. [22]. Our numerical results are also in good agreement with experimental works [6,23], although new experimental measurements for larger values of the interspecies coupling and γ are expected.

It will be quite interesting to study nonsymmetric, e.g. imbalanced two-component Bose mixture, where quasi-magnetic transitions may also take place [21]. Moreover, as has been recently claimed by Naidon and Petrov [24], for unequal interspecies interaction or unequal masses, the mixed phase can form bubbles with a tunable population. This may serve as the next task for the application of present theory, since the ground state physics can be qualitatively understood from the arguments valid for a homogeneous system. However, in the case of real systems in a trap, the phase separation can be suppressed in the inhomogeneous system due to quantum pressure effects [25].

References

1. C.J. Myatt, E.A. Burt, R.W. Ghrist, E.A. Cornell, C.E. Wieman, Production of two overlapping Bose-Einstein condensates by sympathetic cooling. Phys. Rev. Lett. **78**, 586–589 (1997). [Online]. https://doi.org/10.1103/PhysRevLett.78.586
2. D.S. Petrov, Quantum mechanical stabilization of a collapsing Bose-Bose mixture. Phys. Rev. Lett. **115**, 155302 (2015). [Online]. https://doi.org/10.1103/PhysRevLett.115.155302
3. E. Timmermans, Phase separation of Bose-Einstein condensates. Phys. Rev. Lett. **81**, 5718–5721 (1998). [Online]. https://doi.org/10.1103/PhysRevLett.81.5718
4. C.R. Cabrera, L. Tanzi, J. Sanz, B. Naylor, P. Thomas, P. Cheiney, L. Tarruell, Quantum liquid droplets in a mixture of Bose-Einstein condensates. Science **359**(6373), 301–304 (2018). [Online]. https://doi.org/10.1126/science.aao5686

5. S.B. Papp, J.M. Pino, C.E. Wieman, Tunable miscibility in a dual-species Bose-Einstein condensate. Phys. Rev. Lett. **101**, 040402 (2008). [Online]. https://doi.org/10.1103/PhysRevLett.101.040402

6. J.H. Kim, D. Hong, Y. Shin, Observation of two sound modes in a binary superfluid gas. Phys. Rev. A **101**, 061601 (2020). [Online]. https://doi.org/10.1103/PhysRevA.101.061601

7. E. Fava, T. Bienaimé, C. Mordini, G. Colzi, C. Qu, S. Stringari, G. Lamporesi, G. Ferrari, Observation of spin superfluidity in a Bose gas mixture. Phys. Rev. Lett. **120**, 170401 (2018). [Online]. https://doi.org/10.1103/PhysRevLett.120.170401

8. C.R.C. Cabrera, *PhD Thesis: Quantum liquid droplets in a mixture of Bose-Einstein condensates*. Catalonia, Polytechnic University of Catalonia. Institute of Photonic Sciences (2018)

9. L. Wacker, N.B. Jørgensen, D. Birkmose, R. Horchani, W. Ertmer, C. Klempt, N. Winter, J. Sherson, J.J. Arlt, Tunable dual-species Bose-Einstein condensates of ^{39}K and ^{87}Rb. Phys. Rev. A **92**, 053602 (2015). [Online]. https://doi.org/10.1103/PhysRevA.92.053602

10. A. Boudjemâa, Quantum and thermal fluctuations in two-component Bose gases. Phys. Rev. A **97**, 033627 (2018). [Online]. https://doi.org/10.1103/PhysRevA.97.033627

11. J.O. Andersen, Theory of the weakly interacting Bose gas. Rev. Mod. Phys. **76**, 599–639 (2004). [Online]. https://doi.org/10.1103/RevModPhys.76.599

12. I. Stancu, P.M. Stevenson, Second-order corrections to the Gaussian effective potential of $\lambda\phi^4$ theory. Phys. Rev. D **42**, 2710–2725 (1990). [Online]. https://doi.org/10.1103/PhysRevD.42.2710

13. N. Shohno, Low-temperature properties of the interacting Bose system. Prog. Theor. Phys. **31**(4), 553–574 (1964). [Online]. https://doi.org/10.1143/PTP.31.553

14. L. Wen, W.M. Liu, Y. Cai, J.M. Zhang, J. Hu, Controlling phase separation of a two-component Bose-Einstein condensate by confinement. Phys. Rev. A **85**, 043602 (2012). [Online]. https://doi.org/10.1103/PhysRevA.85.043602

15. R.K. Kumar, P. Muruganandam, L. Tomio, A. Gammal, Miscibility in coupled dipolar and non-dipolar Bose–Einstein condensates. J. Phys. Commun. **1**(3), 035012 (2017). [Online]. https://doi.org/10.1088/2399-6528/aa8db5

16. S. Watabe, Hugenholtz-pines theorem for multicomponent Bose-Einstein condensates. Phys. Rev. A **103**, 053307 (2021). [Online]. https://doi.org/10.1103/PhysRevA.103.053307

17. Y.A. Nepomnyashchii, The microscopic theory of a mixture of superfluid liquid. J. Exp. Theor. Phys. **43**, 559 (1976). [Online]. http://jetp.ras.ru/cgi-bin/dn/e_043_03_0559.pdf

18. P. Szépfalusy, I. Kondor, On the dynamics of continuous phase transitions. Ann. Phys. **82**(1), 1–53 (1974). [Online]. https://www.sciencedirect.com/science/article/pii/0003491674903303

19. V.I. Yukalov, Stability and stratification of a quantum liquid mixture. Acta Phys. Pol. A; (Poland) **57**(3) (1980)

20. A. Roy, D. Angom, Thermal suppression of phase separation in condensate mixtures. Phys. Rev. A **92**, 011601 (2015). [Online]. https://doi.org/10.1103/PhysRevA.92.011601

21. M. Ota, S. Giorgini, Thermodynamics of dilute Bose gases: beyond mean-field theory for binary mixtures of Bose-Einstein condensates. Phys. Rev. A **102**, 063303 (2020). [Online]. https://doi.org/10.1103/PhysRevA.102.063303

22. H. Shi, W.-M. Zheng, S.-T. Chui, Phase separation of Bose gases at finite temperature. Phys. Rev. A **61**, 063613 (2000). [Online]. https://doi.org/10.1103/PhysRevA.61.063613

23. K.L. Lee, N.B. Jørgensen, L.J. Wacker, M.G. Skou, K.T. Skalmstang, J.J. Arlt, N.P. Proukakis, Time-of-flight expansion of binary Bose–Einstein condensates at finite temperature. New J. Phys. **20**(5), 053004 (2018). [Online]. https://doi.org/10.1088/1367-2630/aaba39

24. P. Naidon, D.S. Petrov, Mixed bubbles in Bose-Bose mixtures. Phys. Rev. Lett. **126**, 115301 (2021). [Online]. https://doi.org/10.1103/PhysRevLett.126.115301

25. R.W. Pattinson, T.P. Billam, S.A. Gardiner, D.J. McCarron, H.W. Cho, S.L. Cornish, N.G. Parker, N.P. Proukakis, Equilibrium solutions for immiscible two-species Bose-Einstein condensates in perturbed harmonic traps. Phys. Rev. A **87**, 013625 (2013). [Online]. https://doi.org/10.1103/PhysRevA.87.013625

12.1 Introduction

In the present chapter, we shall apply the theory developed in the previous chapter to study ultra-cold quantum liquid droplets qualitatively described in Sects. 2.2 and 5.9. We have mentioned that, in a two-component Bose system, a stable Bose phase-quantum droplet may arise due to the balance between attractive intercomponent ($g_{ab} < 0$) and repulsive intracomponent ($g_{aa} \equiv g_a > 0$, $g_{bb} \equiv g_b > 0$) interactions. Below we quantitatively discuss this point, showing that such a picture will take place when the attractive mean-field term $\mathcal{E}_{MFA} \sim \rho^2$ is compensated by the repulsive LHY energy term $\mathcal{E}_{LHY} \sim \rho^{5/2}$. In other words, the resulting effective interaction $\mathcal{E}_{tot} \sim \tilde{\alpha}\rho^2 + \tilde{\beta}\rho^{5/2}$ with $\tilde{\alpha} < 0$ and $\tilde{\beta} > 0$ will give rise to a self-bound solution: so-called quantum liquid droplet.

12.2 $\mathcal{E}_{LHY}$ and $\mathcal{E}_{MFA}$ at Zero Temperature

Following Petrov [1] we neglect Ω_2 and Ω_4 terms of the thermodynamic potential in Eq. (11.19) and set $\mu_{0a} = \mu_{1a} = \mu_a$, $\mu_{0b} = \mu_{1b} = \mu_b$ to obtain the following expression for the energy of the whole two-component system at zero temperature:

$$\mathcal{E} = \Omega + \mu N \approx \mathcal{E}_{MFA} + \mathcal{E}_{LHY},$$

$$\mathcal{E}_{MFA} = \frac{V}{2}[g_a \rho_{0a}^2 + g_b \rho_{0b}^2 + 2g_{ab}\rho_{0a}\rho_{0b}],$$

$$\mathcal{E}_{LHY} = \frac{1}{2}\sum_{\mathbf{k}}[\omega_1(\mathbf{k}) + \omega_2(\mathbf{k}) + c_t(\mathbf{k})] \tag{12.1}$$

A. Rakhimov and S. Mardonov, *Theory of Quantum Bose Liquids Beyond Bogoliubov Approximation*, Lecture Notes in Physics 1045, https://doi.org/10.1007/978-3-032-05096-0_12

where for equal masses ($m_1 = m_2 \equiv m$)

$$\omega_{1,2}^2 = \frac{\mathbf{k}^4}{4m^2}[\mathbf{k}^2 + 2m(\Delta_a + \Delta_a \pm \sqrt{D_s})], \qquad D_s = 4\Delta_{ab}^2 + (\Delta_a - \Delta_b)^2,$$

(12.2)

(see Eq. (11.34)).

The counter term $c_t(k)$ in Eq. (12.1) can be found easily by using the technique described in Sect. 4.9. As a result we have

$$\mathcal{E}_{LHY} = \frac{V}{4\pi^2} \int_0^\infty k^2 dk \left[\omega_1(\mathbf{k}) + \omega_2(\mathbf{k}) - 2\varepsilon_k - \Delta_a - \Delta_b + \frac{2\Delta_{ab}^2 + \Delta_a^2 + \Delta_b^2}{2\varepsilon_k} \right], \quad (12.3)$$

where $\varepsilon_k = \mathbf{k}^2/2m$ and $\Delta_{a,b}$ are defined in Eq. (11.30). This integral can be successfully evaluated by using e.g. MAPLE[1] and converted into following compact form:

$$\mathcal{E}_{LHY} = \frac{\sqrt{2} V m^{3/2} \Delta_a^{5/2}}{15\pi^2} \tilde{f}(\delta_b, \delta_{ab}),$$

(12.4)

where

$$\tilde{f}(\delta_b, \delta_{ab}) = (1 + \delta_b + \sqrt{d_s})^{5/2} + (1 + \delta_b - \sqrt{d_s})^{5/2},$$

$$d_s = 4\delta_{ab}^2 + (1 - \delta_b)^2, \quad \delta_b = \frac{\Delta_b}{\Delta_a}, \quad \delta_{ab} = \frac{\Delta_{ab}}{\Delta_a}.$$

(12.5)

Now we consider another term in (12.1).

12.2.1 $\mathcal{E}_{MFA}$ and Phase Locking

First, to control the "sign" effects we introduce a widely used notation

$$\delta_g = g_{ab} + \sqrt{g_{aa}g_{bb}},$$

(12.6)

which is allowed to be negative due to the attractive intercomponent interaction $g_{ab} < 0$. Then $\mathcal{E}_{MFA}$ in (12.1) can be presented in following equivalent form:

$$\mathcal{E}_{MFA}(\rho_{0a}, \rho_{0b}) = \frac{V(\rho_{0a}\sqrt{g_a} - \rho_{0b}\sqrt{g_b})^2}{2} + V\delta_g \rho_{0a}\rho_{0b}.$$

(12.7)

It is seen that the mean-field energy consists of two parts: hard and soft modes. The former is always repulsive, the latter may be attractive for $\delta_g < 0$. The stability analysis is based on the fact that a stable droplet can be formed if the hard mode is

---

[1] A more general case with $m_1 \neq m_2$ has been considered in Ref. [2].

not too large, or better, vanishes all. This condition is satisfied by introducing the following constraint[2] :

$$\frac{\rho_{0a}}{\rho_{0b}} = \frac{\sqrt{g_{bb}}}{\sqrt{g_{aa}}} \equiv \alpha = const \tag{12.8}$$

which results in a pure attractive effective interaction

$$\mathcal{E}_{MFA} = V\delta_g \rho_{0a}\rho_{0b}. \tag{12.9}$$

Task 12.1: Find more justification in favor of (12.8) by diagonalazing the quadratic form $\mathcal{E}_{MFA}(\rho_{0a}, \rho_{0b})$ in (12.7).

This constraint means for experimenters that a droplet may arise when the interactions are appropriately tuned and, besides, the number of available condensed atoms in every component must be in the right proportion.

12.3 Droplet Formation

It is understood that it would be quite desirable to proceed in the spirit of OPT by taking into account Ω_2 and Ω_4 terms defined in (11.19), which would force us to solve the Eq. (11.30) with respect to $\Delta_{a,b}$ needed for $\mathcal{E}_{LHY}$. Unfortunately, such a hard task has not been completed yet. On the other hand, we know that the majority of existing BEC-related experiments, e.g. in Ref. [4] could be explained in the Bogolyubov approximation. This allows us to make the following simplifications:

$$\begin{aligned}
\Delta_a &= g_{aa}(\rho_a - \rho_{1a} + \sigma_a) \approx g_{aa}\rho_a, \\
\Delta_b &= g_{bb}(\rho_b - \rho_{1b} + \sigma_b) \approx g_{bb}\rho_b, \\
\Delta_{ab} &= g_{ab}\sqrt{\rho_{0a}\rho_{0b}} + \frac{g_{ab}\rho_{ab}}{2} \approx g_{ab}\sqrt{\rho_a\rho_b}, \\
(\rho_0)_{a,b} &\approx \rho_{a,b},
\end{aligned} \tag{12.10}$$

with the constraint

$$\frac{\rho_a}{\rho_b} = \frac{\sqrt{g_{bb}}}{\sqrt{g_{aa}}} \equiv \alpha = const, \tag{12.11}$$

where, e.g. ρ_a is the number density of the a component, $\rho_a = N_a/V$. Then the total energy functional at zero temperature has the form

$$\mathcal{E} = \frac{V\delta_g \alpha \rho^2}{(1+\alpha)^2} + \frac{\sqrt{2}Vm^{3/2}(g_{aa}\alpha)^{5/2}}{15\pi^2(1+\alpha)^{5/2}}\rho^{5/2}\tilde{f}(\alpha, \delta_{ab}), \tag{12.12}$$

---

[2] Corrections to this approximation are discussed in Ref. [3].

where we used simple relations:

$$\rho_a + \rho_b = \rho, \qquad \rho_a = \frac{\alpha\rho}{1+\alpha}, \qquad \rho_b = \frac{\rho}{1+\alpha},$$
$$\delta_b = \frac{\Delta_b}{\Delta_a} \approx \alpha, \qquad \delta_{ab} = \frac{\Delta_{ab}}{\Delta_a} \approx \frac{g_{ab}}{g_{aa}\sqrt{\alpha}}. \tag{12.13}$$

Therefore, we have revealed that, the energy density of the system of two-component Bose gases with $g_{aa} > 0$, $g_{bb} > 0$, $g_{ab} < 0$ can be presented as

$$\frac{\mathcal{E}}{V} = \tilde{\alpha}\rho^2 + \tilde{\beta}\rho^{5/2}, \tag{12.14}$$

with

$$\tilde{\alpha} = \frac{\alpha\delta_g}{(1+\alpha)^2} < 0, \qquad \tilde{\beta} = \frac{\sqrt{2}Vm^{3/2}(g_{aa}\alpha)^{5/2}}{15\pi^2(1+\alpha)^{5/2}}\tilde{f}(\alpha, \delta_{ab}) > 0. \tag{12.15}$$

It is seen that when the system is nearly at face the collapse (see Fig. 11.1), the energy of quantum fluctuations (the second term in (12.14)) will prevent this collapse to make possible the formation of a superfluid droplet. However, there would be a critical atom number N_c where the repulsive character of the quantum pressure dissociates the system into a gas. Besides, one may intuitively understand from this figure the following picture: For a large atom number $N > N_c$, the droplet is stable. Close to the critical atom number $N \sim N_c$, the droplet becomes metastable and finally, for low atom number $N < N_c$, the droplet dissociates into gas, which is referred to as an LHY gas. This critical atom number may be numerically estimated by analyzing solutions of an appropriate Gross-Pitaevskii equations, which will be performed in the next section.

12.4   Extended Gross-Pitaevskii Equation in the Effective Single-Component Approximation

In general, Gross-Pitaevskii equations are nice tools to study zero temperature properties of microscopical quantum objects such as solitons, kinks, droplets, etc. This simple approach accounts for interactions in the lowest order perturbation in a small gas parameter $\gamma = \rho a_s^3 \ll 1$, where the Bogolyubov approximation with almost a hundred percent of condensed particles is valid. For our quantum droplets, they can be constructed directly from the effective energy functional (12.12), where, in accordance with Petrov's model, LHY energy is a crucial ingredient.

However, examining Eqs. (12.5) and (12.12), one can notice that, the LHY term in (12.12) becomes imaginary. In fact, for $g_{ab}^2 > g_{aa}g_{bb}$ the second term in

$$\tilde{f}(\alpha, \delta_{ab}) = (1 + \delta_b + \sqrt{d_s})^{5/2} + (1 + \delta_b - \sqrt{d_s})^{5/2} \tag{12.16}$$

becomes complex. The root of this evil is hidden in the energy spectrum $\omega_{1,2}$ given by Eq. (12.2): one of branches becomes complex for $|g_{ab}| > \sqrt{g_{aa}g_{bb}}$ in the Bogolyubov approximation, (12.10). This means that, especially in the droplet phase, when $\delta g = g_{ab} + \sqrt{g_{aa}g_{bb}} < 0$, there emerge imaginary excitations, which are the signal of instability, and hence a complex part of the whole energy. To our knowledge, nowadays there are three ways of escaping this catastrophe:

- Neglect the complex part of $\mathcal{E}_{LHY}$ by setting $\mathcal{E}_{LHY} \approx \mathcal{R}e(\mathcal{E}_{LHY})$ [1,5–8]
- Improve Petrovs model in order to obtain appropriate expressions for the dispersions $\omega_{1,2}$, which will be real, regardless of the ratio $g_{ab}^2/g_{aa}g_{bb}$ [9–11].
- Letting alone $\mathcal{E}_{MFA}$, set $g_{ab}^2 = g_{aa}g_{bb}$ in $\mathcal{E}_{LHY}$, i.e. $\mathcal{E}_{MFA} = \mathcal{E}_{MFA}(\delta g \neq 0)$, but $\mathcal{E}_{LHY} \approx \mathcal{E}_{LHY}(\delta g = 0)$, (see e.g. Ref. [3]).

Postponing the detailed discussion of these methods to later (see Sect. 12.6), below for transparency and simplicity, we shall choose the third method working "on the verge of the collapse", limiting ourselves to Bogolyubov approximation (12.10). So, close to the verge, the dispersions and counterterms are simplified as:

$$\omega_1^2 = \frac{k^2[k^2 + 4m(g_{aa}\rho_a + g_{bb}\rho_b)]}{4m^2}, \qquad \omega_2^2 = \frac{k^4}{4m^2},$$

$$c_t(k) = -\frac{k^2}{m} - g_{aa}\rho_a - g_{bb}\rho_b + \frac{m(g_{aa}\rho_a + g_{bb}\rho_b)^2}{k^2}, \tag{12.17}$$

which leads to following energy functional:

$$\mathcal{E}(\rho_a, \rho_b) \approx \mathcal{E}_{MFA} + \frac{8Vm^{3/2}(g_{aa}\rho_a + g_{bb}\rho_b)^{5/2}}{15\pi^2}. \tag{12.18}$$

Now we are on the stage of constructing two-component GP equations, which are actually, nonlinear Schrödinger equations, corresponding to a given Hamiltonian, whose minimal eigenvalue is the ground state energy of the system at zero temperature. In present case the Hamiltonian density of two-component Bose gases with contact interactions and equal masses derived from the Lagrangian (11.1) can be rewritten as:

$$\mathcal{H} = \sum_{i=1,2} \left\{ \varphi_i^\dagger \left[-\frac{\nabla^2}{2m} + \mu_i \right] \varphi_i + \frac{g_i}{2}(\varphi_i^\dagger \varphi_i)^2 \right\} + g_{12}(\varphi_1^\dagger \varphi_1)(\varphi_2^\dagger \varphi_2), \tag{12.19}$$

where for convenience we introduce the following redefinitions: $\psi = \varphi_1$, $\phi = \varphi_2$, $g_{aa} = g_1$, $g_{bb} = g_2$, $g_{ab} = g_{12}$, $\rho_{a,b} = \rho_{1,2}$. This Hamiltonian is rather complicated to be diagonalized. That is finding its eigenvalues is not an easy task. Instead, following Petrov's ideology, we have to replace it with an effective one, $H_{eff}(\varphi_1, \varphi_2)$, so that its lowest eigenvalue will be given by (12.1). Then the stationary GP equations can be obtained by minimization of $H_{eff}(\varphi_1, \varphi_2)$, given by

$$H_{eff}(\varphi_1, \varphi_2) = \int d\mathbf{r} \left[-\sum_{i=1,2} \frac{\varphi_i^\dagger \nabla^2 \varphi_i}{2m} + \mathcal{E}(\varphi_1, \varphi_2) \right], \qquad (12.20)$$

with respect to $[\varphi_1, \varphi_2]$, which are related to the atomic densities as

$$\rho_1 = \varphi_1^\dagger \varphi_1, \qquad \rho_2 = \varphi_2^\dagger \varphi_2, \qquad (12.21)$$

with the normalization $N_{1,2} = \int d\mathbf{r}\rho_{1,2}$. Thus, the procedure $\delta H_{eff}(\varphi_1, \varphi_2)/\delta\varphi_i = 0$, $(i = 1, 2)$ gives

$$\left[-\frac{\nabla^2}{2m} + \frac{\partial \mathcal{E}}{V \partial \rho_i} \right] \varphi_i = i \frac{\partial \varphi_i}{\partial t}. \qquad (12.22)$$

Particularly, using here Eq. (12.7) one obtains

$$i\frac{\partial \varphi_1}{\partial t} = \left[-\frac{\nabla^2}{2m} + \rho_1 g_1 - \rho_2(\sqrt{g_1 g_2} - \delta g) + \frac{4g_1 m^{3/2}(g_1\rho_1 + g_2\rho_2)^{3/2}}{3\pi^2} \right] \varphi_1, \qquad (12.23)$$

and

$$i\frac{\partial \varphi_2}{\partial t} = \left[-\frac{\nabla^2}{2m} + \rho_2 g_2 - \rho_1(\sqrt{g_1 g_2} - \delta g) + \frac{4g_2 m^{3/2}(g_1\rho_1 + g_2\rho_2)^{3/2}}{3\pi^2} \right] \varphi_2, \qquad (12.24)$$

where $\rho_{1,2}$ are given in (12.21).

These two coupled nonlinear equations describe the condensate wave functions of the two-component Bose system with contact interactions when LHY term is taken into account by means of $\mathcal{E}_{LHY}(g_{12}, g_1, g_2) \approx \mathcal{E}_{LHY}(g_{12}^2 = g_1 g_2)$ [12], and the higher terms in the energy functional coming from Ω_2 and Ω_4 in (11.19) as well as the condensate depletions are completely omitted.

On the other hand, a naive question arises: Why is there only one GP equation in most of the droplet-related articles? The answer is that Petrov's GP equation [1] is obtained in a so-called single component approximation under the constraint (12.11). We shall derive it below.

12.4.1 One Component GP Equation

It is easy to show that the constraint (12.11), which guarantees the stability of the self-bound droplet, reduces the number of differential equations (12.23) and (12.24). In fact, the constraint:

$$\frac{\rho_1}{\rho_2} = \frac{\varphi_1^\dagger \varphi_1}{\varphi_2^\dagger \varphi_2} = \frac{\sqrt{g_2}}{\sqrt{g_1}} = \alpha, \qquad (12.25)$$

gives

$$\varphi_1 = \sqrt{\alpha}\varphi_2, \qquad \varphi_1 = \frac{\varphi\sqrt{\alpha}}{\sqrt{1+\alpha}}, \qquad \varphi_2 = \frac{\varphi}{\sqrt{1+\alpha}}, \tag{12.26}$$

where we introduced a unique wave function $\varphi(\mathbf{r})$ with the normalization

$$\int d\mathbf{r}\,\varphi^{\dagger}(\mathbf{r})\varphi(\mathbf{r}) = \int d\mathbf{r}\rho(\mathbf{r}) = N = N_1 + N_2, \tag{12.27}$$

Moreover the constraint simplifies $\mathcal{E}_{MFA}$ such that the total energy becomes

$$\mathcal{E} = V\delta g \rho_1 \rho_2 + \frac{8Vm^{3/2}(g_1\rho_1 + g_2\rho_2)^{5/2}}{15\pi^2} = V\delta g \alpha \rho_2^2 + \frac{8Vm^{3/2}[\alpha(\alpha+1)g_1\rho_2]^{5/2}}{15\pi^2}, \tag{12.28}$$

where $\rho_2 = \rho/(1+\alpha)$.

As to the GP equations, adding up Eqs. (12.23) and (12.24) and using (12.26) we obtain

$$i\frac{\partial\varphi}{\partial t} = -\frac{\nabla^2\varphi}{2m} + \frac{\varphi^3\sqrt{\alpha}\delta g}{1+\alpha} + \frac{4g_1^{5/2}m^{3/2}\alpha^2(1+\alpha-\sqrt{\alpha})\varphi^4}{3\pi^2}. \tag{12.29}$$

This equation leads to the simple single-component GP equation for fully symmetric case, $(\alpha = 1, g_1 = g_2 \equiv g)$:

$$i\frac{\partial\varphi}{\partial t} + \left[\frac{\nabla^2}{2m} - \frac{\delta g|\varphi|^2}{2} - \frac{4g^{5/2}m^{3/2}|\varphi|^3}{3\pi^2}\right]\varphi = 0 \tag{12.30}$$

which has been derived and exploited by Petrov [1] for the first time. This can be reduced to the following stationary equation [12]

$$-\frac{\nabla^2}{2m}\varphi + \frac{\delta g|\varphi|^2\varphi}{2} + \frac{4g^{5/2}m^{3/2}|\varphi|^3}{3\pi^2}\varphi = \epsilon\varphi. \tag{12.31}$$

via standard substitution $\varphi(t,\mathbf{r}) = \exp(it\epsilon)\varphi(\mathbf{r})$, where ϵ defines the collective excitations of droplet states.

12.5 Properties of the Isotropic Droplet

The solutions of (12.31) for a homogeneous system are usually presented in the Gaussian

$$\varphi(r) = \frac{\sqrt{N}e^{-r^2/2\sigma^2}}{\pi^{3/4}\sigma^{3/2}}, \tag{12.32}$$

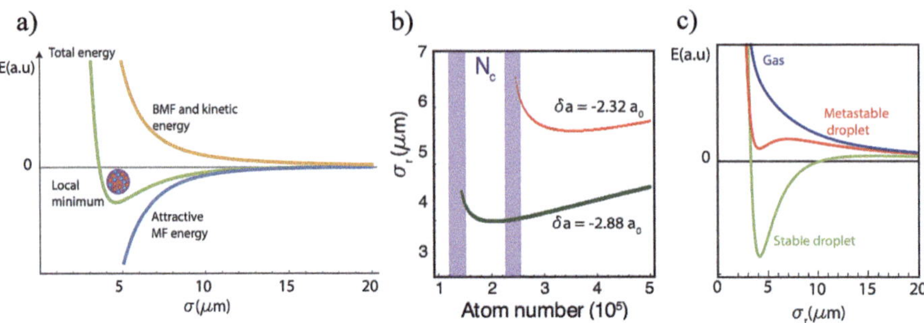

Fig. 12.1 **a** The total energy of the system (12.33) as a function of the variational parameter σ for $N = 5 \times 10^5$, $\delta a = -2.88 a_0$. The local minimum corresponds to the Gaussian-type solution of the GP equation; **b** The variational parameter vs atom number N. It is seen that there is a critical $N = N_c(\delta g)$ showing the boundary of physical solutions; **c** Droplet minima for three different values of N. Reprinted with permission from [13]. © 2018, C. Cabrera. All rights reserved

(or a supergaussin [14]) form, satisfying the condition (12.27), where σ is a variational parameter, which can be interpreted as the width of the droplet. It can be evaluated from the last two equations or by direct minimization of the energy functional

$$\mathcal{E} = \frac{|\nabla\varphi|^2}{2m} - \frac{|\delta g|\varphi^4}{4} + \frac{8g^{5/2}m^{3/2}\varphi^5}{15\pi^2} \tag{12.33}$$

with respect to σ, (see Fig. 12.1a).

Task 12.2: Inserting (12.32) into (12.31) and introducing the spherical system of coordinates derive an explicit equation for σ.

Numerical results for ^{39}K Bose-Bose mixture is presented in Fig. 12.1b for two values of δa: $\delta a = -2.32 a_0$ and $\delta a = -2.88 a_0$ (a_0 is the Bohr radius).

As we have noted in Sect. 12.3 there is a critical value of atoms, N_c, which serves as a boundary of droplet and gas phases (shadowed area in Fig. 12.1b). Numerically N_c can be determined as a point where $\sigma(N)$ diverges. As it is illustrated in Fig. 12.1c for large atom number $N > N_c$, the droplet is stable (green line). Close to the critical value $N \sim N_c$, the droplet becomes metastable (red line). Finally, for low atom number ($N < N_c$), the droplet dissociates into a gas.

Now suppose that, for a given $N > N_c$, an appropriate σ has been found. Then, in accordance with the general rules of quantum mechanics, the radius of the isotropic droplet in the ground state can be calculated from its wave function (12.32) as:

$$R^2 = \int d\mathbf{r} r^2 \varphi^2(r) = \frac{3N\sigma^2}{2}. \tag{12.34}$$

That is $\langle R(\text{droplet})\rangle \sim \sigma$, which displays the physical meaning of σ.

At the end of this section, we briefly outline the excitation spectrum, defined by the eigenvalues ϵ of our nonlinear Schrödinger equation. In general, the spectrum includes very few discrete and continuum modes. It was shown that [1] in the large droplet limit (large R), its spectrum also contains surface modes-ripplons, similar to the modes observed in ^4He. Moreover, the excitation spectrum of a droplet strongly depends on its angular momentum L, which may correspond to different physical effects. For example, the mode with $L = 1$ describes the physical evaluation of the center of mass of the droplet. In the continuous mode, the droplet exhibits self-evaporation by expelling its atoms.

12.6   The Problem of Imaginary Sound Velocity

The essence of this problem is as follows. In Bogolyubov approximation, (12.10) the sound velocities (11.36) has the form:

$$s_\pm = \sqrt{s_\pm^2} \ , s_\pm^2 \equiv s_{1,2}^2 = \frac{g_{aa}\rho_a + g_{bb}\rho_b}{2m} \pm \frac{\sqrt{D_s}}{2m^2}, \tag{12.35}$$

where

$$D_s = 4g_{ab}^2\rho_a\rho_b + (g_{aa}\rho_a - g_{bb}\rho_b)^2. \tag{12.36}$$

Let's consider the symmetric case, introducing following notations:

$$\rho_a = \rho_b = \rho/2, \quad g_{aa} = g_{bb} = g, \quad \rho = \rho_a + \rho_b,$$
$$\delta_g = g + g_{ab}, \quad \alpha = \frac{\delta_g}{g}. \tag{12.37}$$

Then the Eq. (12.35) are simplified as:

$$s_+^2 = \frac{\rho\delta_g}{2m}, \quad s_-^2 = \frac{\rho(2g - \delta_g)}{2m}. \tag{12.38}$$

It is seen that, when the attraction between the two different atoms is dominant, i.e. $\delta_g < 0$ (with $g_{ab} < 0$, $g > 0$), the density sound velocity, s_+, becomes imaginary. This puzzle has caused a lot of discussions in the literature, leading to several attempts, where authors proposed its solution. Below, we outline some of them.

12.6.1 Ota and Astrakharchik

The authors of Ref. [9] proposed to redefine the sound velocities by using the relation between the latter and compressibilities:

$$s_\pm = (m\rho\kappa_\pm)^{-1/2}$$
$$\kappa_\pm = \frac{1}{\rho^2}\left(\frac{\partial^2 E/V}{\partial(\rho_a \pm \rho_b)^2}\right)^{-1} \tag{12.39}$$

where the energy E is given by Eq. (12.12). Then redefined density sound velocity is given by $s_d = s_+ = \sqrt{s_d^2}$ where

$$s_d^2 = \frac{\rho}{2m}\left[\delta_g + 4g\sqrt{\frac{2\gamma}{\pi}}(1 - \frac{g_{ab}}{g})^{5/2}\right] \tag{12.40}$$

Here, since the second term is positive, s_d becomes real above a certain density $\rho\gamma/a_s^3$ for $\delta_g < 0$.

In our opinion, improving the situation in such a way seems not to be selfconsistent, since inclusion of a possible next order term into the energy functional would lead to another redefined expression for the sound velocity.

12.6.2 Bosonic Pairing Theory

Hubbard-Stratanovich transformation, outlined in Chap. 6 has been firstly applied [10] to two-component Bose mixture by Hu and Lin in order to solve the problem of imaginary sound velocity.

In fact, by using the identity

$$\exp\left(-g_{ab}\int dx(\psi^\dagger\psi)(\phi^\dagger\phi)\right) = \int \mathcal{D}\bar{\Delta}\mathcal{D}\bar{\Delta}^\dagger \exp\left(\int dx\left[\frac{\bar{\Delta}\bar{\Delta}^\dagger}{g_{ab}} + \bar{\Delta}^\dagger\phi\psi + \bar{\Delta}\psi^\dagger\phi^\dagger\right]\right) \tag{12.41}$$

in the action (11.4), one can get rid of the fourth-order term, such that the partition function is simplified as

$$Z = \int \mathcal{D}\psi^\dagger\mathcal{D}\psi\mathcal{D}\phi^\dagger\mathcal{D}\phi\, e^{-S_a}e^{-S_b}\int \mathcal{D}\bar{\Delta}\mathcal{D}\bar{\Delta}^\dagger e^{-S_\Delta}, \tag{12.42}$$

with

$$S_\Delta = \int dx\left[-\frac{\bar{\Delta}\bar{\Delta}^\dagger}{g_{ab}} - \bar{\Delta}^\dagger\phi\psi - \bar{\Delta}\psi^\dagger\phi^\dagger\right], \tag{12.43}$$

and e.g. S_a corresponds to the action of the component a. Then after the shift

$$\psi = \sqrt{\rho_{0a}} + \tilde{\psi}, \qquad \phi = \sqrt{\rho_{0b}} + \tilde{\phi}, \qquad \bar{\Delta} = \bar{\Delta}_0 + \tilde{\Delta}, \tag{12.44}$$

the inverse Green function will have the same form as in Eq. (11.15), with

$$\begin{aligned}
X_1 &= -\mu_a + 3g_a\rho_{0a}, & X_2 &= -\mu_a + g_a\rho_{0a}, \\
X_3 &= -\mu_a + 3g_b\rho_{0b}, & X_4 &= -\mu_a + g_b\rho_{0b}, \\
X_5 &= X_6 = -\bar{\Delta}_0.
\end{aligned} \tag{12.45}$$

Note that here and below, the authors used the saddle point approximation, i.e. neglected the fluctuating field $\tilde{\Delta}$, declaring $\bar{\Delta}_0 > 0$ as a variational parameter. Then for e.g. the symmetric case with equal intraspaces interactions, the spectrum of quasiparticles has the form

$$E_-(k) = \sqrt{\varepsilon_k(\varepsilon_k + 2\mu + 2\bar{\Delta}_0)}, \qquad E_+(k) = \sqrt{(\varepsilon_k + 2\bar{\Delta}_0)(\varepsilon_k + 2\mu + 2\bar{\Delta}_0)}. \tag{12.46}$$

It is seen that, in pairing theory, the lower branch $E_-(k)$ is gapless, while the upper branch shows an energy gap: $E_{gap} = E_+(k = 0) = 2\bar{\Delta}_0\sqrt{1 + \mu/\bar{\Delta}_0}$. Hence, the unstable branch in Petrov's theory is automatically removed with the introduction of bosonic pairing via HST.

Here we have to note that the theory has at least two serious shortcomings:
(a) HP relations have not been taken into account;
(b) The chemical potentials were fixed in a rather suspicious way.

12.6.3 Application of Beliaev Theory

Famous Beliaev theory [15], taking into account interaction between quasiparticles, has been applied to cure the present problem of dynamically unstable phonon modes by Gu and Yin recently [16]. It was shown that corrections to the phonon spectrum coming from the interaction between phonon and spin excitations can stabilize the phonon mode. As a result, in contrary to the prediction of the pairing theory, this method predicts both branches of the spectrum as gapless ones.

Unfortunately, the authors could solve the problem only for the equilibrium droplet regime under the zero-pressure condition $P = -\Omega/V = 0$. However, the fate of the problem in the regime of a dimerized gas with $P \neq 0$ remains unclear.

12.6.4 Optimized Perturbation Theory

Can the present theory, described in previous chapters, solve the problem consistently? Below, we show that, in our OPT, such a problem does not appear, at least for moderated values of density and interspecies interactions. We shall illustrate this for the balanced symmetric Bose mixtures, which can be experimentally realized by using two hyperfine states of ^{39}K [17].

To start, coming back to the Chap. 11, we rewrite the Eq. (11.51) as follows:

$$s_+^2 \equiv s_1^2 = \Delta_P/m, \qquad s_-^2 \equiv s_2^2 = \Delta_M/m, \tag{12.47}$$

with $\Delta_{P,M} = (\Delta_a \pm \Delta_{ab})$. For the latters from Eq. (11.30) we have:

$$\Delta_M = (2 - \alpha)g\rho_{0a} + g\sigma_a + \frac{g\rho_{ab}(1 - \alpha)}{2}, \tag{12.48}$$

$$\Delta_P = \alpha g\rho_{0a} + g\sigma_a + \frac{g\rho_{ab}(\alpha - 1)}{2}, \tag{12.49}$$

where

$$\alpha = \frac{\delta_g}{g} = \frac{g_{ab} + g}{g}.$$

(12.50)

It is understood that, the dynamical stability conditions $s_{\pm}^2 \geq 0$, require that $\Delta_P \geq 0$ as well as $\Delta_M \geq 0$.

In Chap. 11 we discussed the case of positive g_{ab}. Now, let $g_{ab} < 0$, or more precisely, $\alpha < 0$. Then, particularly, in the Bilinear approximation, when $\sigma_a = \rho_{ab} = 0$ we have

$$\Delta_P(\text{Bilinear}) = \alpha g \rho_{0a}, \qquad \Delta_M(\text{Bilinear}) = (2 - \alpha) g \rho_{0a}.$$

(12.51)

It is seen that the problem can not be resolved in the Bilinear approximation, since for $\alpha < 0$ we still have

$$s_+^2(\text{Bilinear}) = \Delta_P(\text{Bilinear})/m = \frac{\alpha g \rho_{0a}}{m} < 0.$$

(12.52)

Now let's examine the Eqs. (12.48) and (12.49) in OPT, which takes into account both σ and ρ_{ab}. First, we note that, as it has been proven in Chap. 10, the anomalous density σ at $T = 0$ for the default phase of the condensate ($\theta = 0$ in Eq. (10.3)) is positive i.e. $\sigma(T = 0) > 0$. Secondly, $\rho_{ab}(T = 0)$, given by

$$\rho_{ab} = \frac{m^3 (s_+^3 - s_-^3)}{3\pi^2},$$

(12.53)

(see Eq. (11.50) is negative i.e. $\rho_{ab}(\alpha < 0) < 0$. This is because, at least the real part of s_+ is negligibly small, as it has been pointed out be Petrov [1]. Thus, it is understood that, for $\alpha < 0$ the first two terms in the RHS of Eq. (12.48) are positive, and the third term is negative, but small. Therefore, there would be no problem with s_-^2, since $\Delta_M \geq 0$ in the moderate values of $|\alpha| < 1$.

And what about Δ_P, which determines the sign of s_+^2? Well, in the RHS of Eq. (12.49) one observes that the first term is negative, while the other two are positive. So, the sign of Δ_P, and hence s_+^2, is determined by the interplay of the first and the other two terms in Eq. (12.49). Here we note that, strictly speaking, the densities ρ_{0a}, σ_a and ρ_{ab}, given by Eq. (11.50), depend not only on α, but also on $\gamma = \rho a_s^3$. Thus we come to the following conclusion: In the (α, γ) plane, there would be a region where the first term is comparatively small, i.e.

$$|\alpha g \rho_{0a}| < g \sigma_a + \frac{g \rho_{ab}(\alpha - 1)}{2},$$

(12.54)

so that $s_+^2 \geq 0$. Out of this region, the first term in the RHS of Eq. (12.49) dominates, causing dynamical instability, due to $\Delta_P < 0$. To fix the boundary of such a region more accurately, one has to solve the following equations

$$-2\tilde{s}_1^3 \sqrt{\pi} + \left(6\sqrt{2\gamma n_{0a}}\pi - 3\pi^{3/2}\right)\tilde{s}_1^2 +$$
$$\left(-4\alpha\sqrt{\pi} + 2\sqrt{\pi}\right)\tilde{s}_2^3 + 6\sqrt{2\gamma n_{0a}}\tilde{s}_2^2\pi + 6\alpha\pi^{5/2} = 0,$$

(12.55)

$$- 2\tilde{s}_1^3 \sqrt{\pi} + 6\sqrt{2\gamma n_{0a}}\tilde{s}_1^2 \pi + \left(-6\sqrt{\pi} + 4\alpha\sqrt{\pi}\right)\tilde{s}_2^3 +$$
$$\left(6\sqrt{2\gamma n_{0a}}\pi - 3\pi^{3/2}\right)\tilde{s}_2^2 - 6\pi^{5/2}\gamma\left(-2 + \alpha\right) = 0 \tag{12.56}$$

with respect to dimensionless velocities $\tilde{s}_{1,2} = (s_\pm)ma_s$, $n_{0a} = 1 - n_{1a}$, (n_{1a} is given by Eq. (11.50)) by the conditions $\tilde{s}_{1,2} \geq 0$ and $0 \leq n_{0a} \leq 1$.

The numerical results show [18], e.g. for $\alpha \geq -0.09$ and $10^{-3} < \gamma < 9 \times 10^{-3}$ both sound velocities are real and positive, which supports the stability of the system in this region.

12.7 Summary and Concluding Remarks

- Thus, an interesting new microscopical object-a quantum liquid droplet, similar to Bose-Einstein condensed dilute gas, was first predicted theoretically and then observed experimentally;
- This self-bound cluster of atoms has a surface tension, fixed volume and a very low compressibility;
- The main difference between ordinary liquid droplets and a quantum liquid droplet is in its stabilization mechanism. For example, in a liquid helium droplet, the system is stabilized due to Van der Walls forces, while the quantum droplet to quantum fluctuations. Moreover, unlike the ^4He droplet, it is rather dilute;
- The final conclusion, that the droplet remains really in a liquid phase, is confirmed by the fact that it increases its size when the number of atoms increases. Its density remains practically constant;
- The physical properties of the droplet depend only on the s-wave contact interaction, so that the details of the potential is not seemed to be important, at least for its existence;
- Bose condensed gas, even with repulsive interaction, may exist only inside a confining trap, while a quantum droplet survives even after the removal of the trap. This effect has been observed experimentally in flight images;
- There is a discrepancy in R and N_c between Petrov's theory and experimental observations [4]. The size of the droplet exceeds the theoretical predictions by up to a factor of 3. Furthermore, the critical atom number is a factor of 2 smaller than the theoretical value;
- Although Petrov's theory, being a mismatch of LHY and Bogolyubov approximations, captures the essential features of quantum droplets, it has an intrinsic inconsistency. One of the two gapless Bogolyubov spectra necessarily gets softened and becomes complex. We have shown that these problems find a self-consistent solution within the present OPT, due to taking account of pair correlations in terms of anomalous density σ as well as mixed density ρ_{ab}. In reality, the system, being regardless in droplet or in a gas phase, can be stable when there is a balance between the attractive (g_{ab}) and the repulsive (g_{aa}, g_{bb}) interactions in a two-component mixture. For example, when the attraction dominates, there arise dynamical instabilities leading to collapse. Quantitatively, this tendency is reflected in the specific relations between $\delta_g = \alpha g$ and the densities ($\rho = \gamma/a_s^3$).

Thus we have proven that, beyond Bogolyubov corrections, play an important role in stabilization of the Bose-Bose system with an attractive inter-component interaction both in liquid and gaseous phases, as it was first revealed by D. Petrov [19];

- We haven't touched on the topic of the other types of droplets—dipolar droplets, observed experimentally [20]. Theoretical description of such droplets requires taking into account finite-range interactions, which is beyond the scope of the present lectures.
- Moreover, we didn't consider two dimensional ($2D$) systems, which has also rich physics [21,22].

References

1. D.S. Petrov, Quantum mechanical stabilization of a collapsing Bose-Bose mixture. Phys. Rev. Lett. **115**, 155302 (2015). [Online]. https://doi.org/10.1103/PhysRevLett.115.155302
2. P. Naidon, D.S. Petrov, Mixed bubbles in Bose-Bose mixtures. Phys. Rev. Lett. **126**, 115301 (2021). [Online]. https://doi.org/10.1103/PhysRevLett.126.115301
3. P. Zin, M. Pylak, M. Gajda, Revisiting a stability problem of two-component quantum droplets. Phys. Rev. A **103**, 013312 (2021). [Online]. https://doi.org/10.1103/PhysRevA.103.013312
4. C.R. Cabrera, L. Tanzi, J. Sanz, B. Naylor, P. Thomas, P. Cheiney, L. Tarruell, Quantum liquid droplets in a mixture of Bose-Einstein condensates. Science **359**(6373), 301–304 (2018). [Online]. https://www.science.org/doi/abs/10.1126/science.aao5686
5. A. Cappellaro, T. Macrì, L. Salasnich, Collective modes across the soliton-droplet crossover in binary Bose mixtures. Phys. Rev. A **97**, 053623 (2018). [Online]. https://doi.org/10.1103/PhysRevA.97.053623
6. P. Zin, M. Pylak, Z. Idziaszek, M. Gajda, Self-consistent description of bose–bose droplets: modified gapless hartree–fock–bogoliubov method. New J. Phys. **24**(11), 113038 (2022). [Online]. https://dx.doi.org/10.1088/1367-2630/aca175
7. P. Zin, M. Pylak, M. Gajda, Revisiting a stability problem of two-component quantum droplets. Phys. Rev. A **103**, 013312 (2021). [Online]. https://doi.org/10.1103/PhysRevA.103.013312
8. A. Cappellaro, T. Macrì, G.F. Bertacco, L. Salasnich, Equation of state and self-bound droplet in rabi-coupled bose mixtures. Sci. Rep. **7**(1), 13358 (2017). [Online]. https://doi.org/10.1038/s41598-017-13647-y
9. M. Ota, G.E. Astrakharchik, Beyond Lee-Huang-Yang description of self-bound Bose mixtures. SciPost Phys. **9**, 020 (2020). [Online]. https://scipost.org/10.21468/SciPostPhys.9.2.020
10. H. Hu, X.-J. Liu, Consistent theory of self-bound quantum droplets with bosonic pairing. Phys. Rev. Lett. **125**, 195302 (2020). [Online]. https://doi.org/10.1103/PhysRevLett.125.195302
11. V. Cikojević, K. Dželalija, P. Stipanović, L. Vranješ Markić, J. Boronat, Ultradilute quantum liquid drops. Phys. Rev. B **97**, 140502 (2018). [Online]. https://doi.org/10.1103/PhysRevB.97.140502
12. P. Zin, M. Pylak, M. Gajda, Zero-energy modes of two-component Bose–Bose droplets. New J. Phys. **23**(3), 033022 (2021). [Online]. https://dx.doi.org/10.1088/1367-2630/abe482
13. C.R.C. Cabrera, *PhD Thesis: Quantum liquid droplets in a mixture of Bose-Einstein condensates*. Catalonia, Polytechnic University of Catalonia. Institute of Photonic Sciences (2018)
14. S.R. Otajonov, E.N. Tsoy, F.K. Abdullaev, Modulational instability and quantum droplets in a two-dimensional Bose-Einstein condensate. Phys. Rev. A **106**, 033309 (2022). [Online]. https://doi.org/10.1103/PhysRevA.106.033309

15. S. Beliaev, Application of the methods of quantum field theory to a system of bosons. Sov. Phys. JETP **34**(7), 289–299 (1958). [Online]. http://jetp.ras.ru/cgi-bin/dn/e_007_02_0289.pdf

16. Q. Gu, L. Yin, Phonon stability and sound velocity of quantum droplets in a boson mixture. Phys. Rev. B **102**, 220503 (2020). [Online]. https://doi.org/10.1103/PhysRevB.102.220503

17. G. Semeghini, G. Ferioli, L. Masi, C. Mazzinghi, L. Wolswijk, F. Minardi, M. Modugno, G. Modugno, M. Inguscio, M. Fattori, Self-bound quantum droplets of atomic mixtures in free space. Phys. Rev. Lett. **120**, 235301 (2018). [Online]. https://doi.org/10.1103/PhysRevLett.120.235301

18. A. Rakhimov, S. Tukhtasinova, V.I. Yukalov, Stability of binary Bose mixtures with attractive intercomponent interactions (2025). [Online]. https://arxiv.org/abs/2505.21108

19. D.S. Petrov, *Beyond Mean-Field Effects in Mixtures: Few-Body and Many-Body Aspects*, ed. by M. Inguscio, G. Modugno, Y. Takahashi. IOS Press, Amsterdam, The Netherlands (2025). [Online]. https://ebooks.iospress.nl/doi/10.3254/ENFI250020

20. I. Ferrier-Barbut, H. Kadau, M. Schmitt, M. Wenzel, T. Pfau, Observation of quantum droplets in a strongly dipolar Bose gas. Phys. Rev. Lett. **116**, 215301 (2016). [Online]. https://doi.org/10.1103/PhysRevLett.116.215301

21. A. Tononi, G.E. Astrakharchik, D.S. Petrov, Gas-to-soliton transition of attractive bosons on a spherical surface. AVS Quantum Sci. **6**(2), 023201 (2024). [Online]. https://doi.org/10.1116/5.0190767

22. G. Spada, S. Pilati, S. Giorgini, Quantum droplets in two-dimensional Bose mixtures at finite temperature. Phys. Rev. Lett. **133**, 083401 (2024). [Online]. https://doi.org/10.1103/PhysRevLett.133.083401

13.1 Introduction

We know that, a magnet is an object that produces a magnetic field. This may be an electromagnet or a permanent magnet. The former is made from a coil of wire that acts as a magnet when an electric current passes through it. The magnetic effect stops when one switches off the current. The latter does not need any external source of electric current, and once being magnetized, it creates its own persistent magnetic field. Here, it is interesting to note that an electromagnet can be described in the framework of classical Maxwell equations, while for permanent magnets, one has to resort to quantum mechanics. The point is that magnetization of a permanent magnet is directly caused by the magnetic moments of, say, electrons with spin $s = \hbar/2$, such that the total magnetization is proportional to the vector sum of spins. For example, ^{26}Fe has four unpaired electrons in its $n = 3$ shell. In its turn, spin of a particle is a quantum mechanical quantity by itself, since in classical mechanics Planck's constant is supposed to vanish: $\hbar(Classical) = 0$. So, there is no spin in classical mechanics, and hence, any permanent magnet should be considered as a quantum magnet, since the magnetism related to the orbital moments of electrons is negligibly small.

To describe quantum magnets, one may use the famous Heisenberg Hamiltonian:

$$H_{12} = -\mathcal{J}_{12}(\mathbf{S}_1 \cdot \mathbf{S}_2), \tag{13.1}$$

which is actually an exchange interaction between two spins $\mathbf{S_1}$ and $\mathbf{S_2}$. Here the sign of $\mathcal{J}_{12}$ is utmost important: $\mathcal{J}_{12} > 0$ corresponds to ferromagnetic ordering and $\mathcal{J}_{12} < 0$ to antiferromagnetic one. In terms of spins, this looks as follows: After exposure to a magnetic field, all spins (elementary magnetic moments) in ferromagnetic material are pointed in the same direction (see Fig. 13.1a) while in antiferromagnetic

© The Author(s), under exclusive license to Springer Nature Switzerland AG 2026   193
A. Rakhimov and S. Mardonov, *Theory of Quantum Bose Liquids Beyond
Bogoliubov Approximation*, Lecture Notes in Physics 1045,
https://doi.org/10.1007/978-3-032-05096-0_13

Fig. 13.1 The ground state of ferromagnet (**a**) and antiferromagnet at $T = 0$

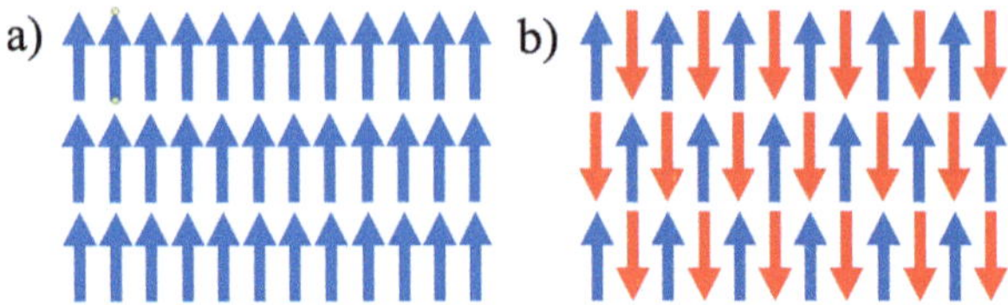

material they are directed oppositely, being equal in strength (Fig. 13.1b). This picture may be realized only in the SSB phase, which starts at a critical temperature $T = T_c^N$. The latter is referred to as the Curie temperature for a ferromagnetic (or the Neel temperature for antiferromagnetic).

In the present lectures, we shall consider only antiferromagnetic materials in the SSB phase. Actually, the situation, illustrated in Fig. 13.1b, corresponds to the ideal case, which takes place at zero temperature when the total magnetization $\mathbf{M}(T = 0) = \sum_i \mathbf{S}_i = 0$. However, at finite temperatures $\mathbf{M}$ may be different from zero, i.e. $\mathbf{M}(T \neq 0, T < T_c^N) \neq 0$. Parallel and anti-parallel spins to the external magnetic field, say to H_z^{ext}, would not compensate each other due to temperature fluctuations and field-induced order. As a result, in reality, an antiferromagnetic behaves like a weak paramagnetic with rather small magnetization. Anyway, it is expected that the lower the temperature, the smaller the fluctuations and therefore the weaker the magnetization.

In the early two thousandths, experimenters from the Tokyo Institute of Technology studied the magnetization of the antiferromagnetic compound $TlCuCl_3$ at different temperatures and magnetic fields. The results were unexpected. It has been found that this sample undergoes three-dimensional ordering in a magnetic field. As it is seen from Fig. 13.2, with lowering the temperature, the magnetization first decreases (as expected) and then suddenly starts to increase at a critical temperature $T = T_c$, which is unusual for an antiferromagnetic material. Moreover, they have observed the existence of a critical magnetic field $H^{ext} = H_c^{ext}$, such that this effect vanishes for $H^{ext} < H_c^{ext}$.

The question arises: Where does this critical field come from? The answer is hidden in the structure of such compounds.

13.2   Dimerized Quantum Magnets (DQM)

In reality, copper salt materials, such as $ACuCl_3$ ($A = Tl, K, NH_4$) have similar dimer structures as it is illustrated, e.g. for $TlCuCl_3$ in Fig. 13.3.

By definition, a dimer is a chemical structure formed from two similar sub-units. In the present case, a pair of Cu^{2+} ions, each with spin $S = 1/2$, coupled antiferromagnetically, form a dimer. The magnetic energy of the whole crystal is very sensitive to the spin states of dimers. The ground state of each dimer corresponds to its singlet state, with $S = S_z = 0$, while the first excited state belongs to the triple with parallel spins ($S = 1; S_z = -1, 0, +1$). Thus, there is a gap $\Delta_g = E_1 - E_0$,

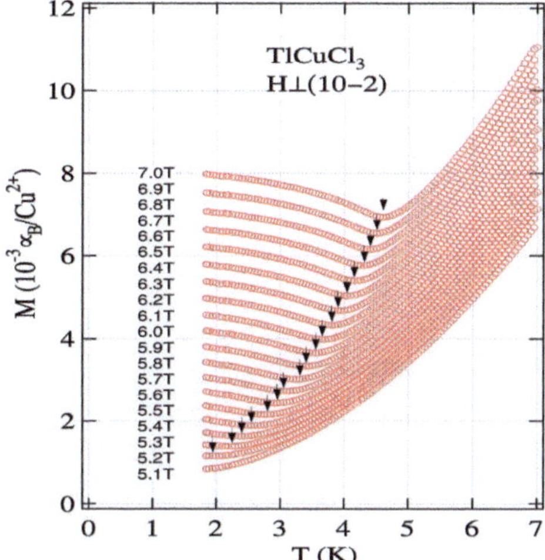

Fig. 13.2 The temperature dependence of magnetizations of $TlCuCl_3$ measured at various magnetic fields $H||b$ (**a**) and $H \perp (1, 0, 2)$ (**b**). Reprinted under CC-BY-4.0 license from [1]. © 2008, F. Yamada et al.

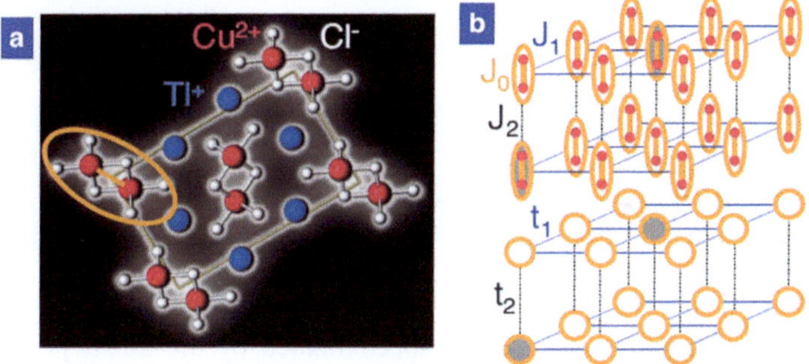

Fig. 13.3 Dimerized quantum copper salt materials. **a** Dimers in $TlCuCl_3$ with $S = 1/2$ from Cu^{2+} ions; **b** Intradimer ($\mathcal{J}_0$) and interdimer ($\mathcal{J}_i$) interactions on a square lattice. Reprinted with permission from [2]. © 2008, Springer Nature Limited. All rights reserved

$e.g.$ $\Delta_g \sim 7.1K$ for $TlCuCl_3$ between the ground state $E = E_0$ and the excited $E = E_1$ states.

Consider now what happens when an external magnetic field H_{ext}^z is imposed as a perturbation. Then a triplet, excited in one dimer, can hop to a neighboring one. A quasiparticle, a magnon, responsible for this process, is referred to in the literature [2] as a triplon.[1] As to the energy spectrum, the energy of the singlet ground state remains unchanged, but the energy of one of the triplet states, say $S = S_z = 1$ will be lowered linearly with the field due to the Zeeman effect. Now the gap becomes

---

[1] We shall discuss the difference between magnons and triplons later.

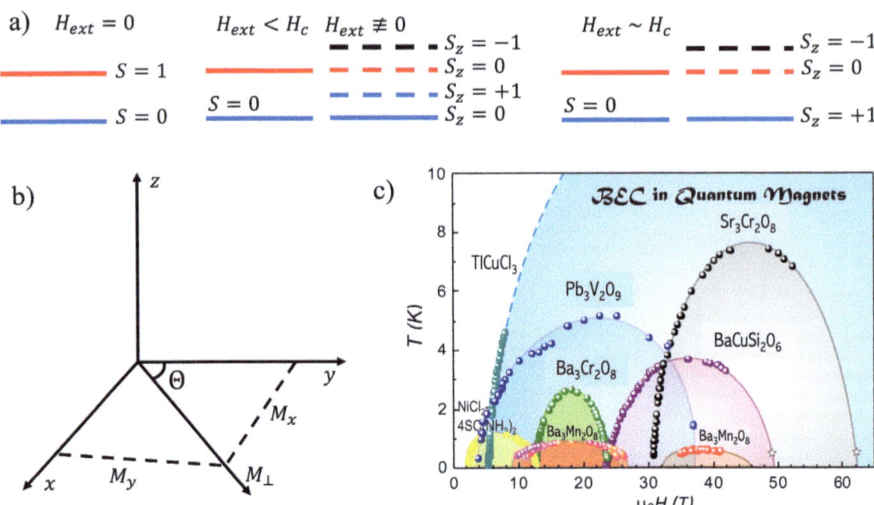

Fig. 13.4 **a** Evolution of the states with the increase in the magnetic field, $H_{ext} = H_c$, the gap is closed and a magnon becomes a triplon; **b** Vector of magnetization; **c** (T, H) phase diagram for several quantum magnets studied in the context of BEC

$$\Delta'_g = E'_1 - E_0 = E_{S_z=+1} - E_{S=0} - g\mu_B H_{ext}, \qquad (13.2)$$

where g is the electron Landé factor $g \approx 2.06$, μ_B is the Bohr magneton $\mu_B = 0.6717 K/T$ and H_{ext} is given in units of Tesla. From Eq. (13.2) it is clear that by choosing H_{ext} one may close the gap. This point defines the critical magnetic field: $\Delta'_g(H = H_c) = 0$.

Therefore, below H_c the ground state is a paramagnet formed by the singlet sea (triplon vacuum) and can be approximated by the direct product of singlet states on each dimer, $|\psi\rangle_i = |S, S_z\rangle$ with $S = 0$, $S_z = 0$. But as soon as the spin gap is closed, $(H \sim H_c)$, the state of a dimer at position $\mathbf{r}$ is well approximated by a coherent superposition of singlet and the $S_z = +1$ triplet states $|\psi\rangle_i = \alpha_i(H)|0, 0\rangle_i + \beta_i(H)|1, +1\rangle_i$ where the amplitudes α_i and β_i depend on the magnetic field H. Further increasing the external field lowers the energy of the triplet $S_z = +1$ states more and more, leading to a finite population of triplons. As it is seen from Fig. 13.2 for $TlCuCl_3$, this happens near $H = H_c \approx 6T$. Since triplons are bosons, there would be a good possibility for them to occupy the same state macroscopically (see illustration in Fig. 13.4a).

13.3   Triplon Condensation and the Chemical Potential

Coming back to the Fig. 13.2b, one may note that the points marked here with arrows correspond to the critical temperatures T_c, where $\partial M/\partial T$ changes its sign. In its turn, the localization of each T_c depends on H_c. Analysis of this dependence has shown the following rather simple behavior $T_c \propto (H - H_c)^{2/3}$, which reminds us of the relation (3.10), describing the critical temperature of Bose condensation of non-interacting particles with the density ρ. Since it is intuitively understood that the triplon density is proportional to the magnetic field, we have received an additional justification in favor of triplon Bose condensation. On the other hand, from Sect. 4.3 we know that this phenomenon is always accompanied by a violation of a certain symmetry spontaneously. So, what kind of SSB does occur here? Actually, in spin language, the condensate corresponds to magnetic order, formed by the transverse components $\langle S_i^x \rangle$ and $\langle S_i^y \rangle$, which spontaneously breaks the rotational $O(2)$ symmetry of the spin Hamiltonian

$$\mathcal{H} = \mathcal{J}_0 \sum_i \mathbf{S}_{1,i}\mathbf{S}_{2,i} + \sum_{mnij} \mathcal{J}_{mnij}\mathbf{S}_{m,i}\mathbf{S}_{n,i} - g\mu_B H \sum_{mi} S^z{}_{m,i}, \tag{13.3}$$

where $\mathcal{J}_0 > 0$, i, j number of dimers, $m, n = 1, 2$ are their magnetic sites, and the external magnetic field is directed along the z axis.

In contrast to BEC of atomic gases, the measurement of the number of particles N and the condensed fraction N_0 for BEC of triplons is rather straightforward. Since magnetization per dimer is proportional to the sum of spin projections ($M \propto \sum S_i$), it is easy to understand that the number of triplons can be evaluated as $N = M/g\mu_B$. To find a relation between the magnetization and the condensed fraction, we make an analogy between the traditional BEC manifest and the order parameter of triplon BEC, given in terms of spin averages as $\langle S_i^x + i S_i^y \rangle$. Now let the Bose field operator of triplons is presented as

$$\psi(\mathbf{r}, t) = e^{i\Theta}\sqrt{\rho_0} + \tilde{\psi}(\mathbf{r}, t), \tag{13.4}$$

where Θ is considered as the phase angle. Then it can be shown that [3] the phase corresponding to the angle of the spin in the XY plane is equal to the phase of the condensate wave function (see Fig. 13.4b).

In other words, the direction of staggered magnetization $\mathbf{M}_\perp$, which emerged due to the triplon condensation, can be presented as

$$M_\perp = |M_x|sin\Theta + |M_y|cos\Theta. \tag{13.5}$$

The magnitude of $M_\perp$ can be evaluated as [4]

$$|M_\perp| = \frac{g\mu_B\sqrt{\rho_0}}{\sqrt{2}}. \tag{13.6}$$

Therefore, by measuring the total magnetization and its staggered component, one may extract $N = V\rho$ and $N_0 = V\rho_0$, per dimer, respectively.

It is understood that in the equilibrium, N should be constant, since the number conservation is an essential condition for the emergence of a stable BEC. In reality,

Table 13.1 Correspondence between a Bose gas and a quantum antiferromagnet

Bose gas	Antiferromagnet
Particles	Spin excitations ($S = 1$ quasiparticles)
Boson number N	Spin component S_z
Charge conservation $U(1)$	Rotational invariance $O(2)$
Condensate wavefunction $\langle \psi(\mathbf{r}) \rangle$	Transverse magnetic order $\langle S_i^x + i S_i^y \rangle$
Chemical potential μ	Magnetic field H
Superfluid density ρ_s	Transverse spin stiffness
Mott insulating state	Magnetization plateau

this is achieved by continuously pumping of triplons into the system by application of a constant external magnetic field. Therefore the quantity

$$\mu(H) = g\mu_B(H - H_c) \tag{13.7}$$

can be interpreted as the chemical potential of $S_z = +1$ triplons. Thus, it is seen that, in contrast to a system of atoms, a triplon gas should be regarded as a $\mu V T$ statistical ensemble, where the chemical potential is fixed for the given H, while the number of particles, N, has to be evaluated by using thermodynamic relations. In Table 13.1 we present correspondence between an atomic Bose gas and a quantum antiferromagnetic [2].

13.3.1 Magnons Versus Triplons

What is the difference between a magnon and a triplon? Well, by definition, a magnon is a quasiparticle corresponding to the excitation in a system of interacting spins. Triplons are auxiliary quasiparticles[2] defined in the bond operator representation [6]. Their main difference is that magnons are actually excitations out of the ground state of the spins, while triplons exist mainly in the equilibrium ground state. Strictly speaking, triplons are conserved particles, while magnons are not, and hence, a stable magnon condensate can not exist. In other words, in BECs of magnons, the BEC is metastable, since magnons are inherently excitations, whereas in quantum magnets, the BEC of triplons occurs in the thermodynamic ground state of the system (see Refs. [6,7] for further discussions). This issue is confused by the fact that the authors of quantum magnet BEC papers sometimes use the terminology "BEC of magnons".

---

[2] Auxiliary quasiparticles arise via an operator transformation of fields, being fictitious particles, related to no real physical object. They may occur, e.g. in Hubbard-Stratonovich transformation, outlined in Sect. 6.5, in bond operator formalism [5] etc.

13.4   Effective Hamiltonian in Bond Operator Formalism

The connection between magnetic systems, gases of itinerant quasiparticles may be clearly established by using one of the mathematical transformations, such as Holstein-Primakoff, Dyson-Maleev, or Matsubara-Matsumoto ones [8]. The choice of a particular method strongly depends on the symmetries of magnetic systems. For the quantum dimerized magnets described by the Hamiltonian (13.3), it is more convenient to use the transformation proposed by Sachdev and Bhatt [5] many years ago in the framework of the bond operator formalism. This leads to the following simple, effective Boson Hamiltonian[3] :

$$\mathcal{H}_{eff} = \int d\mathbf{r}\left[\psi^{\dagger}(\mathbf{r})(\mathcal{K} - \mu)\psi(\mathbf{r}) + \frac{U}{2}(\psi^{\dagger}(\mathbf{r})\psi(\mathbf{r}))^2 \right], \qquad (13.8)$$

where $\psi(\mathbf{r})$ is the Bosonic field operator, U is the strength of contact triplon-triplon interaction, and $\mathcal{K}$ is the kinetic energy operator, which defines the bare dispersion of triplons $\varepsilon(\mathbf{k})$ in momentum space. Since the triplon BEC occurs in solids, we may perform momentum integration over the unit cell of the crystal with the corresponding momenta defined in the first Brillouin zone. Here we set $V = 1$ for the volume of the unit cell.

The general explicit expression for $\varepsilon(\mathbf{k})$ has been obtained in Ref. [3] and looks very complicated. In practical calculations, especially for isotropic quantum magnets, one may use its simplified version such as $\varepsilon(\mathbf{k}) = \mathbf{k}^2/2m$, $\varepsilon(\mathbf{k}) = \sqrt{\Delta_{st}^2 + \mathcal{J}_0^2 \mathbf{k}^2} - \Delta_{st}$ [9, 10] or $\varepsilon_{isotropic}(\mathbf{k}) = J_0[3 - \cos k_x a - \cos k_y a - \cos k_z a]$. Thus summation by moments will be presented as

$$\sum_{\mathbf{k}} f(\varepsilon(k)) = \frac{1}{(2\pi)^3}\begin{cases} 4\pi \int\limits_0^{\infty} k^2 dk f(\varepsilon(k)), & \varepsilon(k) = \frac{k^2}{2m}, \\[2mm] \int\limits_{-\pi}^{\pi} dk_x dk_y dk_z f(\varepsilon(k)), & \varepsilon(k) - \text{isotropic}, \\[2mm] \frac{1}{2}\int\limits_{-\pi}^{\pi} dk_x \int\limits_{-\pi}^{\pi} dk_y \int\limits_{-2\pi}^{2\pi} dk_z \varepsilon(k), & \varepsilon(k) - \begin{cases} \text{realistic}, \\ \text{anisotropic.} \end{cases} \end{cases} \qquad (13.9)$$

where the size of the unit cell a, is set here $a = 1$. Other parameters will be clarified in the next chapter, where we shall study the thermodynamic properties of dimerized quantum magnets by applying OPT to the triplon gas, described by the Hamiltonian (13.8).

13.5   Summary and Concluding Remarks

- BEC can also occur in quantum magnets. Here, in contrast to atomic gases, quasiparticles (magnons, triplons) may undergo condensation. The anomalous

---

[3] The derivation of the Hamiltonian may be found in Appendix E.

behavior of the magnetisation curve of dimerized antiferromagnets may be well explained by the phenomenon of triplon condensation at low temperatures. This phenomenon has its own rather large critical temperature, $T_c \propto 5K$, as well as a critical magnetic field.

- In the present chapter, we have discussed the properties of dimerized quantum magnets mainly on the example of the magnetic insulator $TlCuCl_3$. However, in reality, their list is rather rich, as it was reviewed by Zapf and Jaime [11].
- We have concentrated only on the specific critical magnetic field H_c, where the gap is closed. There exists another critical $H = H_c(sat)$ also, which is related to the saturation effects (see Fig. 13.4c) [2, 11]. For this reason, our Hamiltonian (13.8) is valid only in the region of relatively small magnetic fields with $H \ll H_c(sat)$.

References

1. F. Yamada, T. Ono, H. Tanaka, G. Misguich, M. Oshikawa, T. Sakakibara, Magnetic-field induced Bose–Einstein condensation of magnons and critical behavior in interacting spin dimer system TlCuCl₃. J. Phys. Soc. Jpn. **77**(1), 013701 (2008). [Online]. https://doi.org/10.1143/JPSJ.77.013701
2. T. Giamarchi, C. Rüegg, O. Tchernyshyov, Bose–Einstein condensation in magnetic insulators. Nat. Phys. **4**(3), 198–204 (2008). [Online]. https://doi.org/10.1038/nphys893
3. M. Matsumoto, B. Normand, T.M. Rice, M. Sigrist, Field- and pressure-induced magnetic quantum phase transitions in TlCuCl₃. Phys. Rev. B **69**, 054423 (2004). [Online]. https://doi.org/10.1103/PhysRevB.69.054423
4. H. Tanaka, A. Oosawa, T. Kato, H. Uekusa, Y. Ohashi, K. Kakurai, A. Hoser, Observation of field-induced transverse Néel ordering in the spin gap system TlCuCl₃. J. Phys. Soc. Jpn. **70**(4), 939–942 (2001). [Online]. https://doi.org/10.1143/JPSJ.70.939
5. S. Sachdev, Quantum phase transitions. Phys. World **12**(4), 33 (1999). [Online]. https://doi.org/10.1088/2058-7058/12/4/23
6. V.I. Yukalov, Difference in Bose-Einstein condensation of conserved and unconserved particles. Laser Phys. **22**(7), 1145–1168 (2012). [Online]. https://doi.org/10.1134/S1054660X12070171
7. J.T. Mäkinen, S. Autti, V.B. Eltsov, Magnon Bose–Einstein condensates: from time crystals and quantum chromodynamics to vortex sensing and cosmology. Appl. Phys. Lett. **124**(10), 100502 (2024). [Online]. https://doi.org/10.1063/5.0189649
8. W. Nolting, A. Ramakanth, *Quantum Theory of Magnetism* (Springer, Berlin, Heidelberg, 2009). [Online]. https://books.google.co.uz/books?id=vrcHC9XoHbsC
9. A. Khudoyberdiev, A. Rakhimov, A. Schilling, Bose–Einstein condensation of triplons with a weakly broken U(1) symmetry. New J. Phys. **19**(11), 113002 (2017). [Online]. https://doi.org/10.1088/1367-2630/aa8a2f
10. E.Y. Sherman, P. Lemmens, B. Busse, A. Oosawa, H. Tanaka, Sound attenuation study on the Bose-Einstein condensation of magnons in TlCuCl₃. Phys. Rev. Lett. **91**, 057201 (2003). [Online]. https://doi.org/10.1103/PhysRevLett.91.057201
11. V. Zapf, M. Jaime, C.D. Batista, Bose-Einstein condensation in quantum magnets. Rev. Mod. Phys. **86**, 563–614 (2014). [Online]. https://doi.org/10.1103/RevModPhys.86.563

Thermodynamics of Dimerized Quantum Magnets

14.1 Introduction

In the present chapter, we shall study low temperature $(T < T_{Neel})$ properties of DQMs in the region of $H < H_c(saturation)$ in the framework of optimized perturbation theory, which has been applied to fluids in previous chapters. On first glance, it seems to be no place for OPT in insulating magnets-solids, since it is developed for quantum fluids. Actually, due to the phenomenon of Bose-Einstein condensation of triplons on the one hand, and that of atomic gases on the other hand, these two classes of quantum objects have some similar properties. For example, the behavior of thermodynamic characteristics, such as the heat capacity near the critical temperature, as well as critical exponents [1] are expected to be qualitatively the same both for triplon and atomic gases. However, it should be noted that magnetic materials have their specific features, such as magnetization, magnetic susceptibility, Grüneisen parameter, etc., which are not common for fluids. Thus, applying OPT to the magnets, we shall concentrate mainly on their magnetic properties. For this purpose, we shall exploit the effective Hamiltonian (13.8), which has formally the same structure as the Hamiltonian of Bose gases with the contact interaction used in Chap. 8.

14.2 OPT for the Triplon Gas

We start with the action of the triplon gas, which is similar to the action of an atomic gas (8.41) given by

$$S[\psi, \psi^{\dagger}] = \int_0^{\beta} d\tau \int d\mathbf{r} \left\{ \psi^{\dagger}(\tau, \mathbf{r}) \left[\partial_{\tau} - \hat{K} \right] \psi(\tau, \mathbf{r}) - \mu_0 \rho_0 \right.$$
$$\left. - \mu \tilde{\psi}^{\dagger}(\tau, \mathbf{r}) \tilde{\psi}(\tau, \mathbf{r}) + \frac{U}{2} [\psi^{\dagger}(\tau, \mathbf{r}) \psi(\tau, \mathbf{r})]^2 \right\}, \tag{14.1}$$

© The Author(s), under exclusive license to Springer Nature Switzerland AG 2026
A. Rakhimov and S. Mardonov, *Theory of Quantum Bose Liquids Beyond Bogoliubov Approximation*, Lecture Notes in Physics 1045,
https://doi.org/10.1007/978-3-032-05096-0_14

where the density of the triplon condensate ρ_0 and the fluctuating field $\tilde{\psi}(\tau, \mathbf{r})$ are originated from the Bogolyubov shift $\psi(\tau, \mathbf{r}) = \sqrt{\rho_0} + \tilde{\psi}(\tau, \mathbf{r})$, with the normalization conditions

$$\rho = \rho_0 + \rho_1, \quad N = \rho V = \int d\mathbf{r} \langle \psi^\dagger(\mathbf{r})\psi(\mathbf{r})\rangle,$$

$$N_1 = \rho_1 V = \int d\mathbf{r} \langle \tilde{\psi}^\dagger(\mathbf{r})\tilde{\psi}(\mathbf{r})\rangle. \tag{14.2}$$

The complex fields $\psi^\dagger(\tau, \mathbf{r})$ and $\psi(\tau, \mathbf{r})$ are periodic in τ with the period $\beta = 1/T$. The integration in the coordinate space is taken within the unit crystal cell. Note that, in contrast to the gas of atoms, here the density of particles ρ, and hence their number, depend on temperature, i.e. $\rho = \rho(T)$, while the chemical potential μ is a given constant:

$$\mu(H) = g\mu_B(H - H_c) \tag{14.3}$$

The operator of kinetic energy $\hat{K}$ in (14.1) gives rise to the bare dispersion of triplons $\varepsilon(\mathbf{k})$ as defined in Sect. 13.4.

The application of OPT, introduced in Chap. 7, leads to the following general thermodynamic potential [2,3][1] :

$$\Omega = -T \ln Z = -T \ln \int \mathcal{D}\tilde{\psi}^\dagger \mathcal{D}\tilde{\psi} e^{-S[\psi, \psi^\dagger]} = \Omega_{\text{class}} + \Omega_2 + \Omega_4,$$

$$\Omega_{\text{class}} = -\mu_0\rho_0 + \frac{U\rho_0^2}{2} + \frac{1}{2}\sum_k (E_k - \varepsilon_k) + T\sum_k \ln(1 - e^{-\beta E_k}),$$

$$\Omega_2 = \frac{1}{2}\left[A_2(3U\rho_0 - X_1 - \mu) + A_1(U\rho_0 - X_2 - \mu)\right], \tag{14.4}$$

$$\Omega_4 = \frac{U}{8}\left[3A_1^2 + 2A_1A_2 + 3A_2^2,\right]$$

where $E_k = \sqrt{(\varepsilon(k) + X_1)(\varepsilon(k) + X_2)}$, and $A_{1,2}$ define the normal and anomalous densities as

$$\rho_1 = \frac{A_1 + A_2}{2} = \sum_k \left[\frac{W_k(\varepsilon(k) + X_1/2 + X_2/2)}{E_k} - \frac{1}{2}\right] \equiv \sum_k \rho_{1k},$$

$$\sigma = \frac{A_2 - A_1}{2} = \frac{X_2 - X_1}{2}\sum_k \frac{W_k}{E_k} \equiv \sum_k \sigma_k, \quad A_i = \sum_k \frac{W_k(\varepsilon(k) + X_i)}{E_k}, \tag{14.5}$$

with

$$W_k = \frac{1}{2} + f_B(E_k), \quad f_B(x) = \frac{1}{e^{\beta x} - 1}. \tag{14.6}$$

---

[1] Below we set $V = 1$.

Note that the total differential of Ω is given by

$$d\Omega = -SdT - PdV - Nd\mu - MdH + \frac{\mathcal{C}}{8\pi m a_s^2}da_s, \qquad (14.7)$$

where $a_s = Um/4\pi$ is the s-wave scattering length and m is the triplon effective mass. By S, P, M and $\mathcal{C}$ we have denoted the entropy, pressure, the magnetization and Tan's contact, respectively. Using Eqs. (14.3) and (14.7) one may find the relation between the longitudinal magnetization (i.e. the component parallel to $H = H_z$) and the triplon density as:

$$M = -\frac{\partial\Omega}{\partial H} = -\left(\frac{\partial\Omega}{\partial\mu}\right)\left(\frac{\partial\mu}{\partial H}\right) = \mu_B g \rho. \qquad (14.8)$$

Similarly, differentiating Ω with respect to temperature yields the entropy

$$S = -\left(\frac{\partial\Omega}{\partial T}\right)_H = -\sum_k \ln[1 - \exp(-\beta E_k)] + \beta \sum_k \frac{E_k}{e^{\beta E_k} - 1}. \qquad (14.9)$$

The magnetic heat capacity in constant magnetic field C_H and Tan's contact can be found as

$$C_H = T\left(\frac{\partial S}{\partial T}\right)_H = \beta^2 \sum_k \frac{E_k(E_k - TE_{k,T}')e^{\beta E_k}}{(e^{\beta E_k} - 1)^2,} \qquad (14.10)$$

where $E_{k,T}' = (\partial E_k/\partial T)_H$ and

$$\mathcal{C} = 2U^2 m^2 \left(\frac{\partial\Omega}{\partial U}\right)_{T,\mu}. \qquad (14.11)$$

The Eqs. (14.3)–(14.11) are valid for both normal and condensed phases. The variational parameters $X_{1,2}$, fixed by the principal of minimal sensitivity,

$$\frac{\partial\Omega(X_1, X_2, \rho_0)}{\partial X_1} = 0, \qquad \frac{\partial\Omega(X_1, X_2, \rho_0)}{\partial X_2} = 0, \qquad (14.12)$$

satisfy the following equations:

$$X_1 = -\mu + U(2\rho_1 + 3\rho_0 + \sigma), \qquad X_2 = -\mu + U(2\rho_1 + \rho_0 - \sigma). \qquad (14.13)$$

As to the Lagrange multiplier, μ_0, it may be fixed by the stability condition $\partial\Omega(X_1, X_2, \rho_0)/\partial\rho_0 = 0$, which leads to the equation

$$\mu_0 = U\rho_0 + 2U\rho_1 + U\sigma. \qquad (14.14)$$

Along with a number of triplons, the parameters $X_{1,2}$ strongly depend on temperature.

14.2.1 Normal Phase $T > T_c$

When the temperature exceeds a critical one, $T > T_c$, the condensate fraction as well as the anomalous density vanish, i.e. $\rho_0(T \geq T_c) = \sigma(T \geq T_c) = 0$, so $\rho_1(T \geq T_c) = \rho$. Hence, the basic Eq. (14.13) have the following trivial solutions:

$$X_1(T \geq T_c) = X_2(T \geq T_c) = 2U\rho - \mu. \tag{14.15}$$

The dispersion will be given consequently as

$$E_k(T \geq T_c) \equiv \omega_k = \varepsilon(\mathbf{k}) - (\mu - 2U\rho) \equiv \varepsilon(\mathbf{k}) - \mu_{eff}, \tag{14.16}$$

where we introduced an effective chemical potential $\mu_{eff} = \mu - 2U\rho$. Differentiating both sides of (14.16) with respect to T, such that

$$\frac{d\omega_k}{dT} = 2U\frac{d\rho}{dT}, \tag{14.17}$$

which does not depend on momentum $\mathbf{k}$, and using (14.10) gives the following expression for the heat capacity

$$C_H(T \geq T_c) = \beta^2 \sum_k \frac{\omega_k e^{\beta\omega_k}(\omega_k - 2UT\rho'_T)}{(e^{\beta\omega_k} - 1)^2}. \tag{14.18}$$

The density of triplons can be evaluated as

$$\rho(T \geq T_c) = \rho_1(T \geq T_c) = \sum_k f_B(\omega_k), \tag{14.19}$$

The critical density ρ_c, i.e. the density of particles at critical temperature T_c, is reached when the effective chemical potential vanishes, $\mu_{eff} = 0$ and hence

$$\rho_c = \rho(T = T_c) = \frac{\mu}{2U}. \tag{14.20}$$

Task 14.1. Using Eqs. (14.16)–(14.19) prove that

$$\rho'_T(T \geq T_c) = \frac{d\rho}{dT} = \frac{\beta S_1}{2S_2 - 1}, \tag{14.21}$$

where

$$S_1 = -\beta \sum_k \omega_k f_B^2(\omega_k)e^{\omega_k\beta}, \quad S_2 = -U\beta \sum_k f_B^2(\omega_k)e^{\omega_k\beta}.$$

The thermodynamic potential in (14.4) is simplified as

$$\Omega(T \geq T_c) = -U\rho^2 + T\sum_k \ln(1 - e^{\beta\omega_k}), \tag{14.22}$$

Now using (14.22) in (14.11) one finds Tan's contact

$$\mathcal{C}(T \geq T_c) = 2U^2 m^2 \rho^2 = \frac{2U^2 m^2 M^2}{(g\mu_B)^2}. \tag{14.23}$$

14.2.2 BEC Phase $T \leq T_c$

In Chap. 8, we have proven that, in terms of variational parameters, Hugenholtz-Pines' theorem is equivalent to the condition

$$X_2(T \leq T_c) = 0. \tag{14.24}$$

So, the Eq. (14.13) reads

$$\mu - U(2\rho_1 + \rho_0 - \sigma) = 0. \tag{14.25}$$

Now, eliminating ρ_0 from Eqs. (14.25) and (14.13) one obtains the basic equation

$$\frac{X_1}{2} \equiv \Delta = \mu + 2U(\sigma - \rho_1), \tag{14.26}$$

where

$$\sigma = -\Delta \sum_k \frac{W_k}{E_k}, \tag{14.27}$$

$$\rho_1 = \sum_k \left[\frac{W_k(\varepsilon(\mathbf{k}) + \Delta)}{E_k} - \frac{1}{2} \right], \tag{14.28}$$

and

$$E_k = \sqrt{\varepsilon(\mathbf{k})}\sqrt{\varepsilon(\mathbf{k}) + 2\Delta}. \tag{14.29}$$

In practice the parameters H, H_c, T, U and the function $\varepsilon(\mathbf{k})$ are assumed to be known for a certain DQM. Then one has to solve the Eqs. (14.26)–(14.29) with respect to Δ and evaluate the densities from following equations

$$\rho_0(T \leq T_c) = \frac{\Delta}{U} - \sigma, \qquad \rho(T \leq T_c) = \rho_0 + \rho_1 = \frac{\Delta + \mu}{2U}, \tag{14.30}$$

which can be easily derived from Eqs. (14.25) and (14.26). The Eq. (14.25) with $\rho_0 = \sigma = 0$ gives the same expression for the critical density $\rho_c = \rho(T = T_c) = \mu/2U$ as in (14.20), which proves the self-consistency of the approach.

Due to the condition (14.24) the thermodynamic potential, given by Eq. (14.4) is simplified as

$$\Omega(T \leq T_c) = \frac{U\rho_0^2}{2} - \mu\rho_0 - \frac{U}{2}(2\rho_1^2 + \sigma^2) + \frac{1}{2}\sum_k (E_k - \varepsilon_k - \Delta) + T\sum_k \ln(1 - e^{-\beta E_k}).$$

(14.31)

Now using Eqs. (14.10), (14.11) and (14.31), one can derive the following explicit expressions for the heat capacity and Tan's contact in the BEC phase:

$$C_H(T \leq T_c) = \beta^2 \sum_k \frac{e^{\beta E_k}(E_k^2 - T\varepsilon_k \Delta_T')}{(e^{\beta E_k} - 1)^2},$$

(14.32)

$$\mathcal{C}(T \leq T_c) = C_0 \left\{ \tilde{\rho} - \frac{\tilde{\rho}_1^2 + \tilde{\rho}^2 - \tilde{\sigma}^2}{4} + 2\rho_c \Delta_U' \left[I_3 - 1 - U\tilde{\sigma}\sigma' + \frac{\tilde{\rho}_0}{2} \right.\right.$$
$$\left.\left. + U\rho_1'(2 - \tilde{\rho}_1 - \tilde{\rho}) \right] \right\},$$

(14.33)

$$I_3 = \rho_c^{-1} \sum_k \left(\frac{\varepsilon_k W_k}{E_k} - \frac{1}{2} \right),$$

where $C_0 = m^2\mu^2, \tilde{\rho} = \rho/\rho_c, \tilde{\rho}_1 = \rho_1/\rho_c, \tilde{\rho}_0 = \rho_0/\rho_c, \tilde{\sigma} = \sigma/\rho_c, \rho_1' = (\partial\rho_1/\partial\Delta)$, $\sigma' = (\partial\sigma/\partial\Delta)$.

Task 14.2 Prove following equations:

$$\Delta_T' = \frac{d\Delta}{dT} = \frac{US_4}{2T(2S_5 + 1)}, \qquad \Delta_\mu' = \frac{d\Delta}{d\mu} = \frac{1}{2S_5 + 1},$$

(14.34)

$$\Delta_U' = (\partial\Delta/\partial U) = (\tilde{\sigma} - \tilde{\rho}_1)/[1 + 2U(\rho_1' - \sigma')].$$

$$\rho_1' = \left(\frac{\partial\rho_1}{\partial\Delta}\right) = \sum_k \frac{\varepsilon_k}{E_k^2}\left[\frac{\Delta W_k}{E_k} + \frac{(\varepsilon_k + \Delta)W_k'}{4}\right],$$

(14.35)

$$\sigma' = \left(\frac{\partial\sigma}{\partial\Delta}\right) = -\sum_k \frac{\varepsilon_k}{E_k^2}\left[\frac{\Delta W_k'}{4} + \frac{(\varepsilon_k + \Delta)W_k}{E_k}\right],$$

where

$$S_4 = \sum_k W_k'(\varepsilon_k + 2\Delta), \qquad S_5 = U\sum_k \frac{4W_k + E_k W_k'}{4E_k},$$

(14.36)

$$W_k' = \beta(1 - 4W_k^2) \qquad W_k = \frac{1}{2} + f_B(E_k).$$

Note that, due to the factoring out of C_0, the quantity in curly brackets in (14.33) is dimensionless.

In practical calculations, the Eq. (14.26) may be rewritten as

$$Z = 1 + \tilde{\sigma}(Z) - \tilde{\rho}_1(Z), \tag{14.37}$$

with $Z = \Delta/\mu$. For the condensed phase, after having solved the Eq. (14.37), the total M and staggered magnetizations $M_\perp$ may be evaluated as follows

$$
\begin{aligned}
M(T \le T_c) &= g\mu_B\rho = g\mu_B\rho_c(Z + 1), \\
M_\perp^2(T \le T_c) &= \frac{1}{2}g^2\mu_B^2\rho_0 = \frac{1}{2}g^2\mu_B^2\rho_c(2Z - \tilde{\sigma}),
\end{aligned}
\tag{14.38}
$$

where we used the Eq. (14.30).

14.2.3 Critical Temperature and the Density

The critical temperature T_c is fixed by the condition of continuity of the variational parameters, e.g. $X_2(T = T_c + 0) = X_2(T = T_c - 0)$. So, from Eqs. (14.15) and (14.24) one obtains the condition

$$\mu_{eff}(T = T_c) = \mu - 2g\rho_c = 0, \tag{14.39}$$

which can be represented also as

$$\rho_c = \frac{\mu}{2U}. \tag{14.40}$$

Henceforth, the critical temperature can be evaluated by the equation:

$$\rho_c = \frac{\mu}{2U} = \int \frac{dk_x dk_y dk_z}{(2\pi)^3} \frac{1}{e^{\varepsilon(\mathbf{k})/T_c} - 1}, \tag{14.41}$$

which follows directly from (14.19). Note that, if we use the parabolic dispersion $\varepsilon(k) = \mathbf{k}^2/2m$ here then we obtain well known relation

$$T_c^{\text{parabolic}} = \frac{2\pi}{m}\left(\frac{\mu}{2U\zeta(3/2)}\right)^{2/3}. \tag{14.42}$$

However, when for real quantum magnets we use a more realistic dispersion (13.9), the integration in (14.41) can not be evaluated analytically, and the Eq. (14.42) should be considered as an approximated one, being valid only for qualitative estimations.

14.3 Critical Behavior of C_H and $\mathcal{C}$

It is well known that properties of some thermodynamic quantities in the critical temperature region $T \to T_c \pm 0$ is crucial for the nature of a phase transition. For example, according to Ehrenfest's classification, a discontinuity in the second derivative of Ω at $T = T_c$ with a continuous first derivative indicates that the transition is of the second order [4]. In Chap. 3 we have shown that the heat capacity of an ideal gas has a jump near T_c. In the present section, we study $C_H^{(\pm)} \equiv C_H(T \to T_c \pm 0)$, and $\mathcal{C}^{(\pm)} \equiv \mathcal{C}(T \to T_c \pm 0)$ for the interacting triplon gas making an attempt to evaluate a possible jump by analyzing each region separately.

14.3.1 A. $T \to T_c + 0$ Region

Here $E_k = \omega_k = \varepsilon_k$ and $\mu = 2U\rho$. From Eqs. (14.18) and (14.34) we have

$$C_H^{(+)} = -S_3 + 2U S_1 \rho'_T, \tag{14.43}$$

where $\beta_c = 1/T_c$

$$S_1 = -\beta_c \sum_k \frac{\varepsilon_k e^{\beta_c \varepsilon_k}}{(e^{\beta_c \varepsilon_k} - 1)^2}, \qquad S_2 = -U\beta_c \sum_k \frac{e^{\beta_c \varepsilon_k}}{(e^{\beta_c \varepsilon_k} - 1)^2},$$

$$S_3 = -\beta_c^2 \sum_k \frac{\varepsilon_k^2 e^{\beta_c \varepsilon_k}}{(e^{\beta_c \varepsilon_k} - 1)^2}, \tag{14.44}$$

and

$$\rho'_T = \frac{\beta_c S_1}{2 S_2 - 1}. \tag{14.45}$$

It may be easily shown that

$$\lim_{T \to T_c+0} \rho'_T = 0, \tag{14.46}$$

since in this limit S_2 in the denominator of Eq. (14.45) has an infrared divergence at small k, while the numerator remains finite. In fact, at small momentum, the integrand in S_2 may be rewritten as

$$\lim_{k \to 0} S_{2k} \approx \lim_{k \to 0} \frac{k^2 e^{\beta \varepsilon_k}}{(e^{\beta \varepsilon_k} - 1)^2} \sim \lim_{k \to 0} \frac{k^2}{\beta^2 (k^2/2m)^2} \sim \lim_{k \to 0} \frac{1}{k^2} = \infty. \tag{14.47}$$

Thus from (14.43) we have

$$C_H^{(+)} = -S_3 = \beta_c^2 \sum_k \frac{\varepsilon_k^2 e^{\beta_c \varepsilon_k}}{(e^{\beta_c \varepsilon_k} - 1)^2} \neq \infty. \tag{14.48}$$

As to $\mathcal{C}^{(+)}$, from Eq. (14.23) it can be directly found that $\mathcal{C}^{(+)} = 2(Um\rho_c)^2$.

14.3.2 B. $T \to T_c - 0$ Region

Here $\rho_0 = 0$, $\sigma = 0$ and $\Delta = 0$ and hence $E_k = E_k = \varepsilon_k$. That is, the dispersion is the same on both sides of the critical temperature. For this reason, e.g. the entropy is continuous at $T = T_c$, $S^{(-)} = S^{(+)}$. The heat capacity is given as

$$C_H^{(-)} = \beta_c^2 \sum_k \frac{\varepsilon_k(\varepsilon_k - T_c\Delta_T')e^{\beta_c\varepsilon_k}}{(e^{\beta_c\varepsilon_k} - 1)^2}, \tag{14.49}$$

where we used the relation $\rho_T' = \Delta_T'/2U$. The Δ_T' defined in (14.34) may be rewritten as

$$\Delta_T' = \frac{\beta_c U S_4}{2(2S_5 + 1)},$$

$$S_4 = -4\beta_c \sum_k \frac{\varepsilon_k e^{\beta_c\varepsilon_k}}{(e^{\beta_c\varepsilon_k} - 1)^2}, \tag{14.50}$$

$$S_5 = -U\beta_c \sum_k \frac{T_c + e^{\beta_c\varepsilon_k}(\varepsilon_k - T_c)}{\varepsilon_k(e^{\beta_c\varepsilon_k} - 1)^2}.$$

Now we are in the position of evaluating discontinuity in C_H. The Eqs. (14.48) and (14.49) lead to

$$[\Delta C_H] = C_H^{(-)} - C_H^{(+)} = -\beta_c \sum_k \frac{\varepsilon_k \Delta_T' e^{\beta_c\varepsilon_k}}{(e^{\beta_c\varepsilon_k} - 1)^2} > 0. \tag{14.51}$$

It is seen that C_H has a finite jump near the critical temperature, and therefore the transition is of second order in accordance with the Ehrenfest classification.

What about Tan's contact? Setting in (14.33) $\tilde{\rho}_1 = \tilde{\rho} = 1$, $\sigma = 0$, $\Delta = 0$ and $E_k = \varepsilon_k$ one obtains

$$\mathcal{C}^{(-)} = \frac{m^2\mu^2}{2}[1 + 4(I_3 - 1)\rho_c\Delta_U'] = 2(Um\rho_c)^2, \tag{14.52}$$

where we used $I_3(T = T_c) = \sum_k f_B(\varepsilon_k) = \rho_c$ and $\mu = 2U\rho_c$. Therefore, Tan's contact is continuous at the critical temperature: $\mathcal{C}(T \to T_c^-) = \mathcal{C}(T \to T_c^+) = 2(Um\rho_c)^2$ However, it can be shown that its derivatives with respect to H and T, i.e. $(\partial\mathcal{C}/\partial H)$ and $(\partial\mathcal{C}/\partial T)$ have a cusp at $T = T_c$ [5]. This is illustrated in Fig. 14.1a where $\mathcal{C}(T)$ is presented for three compounds $Ba_3Cr_2O_8$, $Sr_3Cr_2O_8$ and $TlCuCl_3$ calculated in OPT (solid curves) and Bogolyubov approximation (dashed curves):

$$\mathcal{C}^{Bog}(T) = 2(UmM/g\mu_B)^2 \tag{14.53}$$

In Fig. 14.1b, the magnetizations for the same compounds as a function of temperature are also presented for completeness.

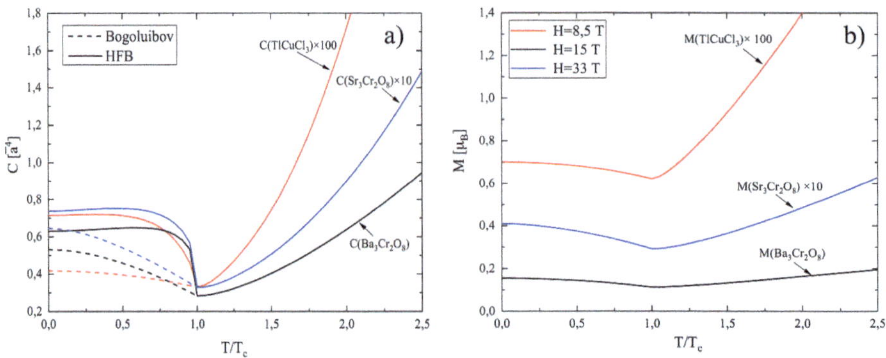

Fig. 14.1 a Contact for spin-gapped magnets in dimensionless units. The solid lines are for the HFB approximation. The dashed lines correspond to the classic approximation given in Eq. (14.53); **b** The magnetizations $M(T)$ per dimer. Note that for a better description of experimental data on $M(T)$ effects of anisotropies [6–9], which have been neglected here, should be included

14.4 Experimental Observation of Tan's Contact for Triplons

In Chap. 5, we have outlined the experimental status of Tan's contact for atomic gases. Unfortunately, the contact for solids, e.g. quantum magnets, where triplon BEC may take place, has not been experimentally studied yet. For this purpose, below we bring a sketch of a possible method of measuring Tan's contact for spin-gapped magnets. We presume that this can be achieved in a similar way to that for atomic gases.

Wild et al. [10] measured $\mathcal{C}$ for ^{85}Rb atoms in a gas with a 60% condensate fraction. The atoms are in the $|F = 2, m_F = -2\rangle$ state, where F is the total atomic spin and m_F is the spin projection. Then, by applying an RF pulse, the atoms are driven from $|2, -2\rangle$ to $|2, -1\rangle$ state. The rate for transferred atoms to the final spin state is given by

$$W = \lim_{\omega \to \infty} \Gamma(\omega) = \mathcal{C}\frac{\Omega_R^2}{4\pi} F(a_s, \omega, m), \qquad (14.54)$$

where Ω_R is the Rabi frequency, $F(a, \omega, m)$ is a given function of scattering length a_s, the frequency ω, and the atomic mass m. Tan's contact has been extracted from this equation after a direct measurement of $\Gamma(\omega)$ as a function of ω.

Now we consider the spin-gapped magnets. Here for $H > H_c$, we have two spin levels with $S_z = 1$ (ground state) and $S_z = -1$ (nearest excited state), as illustrated in Fig. 13.4a. The direct transition between the spin singlet and triplet states is, in principle, forbidden in magnetic dipole transitions. Nonetheless, such transitions have been observed in many spin-gapped systems by means of high-frequency electron spin resonance (ESR) measurements [11–14]. It was shown that these transitions are driven by AC electric fields, and observed electric dipole transitions can be explained by spin-dependent polarization. Very recently, Matsumoto et al. [14] developed a theoretical description of singlet-triplet transitions in spin-gapped magnet KCuCl$_3$ in the framework of spin-wave theory, and proposed practical formulas for the transition probability W, using Fermi's Golden rule (see e.g. Eq. (18) of Ref. [14]). We surmise that, if one uses an effective Hamiltonian similar to (13.8), then the transition probability from Ref. [14] can be related to Tan's contact as in Eq. (14.54), and hence, one will be able to obtain experimental values of C from measurements of W, using proper normalization of observed quantities.

14.5 Summary

In the present chapter, we obtained the grand thermodynamic potential Ω in the framework of OPT, which goes beyond bilinear and Bogolyubov approximations. It can describe properties of dimerized quantum magnets at any temperature below the Neel one, $T < T_{Neel}$. We have proven that the magnetization curves of such magnets display BEC of triplons. The corresponding phase transition is really of a second order in accordance with Ehrenfest classification. We studied analytically Tan's contact and proposed a method for measuring it for spin-gapped dimerized quantum magnets.

In the next two chapters, we shall apply OPT to study specific properties of magnetic materials related to magnetocaloric effects used in the cryogenic technique.

References

1. V. Zapf, M. Jaime, C.D. Batista, Bose-Einstein condensation in quantum magnets. Rev. Mod. Phys. **86**, 563–614 (2014). [Online]. https://link.aps.org/doi/10.1103/RevModPhys.86.563
2. A. Rakhimov, M. Nishonov, B. Tanatar, Joule-Thomson temperature of a triplon system of dimerized quantum magnets. Phys. Lett. A **384**(16), 126313 (2020). [Online]. https://www.sciencedirect.com/science/article/pii/S0375960120301122
3. A. Rakhimov, M. Nishonov, L. Rani, B. Tanatar, Characteristic temperatures of a triplon system of dimerized quantum magnets. Int. J. Modern Phys. B **35**(02), 2150018 (2021). [Online]. https://doi.org/10.1142/S0217979221500181
4. K. Huang, *Introduction to Statistical Physics, Second Edition*, ser. Civil and Environmental Engineering (CRC Press, 2009). [Online]. https://books.google.co.uz/books?id=rKmC3bMEVxIC
5. A. Rakhimov, T. Abdurakhmonov, B. Tanatar, Critical behavior of Tan's contact for bosonic systems with a fixed chemical potential. J. Phys.: Condens. Matter **33**(46), 465401 (2021). [Online]. https://dx.doi.org/10.1088/1361-648X/ac1ec6
6. A. Khudoyberdiev, A. Rakhimov, A. Schilling, Bose–Einstein condensation of triplons with a weakly broken U(1) symmetry. New J. Phys. **19**(11), 113002 (2017). [Online]. https://dx.doi.org/10.1088/1367-2630/aa8a2f
7. J. Sirker, A. Weiße, O.P. Sushkov, Consequences of spin-orbit coupling for the Bose-Einstein condensation of magnons. Europhys. Lett. **68**(2), 275 (2004). [Online]. https://dx.doi.org/10.1209/epl/i2004-10179-4
8. A. Rakhimov, A. Khudoyberdiev, L. Rani, B. Tanatar, Spin-gapped magnets with weak anisotropies I: Constraints on the phase of the condensate wave function. Ann. Phys. **424**, 168361 (2021). [Online]. https://www.sciencedirect.com/science/article/pii/S0003491620302955

9. A. Rakhimov, A. Khudoyberdiev, B. Tanatar, Effects of exchange and weak Dzyaloshinsky–Moriya anisotropies on thermodynamic characteristics of spin-gapped magnets. Int. J. Mod. Phys. B **35**(25), 2150223 (2021). [Online]. https://doi.org/10.1142/S0217979221502234

10. R.J. Wild, P. Makotyn, J.M. Pino, E.A. Cornell, D.S. Jin, Measurements of Tan's contact in an atomic Bose-Einstein condensate. Phys. Rev. Lett. **108**, 145305 (2012). [Online]. https://link.aps.org/doi/10.1103/PhysRevLett.108.145305

11. S. Kimura, K. Kindo, H. Tanaka, ESR study of the spin gap system $KCuCl_3$ using high frequency and high magnetic field. Phys. B: Condens. Matter **346-347**, 15–18 (2004), Proceedings of the 7th International Symposium on Research in High Magnetic Fields. [Online]. https://www.sciencedirect.com/science/article/pii/S0921452604000262

12. S. Kimura, M. Matsumoto, M. Akaki, M. Hagiwara, K. Kindo, H. Tanaka, Electric dipole spin resonance in a quantum spin dimer system driven by magnetoelectric coupling. Phys. Rev. B **97**, 140406 (2018). [Online]. https://link.aps.org/doi/10.1103/PhysRevB.97.140406

13. S. Kimura, M. Matsumoto, H. Tanaka, Electrical switching of the nonreciprocal directional microwave response in a triplon Bose-Einstein condensate. Phys. Rev. Lett. **124**, 217401 (2020). [Online]. https://link.aps.org/doi/10.1103/PhysRevLett.124.217401

14. M. Matsumoto, T. Sakurai, Y. Hirao, H. Ohta, Y. Uwatoko, H. Tanaka, First ESR detection of Higgs amplitude mode and analysis with extended spin-wave theory in dimer system $KCuCl_3$. Appl. Magn. Reson. **52**(4), 523–564 (2021). [Online]. https://doi.org/10.1007/s00723-020-01302-1

Magnetocaloric Effect and Grüneisen Parameter of Quantum Magnets with Spin Gap

15

15.1 Introduction

In Chap. 5 and Appendix B, we discussed methods of cooling dilute Bose gases. As it was pointed out, a gas can be cooled preliminary by the usual vapour compression method, which would be further proceeded by specific methods such as laser and evaporative cooling. However, these methods are not good for solids, particularly for quantum magnets under consideration. In practice, they may be cooled down to $\sim 4K$ by immersion into the liquid helium. Further manipulations are performed by using the magnetocaloric effect.

15.2 Magnetocaloric Effect

Magnetocaloric effect (MCE) is described as the thermal response (heating or cooling) of a magnetic substance when a magnetic field is applied or removed. The MCE is an intrinsic characteristic of magnetic solids, which was originally discovered experimentally in Fe by Warburg in 1881. What is the mechanism of this effect?

Actually, a magnetic substance contains mainly two energy reservoirs, the usual phonon excitations linked to the lattice degrees of freedom and the magnetic excitations linked to the spin degrees of freedom. The spin-lattice coupling often exists in order to ensure the loss-free energy transfer within millisecond time scales. For a usual ferromagnetic (antiferromagnetic) or a paramagnetic substance, an external magnetic field can strongly affect the spin system, but make less influence on the lattice system, that results in the MCE. The MCE can be realized in two fundamental thermodynamic processes, which we illustrate in Fig. 15.1. Here, the initially disordered magnetic moments are aligned by applying a magnetic field isothermally, reducing the disorder of the spin system, thus substantially lowering the magnetic

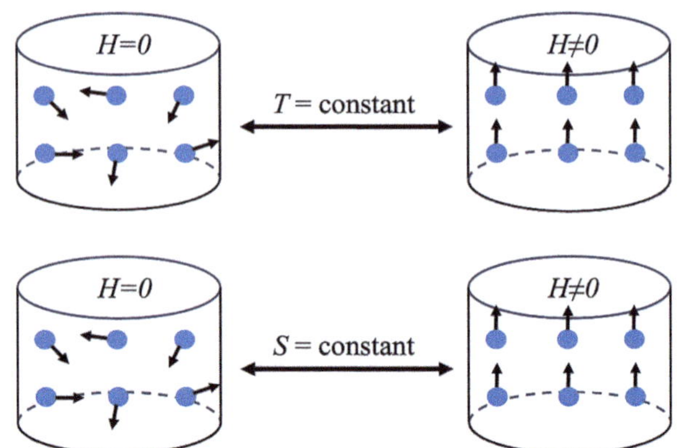

Fig. 15.1 Schematic representation for application of magnetic field H in two basic thermodynamic processes. Total entropy change is non-zero and negative in the isothermal process, and temperature change is non-zero and positive in the adiabatic process

entropy. When the external magnetic field is applied in an adiabatic condition, the entropy of the lattice system must increase in order to compensate for the magnetic entropy reduction, maintaining the total entropy constant. The increase of the lattice entropy represents a rise in the temperature of the material. Therefore, the MCE of a substance is quantified by two parameters, isothermal entropy change, ΔS_T, and adiabatic temperature change, ΔT_S. The reduction of magnetic entropy and the rise of temperature greatly occur near absolute $0\,K$ for a paramagnetic material. While for ferromagnetic (or an antiferromagnetic) material, this occurs at rather higher temperatures, namely, near its spontaneous magnetic ordering temperature (the Curie, or Neel temperature).

Here, a reader who has never thought about the internal mechanism of MCE, may be puzzled by the following paradox: Magnetization rearranges spins in the right order, and the transition *disorder* $\rightarrow$ *order* naturally should reduce the entropy, which is the measure of chaos. Since entropy is qualitatively proportional to the temperature, the sample is expected to cool (and vice versa). So, why in practice do we observe a quite contrary picture, that is, the magnetization process heats up the sample and demagnetization cools it? The explanation is the following. Actually, the total entropy S_{tot} of a magnetic solid sample consists of two parts S_{lat} and S_{spin}, such that $S_{tot} = S_{lat} + S_{spin}$. The former is related to the vibrational degree of freedom of atoms (or ions) near the nodes of the crystal lattice, while the latter is related to the spin degrees of freedom. The S_{lat} is more sensitive to the temperature than S_{spin}, simply put, it is S_{lat} that is mainly responsible for the temperature. Let $S_{tot} = S_{tot}^{(0)}$, $S_{lat} = S_{lat}^{(0)}$, $S_{spin} = S_{spin}^{(0)}$ and $T = T^{(0)}$ in the very beginning. Now we impose an external magnetic field which will reduce $S_{spin}^{(0)}$ down to a $S_{spin} = S_{spin}^{(1)} < S_{spin} = S_{spin}^{(0)}$. What will happen with $S_{lat}^{(0)}$? Well, firstly, since the process is supposed to be adiabatic, S_{tot} remains unchanged: $S_{tot}^{(1)} = S_{tot}^{(0)}$, but S_{lat} will increase to keep the balance in the equation $S_{tot} = S_{lat}^{(1)} + S_{spin}^{(1)}$ i.e. $S_{lat}^{(1)} > S_{lat}^{(0)}$, which results in $T^{(1)} > T^{(0)}$. In this way, adiabatic magnetization heats up the sample! Then we slowly switch off the magnetic field. The temperature fluctuations tend to destroy the spin order increasing S_{spin}, so $S_{spin}^{(2)} > S_{spin}^{(1)}$. This time the system will reduce

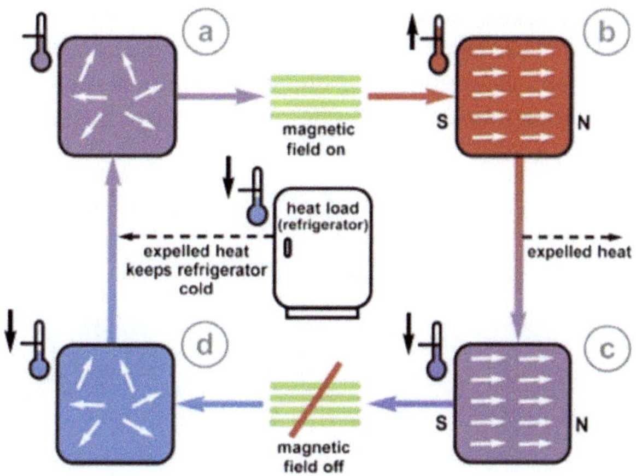

Fig. 15.2 Schematic representation of a magnetic refrigeration cycle which transports heat from a heat load to its surroundings

S_{lat} in order to keep the equivalence $S_{tot}^{(0)} = S_{tot}^{(1)} = S_{tot}^{(2)} = S_{lat}^{(2)} + S_{spin}^{(2)}$, such that $S_{lat}^{(2)} < S_{lat}^{(1)}$. The reduction of S_{lat} will be reflected in lowering the temperature: $T^{(2)} < T^{(1)}$.

You can observe this in home conditions by using a small amount of Gadolinium powders, a toy permanent magnet and a digital thermometer.[1] First, put the powders into a glass ampule and fix its temperature. Then put the ampule on the magnet. You can see the temperature is increasing slowly. After some time, remove the ampule from the magnet and observe the thermometer. You will see that the temperature is going down.

In technical laboratories for magnetic refrigeration based on MCE, a closed process, similar to the Carno cycle is used. A typical one is illustrated in Fig. 15.2. It is seen that this includes the following four stages: Adiabatic magnetization $a \rightarrow b$; Isofield heat extraction $b \rightarrow c$; Adiabatic demagnetization $c \rightarrow d$ and Isofield heat absorption $d \rightarrow a$.

In applied technique, Gadolinium and its alloys are exploited in magnetic refrigeration. There are also some other materials showing a giant magnetic caloric effect due to the existence of a quantum critical point [1]. In present chapter, we show that quantum-dimerized magnets clearly manifest critical points, which can be studied in terms of magnetic Grüneisen parameter Γ_H.

15.2.1 Magnetic Grüneisen Parameter

By definition

$$\Gamma_H = \frac{1}{T}\left(\frac{\partial T}{\partial H}\right)_S.$$

(15.1)

---

[1] They are available on internet shops.

It corresponds to the temperature gradient in the $T(H)$ landscape along an isoentropic (constant entropy S) line and quantifies the cooling or heating of a material when an applied magnetic field is changing while the entropy remains a constant (adiabatic condition). In such a process, the exchange heat is zero:

$$\delta Q = TdS = T\left(\frac{\partial S}{\partial T}\right)_H dT + T\left(\frac{\partial S}{\partial H}\right)_T dH = 0, \tag{15.2}$$

i.e.

$$C_H dT = -T\left(\frac{\partial S}{\partial H}\right)_T dH, \tag{15.3}$$

and hence

$$\Gamma_H = -\frac{1}{C_H}\left(\frac{\partial S}{\partial H}\right)_T, \tag{15.4}$$

where $C_H = T(\partial S/\partial T)_H$ is the specific heat at constant magnetic filed H. Experimentally Γ_H can be directly accessed by measuring the change in temperature at constant entropy upon magnetic field variation using Eq. (15.1). Another equivalent expression for the Grüneisen parameter

$$\Gamma_H = -\frac{1}{C_H}\left(\frac{\partial M}{\partial T}\right)_H, \tag{15.5}$$

can be derived from the grand thermodynamic potential Ω and suitable Maxwell relations (14.7).

Using OPT, developed for quantum dimerized magnets in the previous chapter, one can obtain the following explicit expression for Γ_H

$$\Gamma_H = -\frac{g\mu_B}{C_H}\left(\frac{\partial S}{\partial \mu}\right)_T = \frac{\mu_B g\beta^2}{C_H}\sum_k \frac{E_k E'_{k,\mu} e^{\beta E_k}}{(e^{\beta E_k}-1)^2}, \tag{15.6}$$

by using the Eq. (14.9) for the entropy. In (15.6)

$$E'_{k,\mu} = \frac{\partial E_k}{\partial \mu}. \tag{15.7}$$

Task 15.1: Prove the following equations:

$$E'_{k,\mu}(T \geq T_c) = \frac{\partial \omega_k}{\partial \mu} = 2U\frac{\partial \rho}{\partial \mu} - 1, \qquad \frac{\partial \rho}{\partial \mu} = \frac{S_1}{U(2S_1-1)}, \tag{15.8}$$

and

$$\frac{\partial E_k}{\partial \mu} = \frac{\varepsilon_k}{E_k}\Delta'_\mu, \qquad \Delta'_\mu = \frac{\partial \Delta}{\partial \mu} = \frac{1}{2S_2+1} \tag{15.9}$$

where

$$S_1 = -U\beta \sum_k f_B^2(\omega_k)e^{\omega_k\beta}, \qquad S_2 = U \sum_k \frac{4W_k + E_k W_k'}{4E_k},$$

$$W_k' = \beta(1 - 4W_k^2), \qquad W_k = \frac{1}{2} + f_B(E_k). \tag{15.10}$$

15.3 Critical Properties of Γ_H

We shall study critical properties of the magnetic Grüneisen parameter of dimerized quantum magnets in two aspects: Close to the critical temperature of BEC of triplons and in the region of the quantum critical point (QCP).

15.3.1 Near Critical Temperature T_c

Using Eqs. (15.6)–(15.10) we have

$$\Gamma_H(T \geq T_c) = -\frac{g\mu_B}{C_H}\rho_T', \qquad \Gamma_H(T \leq T_c) = -\frac{g\mu_B\rho_T'}{C_H} = -\frac{g\mu_B\Delta_T'}{2UC_H}, \tag{15.11}$$

where ρ_T' is given in Eqs. (14.45) and (14.50). Now using Eq. (14.46), one may conclude that $\Gamma_H^{(+)} = 0$ and hence, the jump in Γ_H near the critical temperature is nonzero:

$$\Delta\Gamma_H = \Gamma_H^{(-)} - \Gamma_H^{(+)} = -\frac{g\mu_B\Delta_T'}{2UC_H^{(-)}} > 0, \tag{15.12}$$

where Δ_T' and $C_H^{(-)}$ are given by Eqs. (14.50) and (14.49), respectively. From (15.12) one may see that not only C_H but also Grüneisen parameter has a finite jump near the critical temperature.

In Fig. 15.3 we presented Grüneisen parameter vs temperature for two compounds: $Ba_3Cr_2O_8$ and $Sr_3Cr_2O_8$ for various external fields. It is seen that, at the critical temperature, G_H has a jump and changes its sign.

Here it should be noted that the sign change in $\Gamma_H(T)$ has been firstly conjectured by Garst et al. [3] to be universal for systems with magnetically controlled critical points, which is also present at H_c in the triplon systems.

15.3.2 Quantum Critical Point

The behavior of Γ_H at zero temperature and near H_c is also interesting. Particularly, by using scaling analysis, Zhu et al. [4] concluded that the Grüneisen parameter is divergent at any QCP, while Garst and Rosch considered the sign of $\Gamma(H)$ across it [3]. They established following features of the magnetic Grüneisen parameter:

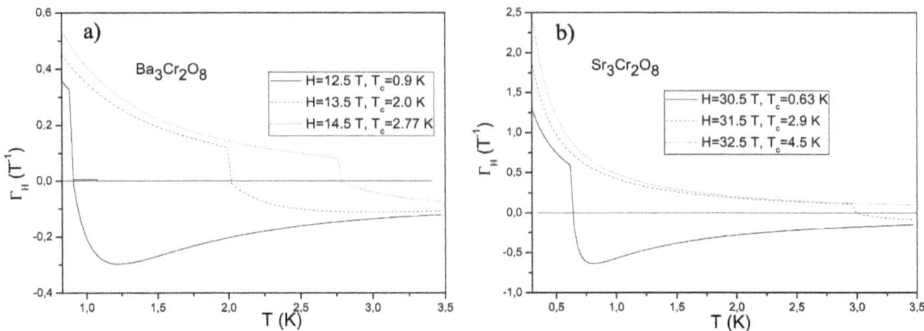

Fig. 15.3 The dependence of the Grüneisen parameter, in units $(1/Tesla)$, on temperature for **a** Ba$_3$Cr$_2$O$_8$ and **b** Sr$_3$Cr$_2$O$_8$ [2]. The input parameters are brought in Table 15.1. Reprinted with permission from [2]. ©1995, American Physical Society. All rights reserved

- Near the QCP, $H \sim H_c$ at $T = 0$, the distance to the QCP can be characterized by a dimensional parameter $r(H) = (H - H_c)/H_c$. Then

$$\Gamma_H(T \to 0, r) = G_r \frac{1}{H - H_c}, \tag{15.13}$$

where $G_r \geq 0$ is a prefactor which can be related to the critical correlation-length exponent ν, the dynamical exponent z, and the dimensionality of the critical fluctuation d as

$$G_r = \nu(d - z). \tag{15.14}$$

For example, for the Ising chain in the transverse magnetic field, one finds $G_r = 1$, while a dilute Bose gas in the symmetry-broken phase shows $G_r = 1/2$ [3].
- The temperature dependence of Γ_H in the critical regime at low temperatures also shows a divergence,

$$\Gamma_H(T, H \to H_c) \sim \frac{1}{T^x}, \tag{15.15}$$

with $x = 1/z\nu$.
- It is predicted that Γ_H exhibits a sign change at the QCP, so that the Grüneisen parameter has a different sign on each side of the phase transition.

The characteristic divergences and the sign change of Γ_H are hallmarks to detect and identify putative quantum critical points.

These properties of Γ_H have been experimentally confirmed by Gegenwart et al., who developed a low-frequency alternating-field technique to measure Γ_H down to low temperatures [5], in order to classify a number of magnetic systems ranging from heavy-fermion compounds to frustrated magnets [6]. They indeed observed a universal scaling of Grüneisen parameter $\Gamma_H = G_r(H - H_c)^{-1}$ with $G_r \approx 0.3$ and $\Gamma_H \sim 1/T^x$, with x varying from 1 to 3 in a certain deviation from the Hertz-Millis theory prediction with $x = 1$ [7].

From simple arguments, the property $\Gamma_H \sim 1/T^x$ can be easily considered for non-interacting Bose systems using $C_H(T \to 0) \sim T^{3/2}$ and $M \sim 1 - (T/T_c)^{3/2}$. From Eq. (15.5), all materials belonging to the non-interacting BEC universality class should therefore obey $\Gamma_H \sim 1/T$, i.e. $x = 1$ [8]. Nevertheless, since the triplon bosonic quasiparticles in the magnetic insulators to be considered here are known as an *interacting* Bose gas [9], a consideration of the effects of interaction on the Grüneisen parameter is of utmost interest. Below, we investigate the properties of Γ_H for such magnets within OPT. We will show that the fractional exponent is $z\nu = 1/2$ (i.e., $x = 2$) and $\Gamma_H \sim 1/(H - H_c)$.

15.4 Low Temperature Expansion

In the present section, we derive analytical expressions in the $T \to 0$ limit. We shall perform the low-temperature expansion as a function of the dimensionless parameter $Tm = \widetilde{T}$. In fact, for the majority of spin gap quantum magnets, the effective triplon mass m is small, e.g. $m \approx 0.02$ K^{-1} for TlCuCl$_3$ [10, 11], so that any power series in the small parameter $\widetilde{T}$ should quickly converge.

In general, the three-dimensional momentum sums and integrals, e.g. in Eq. (14.5), cannot be evaluated analytically. To overcome this difficulty, we use a Debye-like approximation [12] for the low-temperature limit, and replace the Brillouin zone by the Debye sphere with a radius k_D such that for a regular function $f(\mathbf{k})$ one obtains

$$\sum_{\mathbf{k} \in \mathcal{B}} f(\mathbf{k}) = \frac{1}{(2\pi)^3} \int_{\mathcal{B}} f(\mathbf{k}) dk_x dk_y dk_z = \int_{\mathcal{B}} f(\mathbf{q}) dq_x dq_y dq_z$$

$$\approx \frac{\pi}{2} \int_0^{Q_0} q^2 f(q) dq. \tag{15.16}$$

The normalization condition $\sum_{\mathbf{k}} = 1$ yields dimensionless $Q_0 = (6/\pi)^{1/3}$ with $k_D = Q_0 \pi$ and $\mathbf{q} \equiv \mathbf{k}/\pi$. Here we remind that in our units $a_x = a_y = a_z \, sim(V)^{1/3} = 1$. Now, we use the simple symmetric three-dimensional bare dispersion

$$\varepsilon_k = J_0(3 - \cos k_x a - \cos k_y a - \cos k_z a), \tag{15.17}$$

which we will use for numerical calculations as a model relation in gapped quantum magnets [13]. This bare dispersion of triplons enters in Eq. (14.29) for the spectrum of bogolons. At small q, we approximate it as $\varepsilon_q \approx q^2 \pi^2 / 2m = J_0 k^2 / 2$. This approximation can be used at low temperatures, while the effects of the spectrum's nonparabolicity at higher temperatures [14] can be captured by the exact form of ε_k in Eq. (15.17). For bogolon dispersion, we use the long wave approximation:

$$E_q = \sqrt{\varepsilon_q} \sqrt{\varepsilon_q + 2\Delta} \approx \pi c_0 q, \tag{15.18}$$

with the sound velocity at zero temperature $c_0 = \sqrt{\Delta_0/m}$, where $\Delta_0 = \Delta(T = 0)$ and Δ is defined in (14.26). With these approximations for the low-temperature limit, most of the integrals can be evaluated explicitly in terms of logarithmic and poly-logarithmic functions $\mathrm{Li}_s(\tilde{z})$ of the argument $\tilde{z} = \exp(-Q_0 c \pi \beta)$, i.e., as a function $F(T, \tilde{z})$ [15]. Since $\tilde{z}$ decreases quickly with increasing β, we may expand $F(T, \tilde{z})$ in powers of $\tilde{z}$ to extract the leading term.

Now using (14.34), one may find the following low-temperature expansion for Δ_T'

$$\Delta_T' = -\alpha_1 \tilde{T} - \alpha_3 \tilde{T}^3 + O(\tilde{T}^5) \tag{15.19}$$

where

$$\alpha_1 = \frac{2}{3} \frac{U \pi \gamma}{U Q_0 \gamma + \pi c_0}, \quad \tilde{T} = Tm,$$

$$\alpha_3 = \frac{U \pi^2}{45 \gamma^3} \frac{6\pi U Q_0 \gamma + 6\pi^2 c_0 + 5 U \gamma^2}{(U Q_0 \gamma + \pi c_0)^2}. \tag{15.20}$$

and $\gamma \equiv c_0 m$.

Task 15.2 Using Eq. (14.9) derive following low temperature expansions for the entropy:

$$S = \frac{2\pi^2 (\tilde{T})^3}{45 \gamma^3} + O(\tilde{T}^5), \tag{15.21}$$

and hence for the heat capacity:

$$C_H = T \frac{\partial S}{\partial T} \approx \frac{2\pi^2 (\tilde{T})^3}{15 \gamma^3} = \frac{2\pi^2 T^3}{15 c_0^3}, \tag{15.22}$$

Now to obtain low-temperature expansion for the Grüneisen parameter we use the Eq. (14.38) with the relations

$$\Gamma_H = -\frac{1}{C_H} \left(\frac{\partial M}{\partial T} \right)_H = -\frac{g \mu_B}{C_H} \left(\frac{\partial \rho}{\partial T} \right)_H = -\frac{g \mu_B}{2U C_H} \Delta_T'. \tag{15.23}$$

Finally, inserting C_H from (15.22) and Δ_T' from (15.19) we find

$$\Gamma_H = \frac{15 g \mu_B \alpha_1 \gamma^2}{4\pi^2 U} \frac{1}{\tilde{T}^2} + \frac{15 g \mu_B (2\gamma \alpha_3 - \alpha_1^2)}{8U \pi^2 \gamma} + O(\tilde{T}^2). \tag{15.24}$$

This is one of the central results of the present chapter. We will further simplify and discuss it later.

And at last, integrating the relation

$$\Gamma_H = -\frac{1}{C_H} \left(\frac{\partial M}{\partial T} \right)_H, \tag{15.25}$$

and using here (15.24), one may find a low-temperature expansion for the magnetization as

$$M = M(T = 0) - \frac{g \mu_B \alpha_1}{4 U \gamma m} \tilde{T}^2 + O(\tilde{T}^4). \tag{15.26}$$

It is seen that the magnetization varies as $M(T) \sim M(T = 0) - |constant| * T^2$ in the low-temperature limit, in contrast to the condensate fraction of a non-interacting system, which has $\rho_0 \sim \rho_0(T = 0) - |constant|T^{3/2}$ temperature dependence (see Chap. 3).

Now we proceed with the second critical property of Γ_H near the critical point $H \to H_c$, which concerns the statement about $\Gamma \to 1/T^x$, where x has to be determined.

15.5   Divergence of Γ_H Near Quantum Critical Point

In the Sect. 15.1 we have underlined that, in accordance with the renormalization theory [3], Γ_H is expected to diverge near QCP as

$$\Gamma_H(T \to 0, r \to 0) = \frac{G_r}{H_c r}, \qquad r = \frac{H - H_c}{H_c}, \tag{15.27}$$

where the dimensionless constant G_r depends on the individual parameters of a magnetic sample. Combining this with Eq. (15.24) we can represent the latter at low temperatures in the limit $r \to 0$ in the following compact form:

$$\Gamma_H \approx \frac{G_t(H - H_c)}{T^2} + \frac{G_r}{H - H_c}, \tag{15.28}$$

where $G_{r,t}$ are functions of input parameters U, J_0 and the sound velocity at zero temperature $c_0 = c(T = 0)$. The first term in Eq. (15.28) dominates in a fixed magnetic field $H > H_c$ for low temperatures, while the second term dominates in the opposite limit when H approaches the QCP $H \sim H_c$ from above at a fixed low temperature T. Therefore, for numerical evaluation of the coefficients α_1 and α_3 in Eq. (15.20), and hence the coefficients G_r and G_t, we have to clarify the sound velocity $c_0 = \sqrt{\Delta_0/m}$.

15.5.1   Zero Temperature Sound Velocity Near QCP

Let $T \to 0$ and $r = (H - H_c)/H_c \to 0$. Then the Eq. (14.26), with the notation c_0 as $c_0^2 = \Delta(T = 0)/m \equiv \Delta_0/m$ can be simplified as

$$\Delta_0 = \mu + U \sum_k \left(1 - \frac{E_k}{\varepsilon_k}\right), \tag{15.29}$$

with $E_k = \sqrt{\varepsilon_k}\sqrt{\varepsilon_k + 2\Delta_0}$. In the spherical Debye-like approximation, the momentum integration in (15.29) can be taken explicitly even without linear approximation for E_k, resulting in

$$\Delta_0 = \mu + \frac{U}{6\pi^2}\left[8(\Delta_0 m)^{3/2} + \pi^3 Q_0^3 - (\pi^2 Q_0^2 + 4m\Delta_0)^{3/2}\right]. \tag{15.30}$$

Since it is expected that at $r \to 0$ the quantity $m \Delta_0 << 1$ we can expand RHS of (15.30) in powers of $m \Delta_0$ to obtain

$$\Delta_0 = \mu - \frac{U m \Delta_0 Q_0}{\pi} + \mathcal{O}((m \Delta_0)^{3/2}), \tag{15.31}$$

or in terms of r, as

$$\Delta_0 \approx r \Delta_{\mathrm{st}} - \frac{U m \Delta_0 Q_0}{\pi}. \tag{15.32}$$

Thus for small r, the Δ_0 is given by the solution of this linear equation as

$$\Delta \approx \frac{r \Delta_{\mathrm{st}} \pi}{\pi + U Q_0 m}, \tag{15.33}$$

leading to

$$c_0 = \tilde{c} \sqrt{r} + \mathcal{O}(r^{3/2}), \tag{15.34}$$

with

$$\tilde{c} = \left[\frac{\pi \Delta_{\mathrm{st}}}{m(\pi + U Q_0 m)} \right]^{1/2} = \left[\frac{\pi g \mu_B H_c}{m(\pi + U Q_0 m)} \right]^{1/2}. \tag{15.35}$$

15.6   Numerical Results for Realistic Systems

Now inserting (15.34) into (15.24) and making expansion in powers of $\sqrt{r}$ we arrive to the Eq. (15.28) with following explicit expression for $G_{r,t}$

$$G_r = \frac{2}{Q_0 \pi} + \frac{2}{U m Q_0^2} \approx 0.51 + \frac{0.1}{a_s}, \qquad G_t = \left[\frac{\sqrt{5} g \mu_B}{\pi (1 + 4 a_s Q_0)} \right]^2 G_r, \tag{15.36}$$

where $a_s = U m / 4\pi$ is the s-wave scattering length of triplons. From (15.36) it is seen that two quantum magnets have the same G_r if they have the same a_s, which demonstrates universality of G_r, as it is predicted by Garst et al. [3]. Strictly speaking, this universal parameter approaches the value $1/2$ only in the unitary limit $1/a_s \to 0$. However, it should be mentioned that, in real systems, due to a possible anisotropy (see next chapter), the simple isotropic long wave approximation for the dispersion might be violated and hence the Eq. (15.36) would need a correction. Note that, Γ_H and its sign change can be directly measured in a dedicated experiment.

   In Table 15.1 we present the parameters G_r and G_t for three compounds: $Ba_3 Cr_2 O_8$, $Sr_3 Cr_2 O_8$ and $TlCuCl_3$. In practical calculations the set of realistic material parameters g, H_c, U and $J_0 = 1/m$ are optimized to describe experimental magnetization curves [16–18].

Table 15.1 Material parameters used for our numerical calculations. From the input parameters g and H_c, we derived J_0 and U from fitting the experimental data. Δ_{st} corresponds to the energy scale of H_c in Kelvin, while G_r, and G_t come from Eq. (15.36). Note that the lattice constant is chosen to be unity: $\bar{a} = V^{1/3} = 1$. To pass to natural units, following realistic parameters, e.g. for $TlCuCl_3$ may be used: $\bar{a} = (abc \sin\beta)^{1/3}$, $a = 3.98\,\text{Å}$, $b = 14.14\,\text{Å}$, $c = 8.89\text{Å}$, $\beta = 96.32°$ [19]

Material/parameters	g	$H_c(T)$	$J_0(K)$	$U(K)$	$\Delta_{st}(K)$	G_r	G_t
$Ba_3Cr_2O_8$	1.95	12.10	5.045	20	15.85	0.84	0.11
$Sr_3Cr_2O_8$	1.95	30.40	15.86	51.2	39.8	0.9	0.15
$TlCuCl_3$	2.06	5.1	50.	315	7.1	0.72	0.058

15.7 Summary

In the present chapter, we have studied MCE for gapped dimerized quantum magnets, which show a Bose-Einstein condensation of magnetic quasiparticles such as triplons. For this purpose we calculated the Grüneisen parameter, and derived explicit expressions for Γ_H in the limit $T \to T_c$ and $T \to 0$. Near the critical temperature, both the heat capacity and the Grüneisen parameter show a discontinuity with Γ_H changing its sign. Such behavior is expected for systems with a magnetically controlled quantum critical point. In the low-temperature limit near the transition, we find that the Grüneisen parameter diverges as $\Gamma_H \sim T^{-2}$. Approaching the transition field as $H \to H_c$ we find $\Gamma_H \sim G_r(H - H_c)^{-1}$, which is common to a variety of magnetic systems. The parameter G_r reaches its universal value $G_r \to 1/2$ in the unitary limit when the s-wave scattering length greatly exceeds the effective lattice constant, $a_s \gg \bar{a}$. Corresponding experiments to verify these conjectures and compare them with the scaling properties predicted by the full quantum theories of phase transitions are feasible. It should be underlined that the critical properties of the magnetic characteristics of quantum magnets could not be properly explained when one limits oneself to the Bogolyubov approximation.

References

1. B. Wolf, A. Honecker, W. Hofstetter, U. Tutsch, M. Lang, Cooling through quantum criticality and many-body effects in condensed matter and cold gases. Int. J. Mod. Phys. B **28**(26), 1430017 (2014). [Online]. https://doi.org/10.1142/S0217979214300175
2. A. Rakhimov, A. Gazizulina, Z. Narzikulov, A. Schilling, E.Y. Sherman, Magnetocaloric effect and Grüneisen parameter of quantum magnets with a spin gap. Phys. Rev. B **98**, 144416 (2018). [Online]. https://link.aps.org/doi/10.1103/PhysRevB.98.144416
3. M. Garst, A. Rosch, Sign change of the Grüneisen parameter and magnetocaloric effect near quantum critical points. Phys. Rev. B **72**, 205129 (2005). [Online]. https://link.aps.org/doi/10.1103/PhysRevB.72.205129
4. L. Zhu, M. Garst, A. Rosch, Q. Si, Universally diverging Grüneisen parameter and the magnetocaloric effect close to quantum critical points. Phys. Rev. Lett. **91**, 066404 (2003). [Online]. https://link.aps.org/doi/10.1103/PhysRevLett.91.066404

5. Y. Tokiwa, P. Gegenwart, High-resolution alternating-field technique to determine the magnetocaloric effect of metals down to very low temperatures. Rev. Sci. Instrum. **82**(1), 013905 (2011). [Online]. https://doi.org/10.1063/1.3529433

6. P. Gegenwart, Grüneisen parameter studies on heavy fermion quantum criticality. Rep. Prog. Phys. **79**(11), 114502 (2016). [Online]. https://dx.doi.org/10.1088/0034-4885/79/11/114502

7. P. Gegenwart, Classification of materials with divergent magnetic Grüneisen parameter. Philos. Mag. **97**(36), 3415–3427 (2017). [Online]. https://doi.org/10.1080/14786435.2016.1235803

8. H. Nakayama, K. Ando, K. Harii, T. Yoshino, R. Takahashi, Y. Kajiwara, K. Uchida, Y. Fujikawa, E. Saitoh, Geometry dependence on inverse spin Hall effect induced by spin pumping in $Ni_{81}Fe_{19}$/Pt films. Phys. Rev. B **85**, 144408 (2012). [Online]. https://link.aps.org/doi/10.1103/PhysRevB.85.144408

9. V. Zapf, M. Jaime, C.D. Batista, Bose-Einstein condensation in quantum magnets. Rev. Mod. Phys. **86**, 563–614 (2014). [Online]. https://link.aps.org/doi/10.1103/RevModPhys.86.563

10. G. Misguich, M. Oshikawa, Bose–Einstein condensation of magnons in $TlCuCl_3$: phase diagram and specific heat from a self-consistent Hartree–Fock calculation with a realistic dispersion relation. J. Phys. Soc. Jpn. **73**(12), 3429–3434 (2004). [Online]. https://doi.org/10.1143/JPSJ.73.3429

11. R. Dell'Amore, A. Schilling, K. Krämer, U(1) symmetry breaking and violated axial symmetry in $TlCuCl_3$ and other insulating spin systems. Phys. Rev. B **79**, 014438 (2009). [Online]. https://link.aps.org/doi/10.1103/PhysRevB.79.014438

12. V.I. Yukalov, Cold bosons in optical lattices. Laser Phys. **19**(1), 1–110 (2009). [Online]. https://doi.org/10.1134/S1054660X09010010

13. T. Giamarchi, C. Rüegg, O. Tchernyshyov, Bose–Einstein condensation in magnetic insulators. Nat. Phys. **4**(3), 198–204 (2008). [Online]. https://doi.org/10.1038/nphys893

14. E.Y. Sherman, P. Lemmens, B. Busse, A. Oosawa, H. Tanaka, Sound attenuation study on the Bose-Einstein condensation of magnons in $TlCuCl_3$. Phys. Rev. Lett. **91**, 057201 (2003). [Online]. https://link.aps.org/doi/10.1103/PhysRevLett.91.057201

15. J.E. Robinson, Note on the Bose-Einstein integral functions. Phys. Rev. **83**, 678–679 (1951). [Online]. https://link.aps.org/doi/10.1103/PhysRev.83.678

16. A.A. Aczel, Y. Kohama, C. Marcenat, F. Weickert, M. Jaime, O.E. Ayala-Valenzuela, R.D. McDonald, S.D. Selesnic, H.A. Dabkowska, G.M. Luke, Field-induced Bose-Einstein condensation of triplons up to 8K in $Sr_3Cr_2O_8$. Phys. Rev. Lett. **103**, 207203 (2009). [Online]. https://link.aps.org/doi/10.1103/PhysRevLett.103.207203

17. M. Kofu, H. Ueda, H. Nojiri, Y. Oshima, T. Zenmoto, K.C. Rule, S. Gerischer, B. Lake, C.D. Batista, Y. Ueda, S.-H. Lee, Magnetic-field induced phase transitions in a weakly coupled $s = 1/2$ quantum spin dimer system $Ba_3Cr_2O_8$. Phys. Rev. Lett. **102**, 177204 (2009). [Online]. https://link.aps.org/doi/10.1103/PhysRevLett.102.177204

18. Z. Wang, D.L. Quintero-Castro, S. Zherlitsyn, S. Yasin, Y. Skourski, A.T.M.N. Islam, B. Lake, J. Deisenhofer, A. Loidl, Field-induced magnonic liquid in the 3D spin-dimerized antiferromagnet $Sr_3Cr_2O_8$. Phys. Rev. Lett. **116**, 147201 (2016). [Online]. https://link.aps.org/doi/10.1103/PhysRevLett.116.147201

19. F. Yamada, T. Ono, H. Tanaka, G. Misguich, M. Oshikawa, T. Sakakibara, Magnetic-field induced Bose–Einstein condensation of magnons and critical behavior in interacting spin dimer system $TlCuCl_3$. J. Phys. Soc. Jpn. **77**(1), 013701 (2008). [Online]. https://doi.org/10.1143/JPSJ.77.013701

Weak Anisotropy in Spin-Gapped Quantum Magnets

16.1 Introduction

In the last chapters, we have shown that the magnetization curves of quantum dimerized magnets can be successfully described within the concept of triplon BEC in the framework of OPT. Particularly, it was shown that the staggered magnetization is related to the condensate fraction ρ_0 via

$$M_\perp = g\mu_B\sqrt{\rho_0/2} \tag{16.1}$$

Therefore, it is expected that, the function $M_\perp^2(T)$ would behave similarly to $\rho_0(T)$, presented in Fig. 3.3. Namely, the staggered magnetization should dramatically decrease in the limit $T \to T_c - 0$ and vanish at all when $T \to T_c + 0$. However, more precise neutron experiments have revealed that staggered magnetizations of some dimerized spin-gapped magnets remain finite even at $T > T_c$ [1,2]. In fact, as it is seen from Fig. 16.1, $M_\perp^2(T)$ of TlCuCl$_3$ rapidly goes down close to $T \to T_c - 0$, but "survives" even in the region $T > T_c$, not obeying the predictions of the BEC model. How can be this explained?

We have to accept that in such cases, we are dealing with a crossover instead of a clear second-order phase transition. Although there is no strict definition of this term in the literature, one may assume that crossover stands for a smooth transition between two separate phases of matter upon changing its thermodynamic input parameters. In a crossover, the ground state of the system changes radically, but in a very smooth manner, i.e. without any discontinuity in the thermodynamic observables.

A well-known example is BEC-BCS crossover, when an ultracold Fermi gas "crosses over" from a BEC state to a Bardeen-Cooper-Schrieffer (BCS) state without encountering a phase transition [3]. It is interesting that in the BEC-BCS phase

© The Author(s), under exclusive license to Springer Nature Switzerland AG 2026
A. Rakhimov and S. Mardonov, *Theory of Quantum Bose Liquids Beyond Bogoliubov Approximation*, Lecture Notes in Physics 1045,
https://doi.org/10.1007/978-3-032-05096-0_16

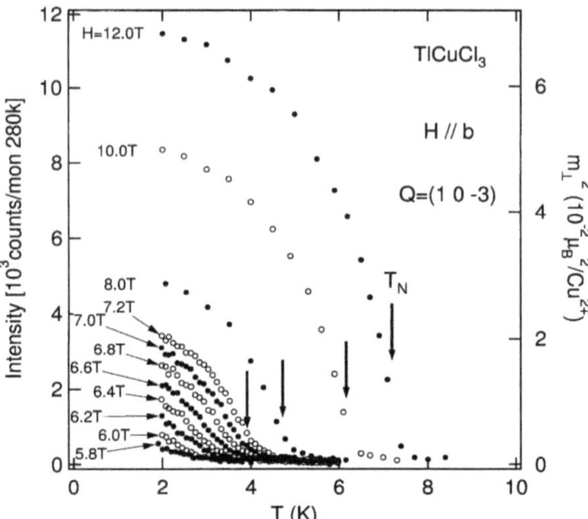

Fig. 16.1 Temperature dependence of the magnetic peak intensity at $(1, 0, -3)$ in various magnetic fields. The square of the transverse magnetization per site $m_\perp^2$ is shown on the right abscissa. Reprinted under CC-BY-4.0 licence from [1]. © 2001, H. Tanaka et al.

diagram, one may find nonzero condensate fractions on both sides of the critical temperature, as it is illustrated in Fig. 16.2 [4].

Coming back to our triplon gas, we note that the pure BEC scenario with the constraint $\rho_0(T \geq T_c) = 0$, exploited in previous chapters, is not directly applicable to the crossover displayed in Fig. 16.1. In fact, in terms of symmetries, the pure Bose-Einstein condensation of triplons corresponds to the spontaneous symmetry breaking of the rotational symmetry in the crystal, while the present crossover may correspond to its explicit violation (see Fig. 3.3a).

The origin of such explicitly broken symmetry is hidden in crystalline anisotropies, spin-orbit coupling and dipole interactions. Even when they are weak, the anisotropies become important at low temperatures and modify physical properties [5,6].

In general, the exchange interaction between two moments has the form $\Sigma_{ij} S_r^i T^{ij} S_{r+e_\nu}^j$ [8] where $\mathbf{T} = (1/3)Tr(\mathbf{T})I + \mathbf{T}_{as} + \mathbf{T}_{EM}$. Here the first term leads to the usual isotropic exchange coupling, $\mathbf{T}_{as}$ is an antisymmetric tensor that describes the Dzyaloshinsky-Moriya (DM) interaction $\mathbf{D} \cdot [\mathbf{S}_r \mathbf{S}_{r+e_\nu}]$, where $\mathbf{D}$ is the DM vector. The last term contains the so-called symmetric exchange anisotropy (EA) and has contributions from the classical dipole-dipole interaction between magnetic moments. Clearly, such interactions break axial symmetry not spontaneously but explicitly. Therefore, when the system of triplons is invariant under rotational symmetry, one deals with spontaneous symmetry breaking and hence pure BEC is termed as isotropic. On the other hand, when DM or EA interactions exist, one has to deal with an explicitly broken axial symmetry where, strictly speaking, no BEC can take place [9]. Nevertheless, due to the weakness of the observed anisotropy, we shall apply the BEC model, calling it the anisotropic case, and study its consequences for physical observables.

Thus, in the present chapter, we shall study the low-temperature properties of spin-gapped magnets with DM and exchange anisotropies in the concept of triplon

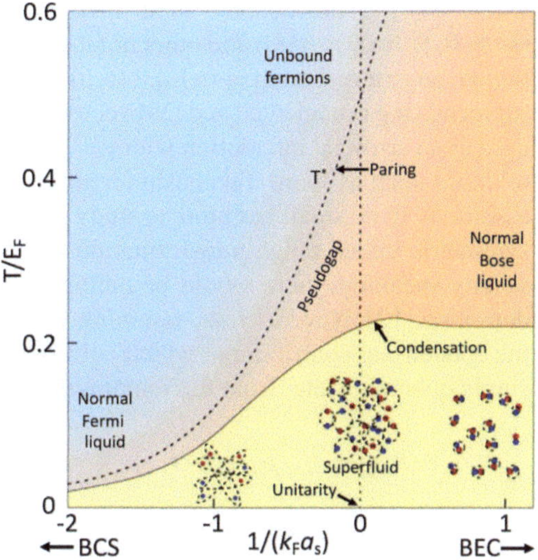

Fig. 16.2 Qualitative phase diagram of the BCS to BEC crossover as a function of the temperature T/T_F and the dimensionless coupling $1/(k_F a_s)$, where k_F is the Fermi momentum and a_s the scattering length. The picture shows schematically the evolution of the ground state from the BCS limit with large, spatially overlapping Cooper pairs to the BEC limit with tightly bound molecules. Reprinted with permission from [7]. © 2011, Springer Nature Limited. All rights reserved

BEC. For this purpose, we propose an extended Hamiltonian and develop an appropriate optimized perturbation theory. We show that the present approach can describe experimental data in Fig. 16.1 successfully.

16.2   The Hamiltonian and the Thermodynamic Potential Including EA and DM Interactions

The bond operator Hamiltonian corresponding to exchange and DM interactions have been derived in Refs. [6, 10]. Together with our "isotropic" Hamiltonian (13.8) the total Hamiltonian has the form:

$$H_{\text{tot}} = H_{\text{iso}} + H_{\text{aniso}}, \tag{16.2}$$

$$H_{\text{iso}} = \int d\mathbf{r} \left[\psi^+(\mathbf{r})(\hat{K} - \mu)\psi(\mathbf{r}) + \frac{U}{2}(\psi^+(\mathbf{r})\psi(\mathbf{r}))^2 \right], \tag{16.3}$$

$$H_{\text{aniso}} = H_{EA} + H_{DM}, \tag{16.4}$$

$$H_{EA} = \frac{\gamma}{2} \int d\mathbf{r} \left[\psi^+(\mathbf{r})\psi^+(\mathbf{r}) + \psi(\mathbf{r})\psi(\mathbf{r}) \right], \tag{16.5}$$

$$H_{DM} = i\gamma' \int d\mathbf{r} \left[\psi(\mathbf{r}) - \psi^+(\mathbf{r}) \right], \tag{16.6}$$

where $\psi(r)$ is the bosonic field operator, U, γ, γ' are the interaction strengths ($U \geq 0$, $\gamma \geq 0$, $\gamma' \geq 0$) and other notations are given in Chap. 14. The linear Hamiltonian, an external source, in Eq. (16.6) corresponds to a simple case when singlet-triplet mixing is neglected and DM vector is chosen as $D \parallel x$ and $H \parallel z$ [6]. It is often used as an artificial interaction with $\gamma' \to 0$ in quantum field theories to derive exact relations such as Ward-Takahashi identities [11]. However, in the present work, we assume γ' to be small but finite to study its physical consequences. Note that, H_{iso} is symmetric under gauge transformation $\psi \to e^{i\phi}\psi$, while H_{aniso} is not. Therefore, strictly speaking, there would be neither a Goldstone mode nor a pure Bose condensation [12]. Nevertheless, assuming $\gamma/U \ll 1$ and $\gamma'/U \ll 1$, one may separate the condensate contribution, which corresponds to the macroscopic occupation of a single quantum state, from the remaining part of the Bose field operator:

$$\psi(\mathbf{r}, t) = \eta\sqrt{\rho_0} + \tilde{\psi}(\mathbf{r}, t), \qquad \psi^{+}(\mathbf{r}, t) = \bar{\eta}\sqrt{\rho_0} + \tilde{\psi}^{\dagger}(\mathbf{r}, t). \tag{16.7}$$

Here it should be noted that, in contrast to previous cases, the phase of the condensate has been chosen arbitrarily, as in Chap. 10.

The thermodynamic potential will be evaluated in a usual way, $\Omega = -T \ln Z$, with

$$Z = \int \mathcal{D}\tilde{\psi}\mathcal{D}\tilde{\psi}^{\dagger} \exp\left\{-S[\psi, \psi^{\dagger}]\right\} \tag{16.8}$$

and

$$\begin{aligned} S[\psi, \psi^{\dagger}] = \int_0^{\beta} d\tau d\mathbf{r} &\left\{\psi^{\dagger}\left[\frac{\partial}{\partial\tau} - \hat{K} - \mu\right]\psi + \frac{U}{2}\left(\psi^{\dagger}\psi\right)^2 \right. \\ &\left. + \frac{\gamma}{2}\left(\psi\psi + \psi^{\dagger}\psi^{\dagger}\right) + i\gamma'\left(\psi - \psi^{\dagger}\right)\right\}. \end{aligned} \tag{16.9}$$

We apply OPT to Eqs. (16.8) and (16.9) in the same manner as in Chap. 13, making the following substitutions: $U \to \delta U$, $\gamma \to \delta\gamma$, $\gamma' \to \delta\gamma'$. Then one may formally obtain the same expressions for the Green function, the dispersion and the densities as in Eqs. (8.45), (8.46) and (14.5), respectively in the first order of δ.

Now, letting a reader make some algebraic calculations, we present below the final expression for the grand thermodynamic potential:

$$\Omega = \Omega_{\mathrm{iso}} + \Omega_{EA} + \Omega_{DM}, \tag{16.10}$$

$$\Omega_{EA} = \frac{\gamma\rho_0}{2}(\eta^2 + \bar{\eta}^2) + \frac{\gamma}{2}(B - A), \tag{16.11}$$

$$\Omega_{DM} = -i\gamma'(\bar{\eta} - \eta)\sqrt{\rho_0}, \tag{16.12}$$

where Ω_{iso} corresponding to the pure BEC is given by (10.15) and $\eta = e^{i\theta}, \bar{\eta} = e^{-i\theta}$ with θ-the phase angle of the condensate wave function.

Task 16.1 Derive Eqs. (16.10)–(16.12) from (16.8)–(16.9).

Minimization Ω with respect to ρ_0 leads to the equation

$$\frac{\partial \Omega}{\partial \rho_0} = \cos 2\theta (U\sigma + \gamma) + U(\rho_0 + 2\rho_1) - \mu - \frac{\gamma' \sin \Theta}{\sqrt{\rho_0}} = 0, \qquad (16.13)$$

Further, minimization of Ω with respect to the variational parameters $X_{1,2}$ gives

$$X_1 = 2U\rho + U\sigma - \mu + \frac{U\rho_0(\eta^2 + \bar{\eta}^2)}{2} + \gamma + \mathcal{O}(\gamma'^2), \qquad (16.14)$$

$$X_2 = 2U\rho - U\sigma - \mu - \frac{U\rho_0(\eta^2 + \bar{\eta}^2)}{2} - \gamma + \mathcal{O}(\gamma'^2). \qquad (16.15)$$

The inverse Green function is given in terms of $X_{1,2}$ as

$$D^{-1}(\omega_n, \mathbf{k}) = \begin{pmatrix} \varepsilon_k + X_1 & \omega_n \\ -\omega_n & \varepsilon_k + X_2 \end{pmatrix}, \qquad (16.16)$$

leading to the energy dispersion

$$E_k = \sqrt{\varepsilon_k + X_1}\sqrt{\varepsilon_k + X_2}, \qquad (16.17)$$

and the densities as

$$\rho_1 = \int \langle \tilde{\psi}^\dagger(\mathbf{r}) \tilde{\psi}(\mathbf{r}) \rangle d\mathbf{r} = \sum_k \left[\frac{W_k(\varepsilon_k + X_1/2 + X_2/2)}{E_k} - \frac{1}{2} \right], \qquad (16.18)$$

$$\sigma = \int \langle \tilde{\psi}(\mathbf{r}) \tilde{\psi}(\mathbf{r}) \rangle d\mathbf{r} = \frac{(X_2 - X_1)}{2} \sum_k \frac{W_k}{E_k} \equiv \frac{(X_2 - X_1)}{2U} \tilde{S},$$

$$\tilde{S} = \sum_k \frac{U W_k}{E_k}. \qquad (16.19)$$

Here still the total density of triplons per dimer is the sum of condensed and uncondensed fractions:

$$\rho = \frac{N}{V} = \rho_0 + \rho_1 \qquad (16.20)$$

which defines the uniform magnetization per dimer $M = g\mu_B \rho$. In the next chapters, we shall consider the effects of EA and DM interactions separately.

16.3 Effects of EA Interaction, $\gamma \neq 0$, $\gamma' = 0$

In the present section, we show that the presence of exchange interaction leads to the modification of anomaleous density as well as the critical temperature.

16.3.1 Violation of the Phase Invariance Due to Anisotropies

In Chap. 10 we have shown that physical observables of the pure condensate, caused by the spontaneous breaking, do not depend on the phase angle θ. However, in the present case, the symmetry is broken explicitly. Particularly, it is naturally expected that the phase invariance with respect to, say, $\theta \to \theta + \pi/2$, must be broken also. Below we show that even the presence of the exchange anisotropy makes, e.g. the energy dispersion phase dependent.

Actually, let's rewrite Eqs. (16.14) and (16.15) as follows

$$X_1 = U\tilde{\sigma} + U\rho_0\eta^2 + 2U\rho - \mu, \quad X_2 = -U\tilde{\sigma} - U\rho_0\eta^2 + 2U\rho - \mu, \quad (16.21)$$

where we set for the simplicity $\bar{\eta}^2 = \eta^2$ and introduced $\tilde{\sigma} = \sigma + \tilde{\gamma}$, $\tilde{\gamma} = \gamma/U$. Inserting (16.21) into (16.19) leads to the following equation with respect to $\tilde{\sigma}$

$$\tilde{\sigma}(1 + \tilde{S}) - \tilde{\gamma} + \tilde{S}\rho_0\eta^2 = 0, \qquad (16.22)$$

whose formal solution can be presented as

$$\tilde{\sigma} = \tilde{\sigma}_1 + \eta^2\tilde{\sigma}_2, \quad \tilde{\sigma}_1 = \tilde{\gamma}/(1 + \tilde{S}), \quad \tilde{\sigma}_2 = -\tilde{S}\rho_0/(1 + \tilde{S}), \qquad (16.23)$$

where the dependence on the phase is explicitly highlighted. Inserting back these equations to (16.21) we have

$$X_1 = a + b\eta^2, \quad X_2 = a - b\eta^2 - 2U\tilde{\sigma}_1, \qquad (16.24)$$

where $a = U(\tilde{\sigma}_1 + 2\rho) - \mu$, and $b = U(\tilde{\sigma}_2 + \rho_0)$. Now inserting (16.24) into the Eq. (16.17), we come to the following final expression for the energy dispersion

$$E_k^2(\gamma' = 0) = (\varepsilon_k - \mu + 2U\rho)^2 - \frac{U^2(\rho_0^2 + \tilde{\gamma}^2)}{(1 + \tilde{S})^2} - \frac{2U^2\rho_0\tilde{\gamma}}{(1 + \tilde{S})^2}\eta^2. \qquad (16.25)$$

Thus, we have revealed explicit dependence of the dispersion on the phase. Particularly, this dependence vanishes in the isotropic case when $\tilde{\gamma} = \gamma/U \to 0$.

What phase angle will the system of triplons with exchange anisotropy prefer? Strictly speaking, to find the correct answer to this question, one has to study numerically the dependence of the free energy $F = E - TS = \Omega + \mu N$ on the phase $\eta = \exp(i\theta)$ to find the minimum of the function $F(\theta)$. Postponing this work for future studies, below we shall make a choice $\eta^2 = +1$, following Refs. [12,13]. The plausibility of such a choice is based on Eq. (16.25): It is seen that the multiplier before η^2 is negative and hence $E_k(\eta^2 = 1) < E_k(\eta^2 = -1)$. So, the choice $\eta^2 = 1$, i.e. $\theta = \pi n$ $(n = 0, 1, 2\dots)$ for the case of $\gamma \neq 0$, $\gamma' = 0$ has been, more or less, justified. In a similar way one can show [6,14] that the presence of DM interaction $(\gamma' \neq 0)$ requires the choice $\eta = i$, i.e. $\theta = \pi/2 + 2\pi n$, which will be exploited in Sect. 16.4.

16.3.2 HP Relation for a Quasicondensate

Due to the presence of the term H_{EA} in equation (16.6) the Hamiltonian is no more symmetric under the transformation $\psi \to \exp(i\alpha)\psi$, which excludes the emergence of a pure BEC. However, assuming this symmetry violation to be small, one may discuss the emergence of a quasicondensate, which "becomes" pure, when this term vanishes ($\gamma \to 0$).

So, let $\gamma' = 0$, $\gamma/U << 1$ and $\eta^2 = +1$. Then the Eqs. (16.13), (16.14) and (16.15) are simplified as

$$- \mu_0 + U(2\rho_1 + \rho_0 + \tilde{\gamma} + \sigma) = 0, \tag{16.26}$$

$$X_1 = -\mu_1 + U(2\rho + \tilde{\gamma} + \rho_0 + \sigma), \tag{16.27}$$

$$X_2 = -\mu_1 + U(2\rho - \tilde{\gamma} - \rho_0 - \sigma), \tag{16.28}$$

where we have introduced two chemical potentials, μ_0 and

$$\mu_1 = g\mu_B(H_{ext} - H_c) \tag{16.29}$$

in accordance with Yukalov's prescription (see Chap. 8). From the Eq. (16.26) we may conclude that, in contrast to the case with $\gamma' \neq 0$, in the present case with the only exchange anisotropy ($\gamma \neq 0$, $\gamma' = 0$) we are allowed to have $\rho_0 = 0$ solution also. In other words, there may exist a critical temperature $T = T_c$, which controls the transition (not a crossover!) from a condensate phase with $\rho_0(T < T_c) \neq 0$ to the normal phase with $\rho_0(T \geq T_c) = 0$. Here, to put it more accurately, it would be correct to use the term quasi-condensate, since the necessary condition for Bose-Einstein condensation-SSB is not fulfilled due to the term (16.5) in the Hamiltonian. So, what about the Hugenholtz-Pines theorem? Can we exploit it? The rigorous answer is certainly negative for the following reasons:

- The declared critical temperature has nothing to do with SSB, which is the logical prerequisite in the accurate derivation of HP relations [15].
- Application of the theorem leads to the appearance of the Goldstone mode with $E_k(k \to 0) \approx ck$. However, the ESR measurements [16] have shown a direct singlet-triplet transition, which indicates to the gap in the energy dispersion E_k. Consequently, $E_k^{exper}(k \to 0) \neq 0$.

Nevertheless, bearing in mind that, $U(1)$ symmetry is broken weakly, since $\tilde{\gamma} = \gamma/U \ll 1$, we propose following modified version of HP relations:

$$\frac{\Sigma_n - \Sigma_{an}}{\mu_1} = 1 + 2\tilde{\gamma}c_\gamma + \mathcal{O}(\tilde{\gamma}^2) \tag{16.30}$$

which is exact in the SSB phase when $\gamma = U\tilde{\gamma} = 0$. In (16.30) c_γ may be considered as a linear dimensionless coefficient in the expansion of $(\Sigma_n - \Sigma_{an})/\mu_1$ in powers of γ/U which can be fixed, e.g. by fitting the gap in the energy spectrum observed experimentally at small momentum transfer. Now we discuss $T < T_c$ and $T > T_c$ phases separately.

16.3.3 Condensed Phase $T < T_c$

Our modified HP relation is equivalent to the constraint:

$$X_2 = 2\mu_1\tilde{\gamma}c_\gamma \tag{16.31}$$

where we used the general relations (4.84). Now setting in Eq. (16.28) $\rho = \rho_0 + \rho_1$ and using (16.31) we have

$$\rho_0 = \frac{\mu_1}{U} - 2\rho_1 + \sigma + \tilde{\gamma}(1 + \frac{2\mu_1 c_\gamma}{U}) \tag{16.32}$$

which being inserted to (16.27) leads to the following main equation for X_1

$$X_1 = 2\mu_1 + 4U(\sigma - \rho_1) + 2\tilde{\gamma}(2U + 3\mu_1 c_\gamma) \tag{16.33}$$

where

$$\rho_1 = \sum_k \left[\frac{W_k(\varepsilon_k + X_1/2 + \mu_1\tilde{\gamma}c_\gamma)}{E_k} - \frac{1}{2} \right], \tag{16.34}$$

$$\sigma = \frac{2\mu_1\tilde{\gamma}c_\gamma - X_1}{2} \sum_k \frac{W_k}{E_k}, \tag{16.35}$$

$$E_k = \sqrt{\varepsilon_k + X_1}\sqrt{\varepsilon_k + 2\mu_1\tilde{\gamma}c_\gamma} . \tag{16.36}$$

In practical calculations, one has to start with the input parameters μ_1, the coupling constant U, the temperature T, as well as the explicit expression for the bare dispersion ε_k [12] and solve the set of algebraic equations (16.33)–(16.36) with respect to X_1 and then evaluate the condensate fraction from Eq. (16.32). As to the total density of triplons, it may be found from the normalization condition $\rho = \rho_0 + \rho_1$. As to the parameter c_γ, it should be guessed and optimized, for instance, to describe the experimental energy dispersion E_k.

16.3.4 Normal Phase $T > T_c$

In the normal phase, $\rho_0 = 0$, $\rho_1 = \rho$, and $T > T_c$, the energy dispersion has a gap even for $\gamma = 0$ and the Eq. (16.31) is no longer valid. However the main Eqs. (16.27) and (16.28) with $\rho_0 = 0$, $\mu = \mu_1$

$$X_1 = -\mu + U[2\rho + \sigma + \tilde{\gamma}], \tag{16.37}$$

$$X_2 = -\mu + U[2\rho - \sigma - \tilde{\gamma}], \tag{16.38}$$

make sense. The energy dispersion, given by

$$E_k = \sqrt{(\varepsilon_k - \mu_{eff})^2 - U^2(\sigma + \tilde{\gamma})^2}, \qquad \mu_{eff} = \mu - 2U\rho, \tag{16.39}$$

still includes the anomalous density σ:

$$\sigma = \frac{(X_2 - X_1)}{2} \sum_k \frac{f_B(E_k)}{E_k}, \tag{16.40}$$

which can be formally presented as

$$\sigma(T \geq T_c) = -\frac{\tilde{\gamma}\tilde{S}}{1 + \tilde{S}}, \qquad \tilde{S} = U \sum_k \frac{f_B(E_k)}{E_k}. \tag{16.41}$$

The last equation confirms that $\sigma(T \geq T_c, \gamma = U\tilde{\gamma} = 0) = 0$, but at the same time makes it possible to assume that the anomalous density may survive even in the normal phase due to the presence of exchange anisotropy, i.e. $\sigma(T \geq T_c, \gamma \neq 0) \neq 0$. To analyze this point, we consider the normal density also, which is given by

$$\rho(T \geq T_c) = \sum_k \left[\frac{f_B(E_k)(\varepsilon_k + X_1/2 + X_2/2)}{E_k} \right] = \sum_k \left[\frac{f_B(E_k)(\varepsilon_k - \mu + 2U\rho)}{E_k} \right]. \tag{16.42}$$

It is understood that, actually, $\sigma(T \geq T_c)$ and $\rho(T \geq T_c)$ could be evaluated as numerical solutions of the nonlinear algebraic equations (16.39)–(16.42).

We present in Fig. 16.3a the density of condensed particles ρ_0 (solid line) and the absolute value of the anomalous density $|\sigma|$ (dashed line) versus the reduced temperature $t = T/T_c$, ($t = 0 \div 1.2$) for the quantum dimerized magnet TlCuCl$_3$. It is seen that $|\sigma|$ is comparable with ρ_0 at all temperatures. Another interesting fact, which is demonstrated in Fig. 16.3b is that the anomalous density survives, although on a small level, even above the critical temperature where it vanishes asymptotically due to the factor $f_B(E_k) = 1/(\exp(E_k/T) - 1)$ in Eq. (16.41). For example, $\sigma(t = 0)/\sigma(t = 1) \approx 100$. Note that without the exchange anisotropy $\sigma(\gamma = 0)|_{T \geq T_c} = 0$. A similar phase with $\rho_0 = 0$ and $\sigma \neq 0$ has been reported by Cooper et al. [17] within a lowest-order auxiliary field formalism. This approach predicts the existence of two critical temperatures, one T_c, where $\rho_0 = 0$, $\sigma \neq 0$ and another one T^*, where $\rho_0 = 0$, $\sigma = 0$, with $T^* > T_c$. This exotic state in the region $T_c < T < T^*$ has not been experimentally observed yet, but it is predicted to exhibit a modified dispersion relation. The question about observing such a phase still remains open.

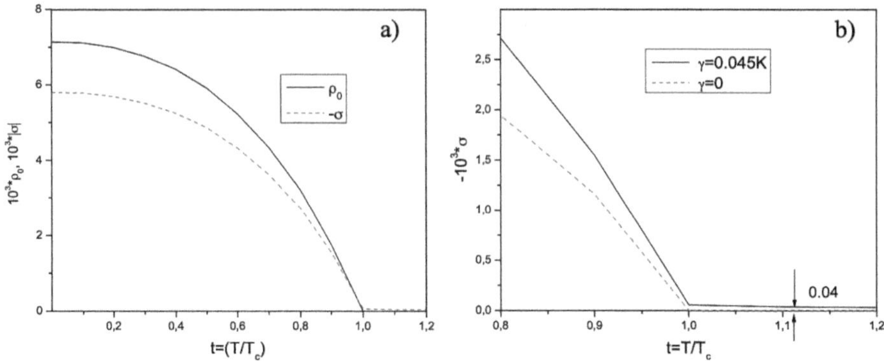

Fig. 16.3 a The condensed (solid line) and the absolute value of anomalous densities (dashed line) at $H_{ext} = 7T$; **b** The behavior of $|\sigma(t)|$ near the critical temperature with (solid line) and without (dashed line) exchange anisotropy. The input parameters are $U = 367.5\,$K, $g = 2.06$, $c_\gamma = 1.67U/\mu$

16.3.5 The Shift of the Critical Temperature Due to Exchange Anisotropy

Let's denote the critical temperature of pure BEC transition as $T_c^{(0)}$, and that of the quasicondensate in the presence of EA interaction as T_c. By definition $T_c(\gamma = 0) = T_c^{(0)}$, where $T_c^{(0)}$ is given by (14.41). As to T_c, it can be determined from the continuity condition of $X_{1,2}$ e.g. $X_2(T = T_c - 0) = X_2(T = T_c + 0)$. Thus the Eqs. (16.31) and (16.38) lead to

$$2\mu_1\tilde{\gamma}c_\gamma = -\mu_1 + U(2\rho_c - \sigma_c - \tilde{\gamma}), \tag{16.43}$$

where

$$\rho_c = \rho(T = T_c) = \sum_k \left[\frac{f_B(E_k^c)}{E_k^c}(\varepsilon_k + X_1^c/2 + \mu\tilde{\gamma}c_\gamma)\right], \tag{16.44}$$

with

$$E_k^c = \sqrt{(\varepsilon_k + X_1^c)(\varepsilon_k + 2\mu_1\tilde{\gamma}c_\gamma)}, \qquad f_B(E_k^c) = \frac{1}{e^{E_k^c/T_c} - 1},$$
$$X_1^c = X_1(T = T_c) = 2U\sigma_c + 2U\tilde{\gamma}(1 + \frac{\mu}{U}c_\gamma) \tag{16.45}$$

and

$$\sigma_c = \sigma(T = T_c) = -\frac{\tilde{\gamma}\tilde{S}_c}{1 + \tilde{S}_c}, \qquad \tilde{S}_c = U\sum_k \frac{f_B(E_k^c)}{E_k^c}. \tag{16.46}$$

The critical temperature is numerically evaluated from the two coupled nonlinear Eqs. (16.43) and (16.44) with respect to T_c and σ_c. Clearly, T_c will be a function of H_{ext}, which fixes the chemical potential as the input parameter: $\mu = g\mu_B(H_{ext} - H_c)$.

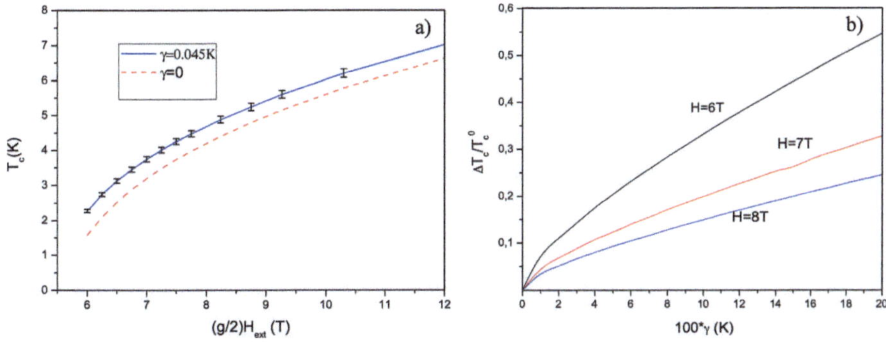

Fig. 16.4 Phase diagram normalized by the g-factor (**a**) and the shift of the critical temperature, $(T_c(\gamma) - T_c(\gamma = 0))/T_c(\gamma = 0)$ due to the exchange anisotropy (**b**) for TlCuCl$_3$. The experimental data are taken from [18]. The input parameters are the same as in Fig. 16.3. Reprinted under CC-BY-4.0 [18]. ©2008, F. Yamada et al.

The phase boundary $T_c(H_{ext})$, is presented in Fig. 16.4a. Actually, it contains information about the critical exponent ϕ, defined as $T_c \sim (H_{ext} - H_c)^\phi$ or more precisely as $T_c = C\,(\mu/U)^\phi$, where the constant C and ϕ are fitting parameters. Note that, for the case of a homogeneous ideal gas with the quadratic dispersion $\phi = 2/3$. From Fig. 16.4a, we have found that $C = 47.4\,K$ and $\phi = 0.53$ (solid line) and $C = 63.2\,K$ and $\phi = 0.62$ (dashed line) for the $\gamma = 0.045\,K$ and $\gamma = 0$ cases respectively. This means that the inclusion of a finite exchange anisotropy reduces the exponent ϕ, and one does not need to expect $\phi = 2/3$ as it has been debated in the literature [19–21]. The presence of interparticle interaction as well as using a more realistic dispersion than a simple quadratic one, leads to a shift of the critical temperature, especially at high temperatures $T > 2K$ [22]. Here we note that if we restrict to fit ϕ in the range of $0 \leq T \leq 1.5\,K$ (not drawn in Fig. 16.4a), the solid line in Fig. 16.4a may also be well fitted by $\phi \approx 2/3$.

From Fig. 16.4a we can state that the exchange anisotropy term H_{EA} given in (16.5) leads to an increase in the critical temperature at a given magnetic field. To study this issue in more detail, we present in Fig. 16.4b the shift of the critical temperature due to the anisotropy $\Delta T/T_c^0$ vs γ for various values of H_{ext}. It is seen that

- ΔT_c increases with the increase of γ;
- For a moderate value of gamma $\gamma \sim 0.04\,K$ the shift is nearly 10% at $H_{ext} = 7T$;
- With increasing the magnetic field, the upward shift in the critical temperature decreases. In terms of the whole Hamiltonian, this may be explained as follows: when γ is fixed the increase of H_{ext} leads to the increase of the term with μ. So, this term becomes much larger than the term H_{EA}, and hence the influence of EA interaction becomes negligible.

A similar dependence of the shift on γ and H_{ext} has been predicted by Dell'Amore et al. [23].

Therefore, one concludes that the exchange anisotropy does not change the form of the transition qualitatively. Its main effect on magnetization curves is a modification of the transition temperature, leaving the behavior of the staggered magnetization unchanged [24]. On the other hand, as it has been revealed at the beginning of the present chapter, a crossover-like behavior of the staggered magnetization can be explained as the consequence of DM interaction. So, we move on to study its effects.

16.4　Effects of DM Anisotropy, $\gamma = 0$, $\gamma' \neq 0$

First, we note that the last term of Eq. (16.13) includes ρ_0 in the denominator, prohibiting ρ_0 from reaching zero. Therefore, strictly speaking, $\rho_0(T) \neq 0$ for any T. Physically, this means that there would be no pure BEC transition as long as the DM interaction is present. Thus, we have proven that the presence of Dzyaloshinsky-Moriya anisotropy converts a second-order phase transition into a crossover.

To make a quantitative analysis, we have to concentrate on the main equations (16.13)–(16.15). It is understood that now we have to solve the coupled three equations with respect to the variational parameters ρ_0, X_1 and X_2. In fact, in previous sections, dealing with $\gamma' = 0$ case, we are practically left with only one equation with respect to X_1. Such simplification has been provided to us due to the Yukalov prescription and Hugenholtz-Pines relation. However, the presence of DM anisotropy ($\gamma' \neq 0$) changes the situation dramatically: There is no constraint for our three parameters, but $\rho_0(T) \neq 0$, which, on the contrary, complicates the problem. In this case, as it has been shown in Refs. [14,25] the phase angle $\theta = \pi/2 + 2\pi n$, that is $\eta = i$ seems to be preferable.

So, accepting the choice $\eta = i$ we present the Eqs. (16.13)–(16.15) as follows

$$-U\sigma + U(\rho_0 + 2\rho_1) - \mu - \frac{\gamma'}{\sqrt{\rho_0}} = 0, \tag{16.47}$$

$$X_1 = 2U\rho + U\sigma - \mu - U\rho_0, \tag{16.48}$$

$$X_2 = 2U\rho - U\sigma - \mu + U\rho_0. \tag{16.49}$$

Excluding μ from the first equation these can be rewritten in the following equivalent form

$$\mu = -U\sigma + U(\rho_0 + 2\rho_1) - \frac{\gamma'}{\sqrt{\rho_0}}, \tag{16.50}$$

$$X_1 = 2U\sigma + \frac{\gamma'}{\sqrt{\rho_0}}, \tag{16.51}$$

$$X_2 = 2U\rho_0 + \frac{\gamma'}{\sqrt{\rho_0}}. \tag{16.52}$$

For practical calculations, the last three equations are simplified in the following dimensionless form

$$r_0^3 + Pr_0 + Q = 0, \tag{16.53}$$

$$Z_1 - \frac{\tilde{\sigma}}{2} - \tilde{\gamma} + \frac{Q}{4r_0} = 0, \tag{16.54}$$

$$Z_2 - \frac{r_0^2}{2} + \frac{Q}{4r_0} = 0, \tag{16.55}$$

where we have introduced $P = -\tilde{\sigma} + 2(\tilde{\rho}_1 - 1 - \tilde{\gamma})$, $Q = -2\tilde{\gamma}'/\sqrt{\rho_{c0}}$, $r_0^2 = \rho_0/\rho_{c0}$, $\tilde{\sigma} = \sigma/\rho_{c0}$, $\tilde{\rho}_1 = \rho_1/\rho_{c0}$, $\tilde{\gamma} = \gamma/\mu$, $\tilde{\gamma}' = \gamma'/\mu$, in which ρ_{c0} is the critical density of pure BEC, $\rho_{c0} = \mu/2U$. In general, one has to solve these three coupled nonlinear algebraic equations for the unknown quantities $X_{1,2} = Z_{1,2}/2\mu$, and $r_0 = \sqrt{\rho_0/\rho_{c0}}$ at a given temperature and magnetic field. Clearly, in such cases, it is important to guess the initial values of $X_1(T)$ and $X_2(T)$, since the solutions are not unique. For this purpose, it will be convenient to start from a higher temperature, say $T \approx 15K$, where $\sigma \approx 0$, and hence Eqs. (16.54) and (16.55) are simplified to

$$Z_1 = \frac{\gamma}{\mu} - \frac{Q}{4r_0}, \quad Z_2 = \frac{r_0^2}{2} - \frac{Q}{4r_0}, \tag{16.56}$$

Actually, before speculating about low-high temperatures, it would be advisable to define a critical temperature of a crossover T_c somehow.

16.5   Critical Temperature of a Crossover

Let's assume that EA and DM interactions are turned off. Then we come back to a pure BEC transition, whose critical temperature T_c^0 is defined from the condition $\rho_0(T = T_c^0) = 0$ and given by Eq. (14.41). Now, when we "turn on" those interactions, the Eq. (16.53) has no zero solution and we have to seek another way for defining a proper T_c. For this purpose, we come back to the Figs. 13.2 and note that experimental magnetization curve $M(T)$ has a local minimum for any $H_{ext} > H_c$. Therefore, bearing in mind that $M(T) \sim \rho(T)$, we can define critical temperature from following equations

$$(d\rho/dT)|_{T=T_c} = 0, \quad (d^2\rho/dT^2)|_{T=T_c} \geq 0 \tag{16.57}$$

i.e. as the point $T = T_c(H_{ext})$, where the total number of triplons at fixed magnetic field reaches its minimum.[1] Naturally, this T_c is not expected to coincide with T_c^0, since the presence of anisotropies causes a shift, which we denote as

$$\frac{\Delta T_c}{T_c^0} \equiv \frac{T_c - T_c^0}{T_c^0} \tag{16.58}$$

---

[1] See Ref. [26] for an alternative definition.

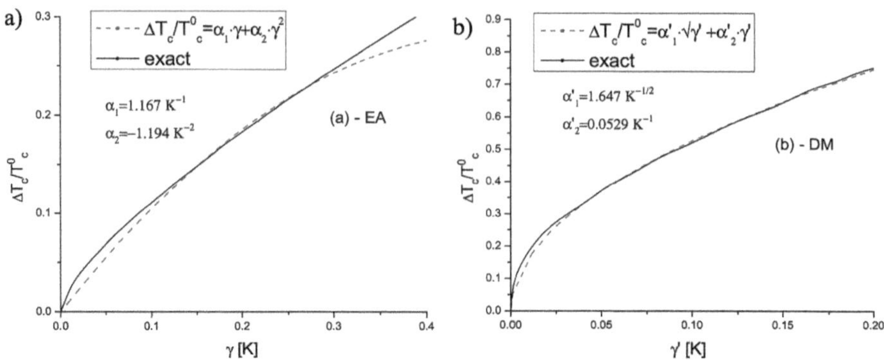

Fig. 16.5 The shift in critical temperature due to EA (**a**) and DM (**b**) interactions (solid curves). Dashed curves are phenomenological fits. The input parameters are $g = 2.06$, $U = 315\,\text{K}$ and $H_{ext} = 8.5\,\text{T}$, which are favorable for the compound $TlCuCl_3$

In Fig. 16.5a, b, we present the shifts due to EA and DM interactions, respectively. For weak anisotropies these can be approximated in powers of γ/U and $\sqrt{\gamma'/U}$ as $\Delta T_c/T_c^0(\gamma) \approx a_1(\gamma/U) + a_2(\gamma/U)^2$ and $\Delta T_c/T_c^0(\gamma') \approx a_1'\sqrt{(\gamma'/U)} + a_2'(\gamma'/U)$ for the cases of EA and DM interactions, respectively. Clearly, the optimized parameters a_i and a_i' depend also on the external magnetic field H. Particularly, for $TlCuCl_3$ with $U = 315\,\text{K}$ at $H = 8.5\,\text{T}$ we obtained $a_1/U = 1.167\,\text{K}^{-1}$, $a_2/U^2 = -1.194\,\text{K}^{-2}$, $a_1'/\sqrt{U} = 1.647\,\text{K}^{-1/2}$ and $a_2'/U = 0.053\,\text{K}^{-1}$, as illustrated in Fig. 16.5.

Firstly, one may note that in both cases $\Delta T_c \geq 0$, which means that the presence of the anisotropies shifts the critical temperature of the BEC transition (or a crossover in the case of DM anisotropy) toward higher values. Secondly, it is seen that the influence of anisotropy is not negligibly small at moderate values of the intensities. For instance, DM interaction with $\gamma' \approx 0.1\,\text{K}$ modifies T_c with $\Delta T_c/T_c^0(\gamma' = 0.1\,\text{K}) \sim 50\%$. Thirdly, DM anisotropy modifies the critical temperature more strongly than EA anisotropy. For example, for the equal values of intensities, say, $\gamma = \gamma' \approx 0.1\,\text{K}$, the shift due to DM interaction is nearly five times larger than due to EA interaction. Thus, the critical temperature is more sensitive to DM interaction than to EA.

16.6   Magnetization Curves for $TlCuCl_3$

In the previous section, we studied the effect of anisotropies on thermodynamic quantities. Particularly, we have shown that in contrast to EA interaction, DM interaction modifies their behavior dramatically. It smears the BEC transition to a crossover and changes the sign of the anomalous density. Clearly, the significance or measurability of such effects depends on their interaction strengths γ and γ'. Evidently, unless we have realistic values for these parameters for real materials, our studies will remain purely academic.

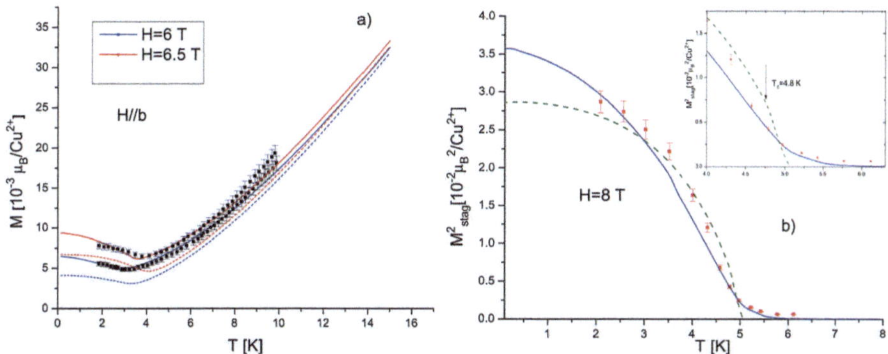

Fig. 16.6 Uniform (**a**) and staggered (**b**) magnetizations for TlCuCl$_3$, $H//b$. Solid and dashed lines correspond to the present approximation and approximation in Ref. [6], respectively. Experimental data are taken from Ref. [1]. The optimized parameters are $\gamma = 0.05$ K, $\gamma' = 0.0201$ K and $U = 367$ K. Diamagnetic and other contributions to total experimental magnetizations are taken into account following the ansatz by Dell'amore et al. [30]. Reprinted under CC-BY-4.0 [1]. ©2001, H. Tanaka et al.

As we have noted in previous chapters, among the 3D quantum dimerized magnets with a spin gap, the compound TlCuCl$_3$ seems to be the most experimentally studied [19, 27–29].

The observation of a finite $M_\perp$ at $T \geq T_c$ [1] uniquely indicates the presence of DM interaction with a finite γ'. Thus, using existing experimental data on the magnetization of TlCuCl$_3$, we can make an attempt to obtain optimal values of input parameters of the present approach.[2]

The result for $H//b$ is as follows: $g = 2.06$, $U = 367$ K, $\gamma = 0.05$ K and $\gamma' = 0.0201$ K. The magnetizations M and $M_\perp$ corresponding to this set of parameters are depicted in Fig. 16.6a, b, respectively. It is seen that the inclusion of DM anisotropy gives a good description of the staggered magnetization, especially at higher temperatures (see inset of Fig. 16.6b). Moreover, taking into account the anomalous density σ leads to a better description of M, e.g., at low temperatures, compared with the Bilinear approximation exploited in Ref. [25].

Remarkably, the experimental fit of parameters can be reached with rather small values of anisotropies, namely $\gamma/U = 1.36 \times 10^{-4}$ and $\gamma'/U = 5.47 \times 10^{-5}$. Thus, we have found that the experimental data on magnetization of TlCuCl$_3$ can be well described by the present approach, based on variational perturbation theory, i.e. OPT.

16.7 Summary

In the present chapter, we have studied the effects of lattice anisotropies on thermodynamic characteristics of spin-gapped quantum magnets for $H_c \leq H < H_{Saturation}$ by applying OPT, developed in previous chapters. This nonperturbative approach

---

[2] A reader can find the details of calculations in Ref. [12, 24].

takes into account the anomalous density and both EA and DM interactions more accurately than it is done, e.g., in the Bilinear approximation. We derived explicit expressions for some thermodynamic quantities, which include the self-energies X_1 and X_2, and the condensate fraction ρ_0. Analysis of the coupled equations with respect to these three quantities show that at high temperatures $T \gg T_c$, the self-energies $X_{1,2}$ are not significantly affected by EA and DM interactions. Meanwhile, the latter strongly modifies the condensate fraction, converting the BEC transition into a crossover.

At low temperatures, the DM interaction increases ρ_0, but leads to rather small values of X_1, compared with the isotropic case. As a result, the energy dispersion $E_k = \sqrt{(\varepsilon_k + X_1)(\varepsilon_k + X_2)}$, develops a linear dependence at small momentum, in accordance with experimental measurements.

In contrast to EA interaction, the presence of DM interaction, even in the simple linear form in the Hamiltonian, smears the BEC phase transition into a crossover and modifies the anomalous density, changing its sign. Particularly, it is expected that, the usual "λ-shape" of the heat capacity disappears due to strong DM interactions.

We have found optimal input parameters of the Hamiltonian for the compound TlCuCl$_3$, which describes experimental data on magnetizations, at least for Hb, quite well. This set of parameters leads to a linear dispersion of energy of quasiparticles but predicts a fairly small value of an anisotropy gap, estimated by ESR measurements.

References

1. H. Tanaka, A. Oosawa, T. Kato, H. Uekusa, Y. Ohashi, K. Kakurai, A. Hoser, Observation of field-induced transverse Néel ordering in the spin gap system TlCuCl$_3$. J. Phys. Soc. Jpn. **70**(4), 939–942 (2001). [Online]. https://doi.org/10.1143/JPSJ.70.939
2. B. Thielemann, C. Rüegg, K. Kiefer, H.M. Rønnow, B. Normand, P. Bouillot, C. Kollath, E. Orignac, R. Citro, T. Giamarchi, A.M. Läuchli, D. Biner, K.W. Krämer, F. Wolff-Fabris, V.S. Zapf, M. Jaime, J. Stahn, N.B. Christensen, B. Grenier, D.F. McMorrow, J. Mesot, Field-controlled magnetic order in the quantum spin-ladder system Hpip$_2$CuBr$_4$. Phys. Rev. B **79**, 020408 (2009). [Online]. https://link.aps.org/doi/10.1103/PhysRevB.79.020408
3. Y. Ohashi, H. Tajima, P. van Wyk, BCS–BEC crossover in cold atomic and in nuclear systems. Prog. Part. Nucl. Phys. **111**, 103739 (2020). [Online]. https://www.sciencedirect.com/science/article/pii/S0146641019300742
4. W. Zwerger, *The BCS-BEC Crossover and the Unitary Fermi Gas*, ser. Lecture Notes in Physics (Springer, Berlin, Heidelberg, 2011). [Online]. https://books.google.co.uz/books?id=VjdUd5JbRooC
5. R. Chitra, T. Giamarchi, Critical properties of gapped spin-chains and ladders in a magnetic field. Phys. Rev. B **55**, 5816–5826 (1997). [Online]. https://link.aps.org/doi/10.1103/PhysRevB.55.5816
6. J. Sirker, A. Weiße, O.P. Sushkov, The field-induced magnetic ordering transition in TlCuCl$_3$. J. Phys. Soc. Jpn. **74**(Suppl), 129–134 (2005). [Online]. https://doi.org/10.1143/JPSJS.74S.129
7. M. Randeria, W. Zwerger, M. Zwierlein, *The BCS–BEC Crossover and the Unitary Fermi Gas* (Springer, Berlin, Heidelberg, 2012), pp. 1–32. [Online]. https://doi.org/10.1007/978-3-642-21978-8_1
8. S.E. Sebastian, P. Tanedo, P.A. Goddard, S.-C. Lee, A. Wilson, S. Kim, S. Cox, R.D. McDonald, S. Hill, N. Harrison, C.D. Batista, I.R. Fisher, Role of anisotropy in the spin-dimer com-

pound $BaCuSi_2O_6$. Phys. Rev. B **74**, 180401 (2006). [Online]. https://link.aps.org/doi/10.1103/PhysRevB.74.180401

9. V.I. Yukalov, Basics of Bose-Einstein condensation. Phys. Part. Nucl. **42**(3), 460–513 (2011). [Online]. https://doi.org/10.1134/S1063779611030063

10. S. Miyahara, J.-B. Fouet, S.R. Manmana, R.M. Noack, H. Mayaffre, I. Sheikin, C. Berthier, F. Mila, Uniform and staggered magnetizations induced by Dzyaloshinskii-Moriya interactions in isolated and coupled spin-1/2 dimers in a magnetic field. Phys. Rev. B **75**, 184402 (2007). [Online]. https://link.aps.org/doi/10.1103/PhysRevB.75.184402

11. H. Enomoto, M. Okumura, Y. Yamanaka, Goldstone theorem, Hugenholtz–Pines theorem, and Ward–Takahashi relation in finite volume Bose–Einstein condensed gases. Ann. Phys. **321**(8), 1892–1917 (2006). [Online]. https://www.sciencedirect.com/science/article/pii/S0003491605002940

12. A. Khudoyberdiev, A. Rakhimov, A. Schilling, Bose–Einstein condensation of triplons with a weakly broken U(1) symmetry. New J. Phys. **19**(11), 113002 (2017). [Online]. https://dx.doi.org/10.1088/1367-2630/aa8a2f

13. J. Sirker, A. Weiße, O.P. Sushkov, Consequences of spin-orbit coupling for the Bose-Einstein condensation of magnons. Europhys. Lett. **68**(2), 275 (2004). [Online]. https://dx.doi.org/10.1209/epl/i2004-10179-4

14. A. Rakhimov, A. Khudoyberdiev, L. Rani, B. Tanatar, Spin-gapped magnets with weak anisotropies I: constraints on the phase of the condensate wave function. Ann. Phys. **424**, 168361 (2021). [Online]. https://www.sciencedirect.com/science/article/pii/S0003491620302955

15. S. Watabe, Hugenholtz-pines theorem for multicomponent Bose-Einstein condensates. Phys. Rev. A **103**, 053307 (2021). [Online]. https://link.aps.org/doi/10.1103/PhysRevA.103.053307

16. E. Čižmár, M. Ozerov, J. Wosnitza, B. Thielemann, K. W. Krämer, C.Rüegg, O. Piovesana, M. Klanjšek, M. Horvatić, C. Berthier, S.A. Zvyagin, Anisotropy of magnetic interactions in the spin-ladder compound $C_5H_{12}N_2CuBr_4$. Phys. Rev. B **82**, 054431 (2010). [Online]. https://link.aps.org/doi/10.1103/PhysRevB.82.054431

17. F. Cooper, B. Mihaila, J.F. Dawson, C.-C. Chien, E. Timmermans, Auxiliary-field approach to dilute Bose gases with tunable interactions. Phys. Rev. A **83**, 053622 (2011). [Online]. https://link.aps.org/doi/10.1103/PhysRevA.83.053622

18. F. Yamada, T. Ono, H. Tanaka, G. Misguich, M. Oshikawa, T. Sakakibara, Magnetic-field induced Bose–Einstein condensation of magnons and critical behavior in interacting spin dimer system $TlCuCl_3$. J. Phys. Soc. Jpn. **77**(1), 013701 (2008). [Online]. https://doi.org/10.1143/JPSJ.77.013701

19. E.Y. Sherman, P. Lemmens, B. Busse, A. Oosawa, H. Tanaka, Sound attenuation study on the Bose-Einstein condensation of magnons in $TlCuCl_3$. Phys. Rev. Lett. **91**, 057201 (2003). [Online]. https://link.aps.org/doi/10.1103/PhysRevLett.91.057201

20. G. Misguich, M. Oshikawa, Bose–Einstein condensation of magnons in $TlCuCl_3$: phase diagram and specific heat from a self-consistent Hartree–Fock calculation with a realistic dispersion relation. J. Phys. Soc. Jpn. **73**(12), 3429–3434 (2004). [Online]. https://doi.org/10.1143/JPSJ.73.3429

21. F. Yamada, H. Tanaka, T. Ono, H. Nojiri, Transition from Bose glass to a condensate of triplons in $Tl_{1-x}K_xCuCl_3$. Phys. Rev. B **83**, 020409 (2011). [Online]. https://link.aps.org/doi/10.1103/PhysRevB.83.020409

22. B. Kastening, Bose-Einstein condensation temperature of a homogeneous weakly interacting Bose gas in variational perturbation theory through six loops. Phys. Rev. A **68**, 061601 (2003). [Online]. https://link.aps.org/doi/10.1103/PhysRevA.68.061601

23. R. Dell'Amore, A. Schilling, K. Krämer, U(1) symmetry breaking and violated axial symmetry in $TlCuCl_3$ and other insulating spin systems. Phys. Rev. B **79**, 014438 (2009). [Online]. https://link.aps.org/doi/10.1103/PhysRevB.79.014438

24. A. Rakhimov, A. Khudoyberdiev, B. Tanatar, Effects of exchange and weak Dzyaloshinsky–Moriya anisotropies on thermodynamic characteristics of spin-gapped magnets. Int. J. Mod. Phys. B **35**(25), 2150223 (2021). [Online]. https://doi.org/10.1142/S0217979221502234

25. J. Sirker, A. Weiße, O.P. Sushkov, Consequences of spin-orbit coupling for the Bose-Einstein condensation of magnons. Eur. Lett. **68**(2), 275 (2004). [Online]. https://dx.doi.org/10.1209/epl/i2004-10179-4

26. A. Rakhimov, A. Khudoyberdiev, Z. Narzikulov, B. Tanatar, Defining a critical temperature of a crossover from BEC to the normal phase. Mod. Phys. Lett. B **37**(03), 2250206 (2023). [Online]. https://doi.org/10.1142/S0217984922502062

27. M. Matsumoto, B. Normand, T.M. Rice, M. Sigrist, Field- and pressure-induced magnetic quantum phase transitions in $TlCuCl_3$. Phys. Rev. B **69**, 054423 (2004). [Online]. https://link.aps.org/doi/10.1103/PhysRevB.69.054423

28. V.N. Glazkov, A.I. Smirnov, H. Tanaka, A. Oosawa, Spin-resonance modes of the spin-gap magnet $TlCuCl_3$. Phys. Rev. B **69**, 184410 (2004). [Online]. https://link.aps.org/doi/10.1103/PhysRevB.69.184410

29. N. Cavadini, G. Heigold, W. Henggeler, A. Furrer, H.-U. Güdel, K. Krämer, H. Mutka, Magnetic excitations in the quantum spin system $TlCuCl_3$. Phys. Rev. B **63**, 172414 (2001). [Online]. https://link.aps.org/doi/10.1103/PhysRevB.63.172414

30. R. Dell'Amore, A. Schilling, K. Krämer, Fraction of Bose-Einstein condensed triplons in $TlCuCl_3$ from magnetization data. Phys. Rev. B **78**, 224403 (2008). [Online]. https://link.aps.org/doi/10.1103/PhysRevB.78.224403

Optimized Perturbation Theory in δ^2 Order

17

17.1 Introduction

In previous chapters, we have studied BEC of weakly interacting atomic gases and triplons within OPT, limiting ourselves to the linear order of the auxiliary parameter-δ. This approach, being equivalent to the HFB approximation, turns out to be a powerful tool to describe the physical properties of at least homogeneous Bose systems. One of the main reasons for such success is that in majority of experiments with ultracold gases deal with weakly interacting atoms: $\gamma = \rho a_s^3 \sim 10^{-8} \div 10^{-4}$. In reality, experimenters may create BECs with stronger interactions, that is, large γ, by means of the Feshbach resonance technique [1] Besides, we have to note that superfluid helium, consisting of $\sim 10\%$ of condensed fraction, is a strongly interacting system. At saturated vapor pressure one has $a_s \approx 2.2\text{Å}$, $\rho \approx 0.022\text{Å}^{-3}$, and hence $\gamma \approx 0.215$ [2].

Does it make sense to make an attempt to study superfluid helium within the HFB approximation? Clearly, it does not. As we pointed out in Chap. 8, HFB works properly for $\gamma \leq 0.1$. This failure can be revealed, e.g. by comparison of HFB predictions with Monte-Carlo ones, which are supposed to be much more powerful and accurate than most analytical theories. In fact, for zero temperature expansion of the condensate fraction, HFB predicts (see and Chap. 8):

$$
\begin{aligned}
n_0(HFB) &= 1 - \frac{8\sqrt{\gamma}}{3\sqrt{\pi}} - \frac{64\gamma}{3\pi} - \frac{640\gamma^{3/2}}{9\pi^{3/2}} + \mathcal{O}(\gamma^2) \\
&\approx 1 - \frac{8\sqrt{\gamma}}{3\sqrt{\pi}} - 6.8\gamma - 12.8\gamma^{3/2} + \mathcal{O}(\gamma^2),
\end{aligned}
\tag{17.1}
$$

© The Author(s), under exclusive license to Springer Nature Switzerland AG 2026

A. Rakhimov and S. Mardonov, *Theory of Quantum Bose Liquids Beyond Bogoliubov Approximation*, Lecture Notes in Physics 1045,

https://doi.org/10.1007/978-3-032-05096-0_17

while Monte-Carlo calculations predict [3]

$$n_0(MC) = 1 - \frac{8\sqrt{\gamma}}{3\sqrt{\pi}} - 5.5\gamma + 7.86\gamma^{3/2} + \mathcal{O}(\gamma^2). \tag{17.2}$$

Therefore, going beyond HFB approximation, that is, developing OPT for the next orders of δ is of the utmost desirable. In the present chapter, we make an attempt to obtain next-order corrections to the free energy analytically. For simplicity, we limit ourselves to zero temperature only.

17.2 Zero Temperature Effective Action and Partition Function

It is intuitively clear that the $T = 0$ case is simpler than the $T \neq 0$ one. In fact, it is understood that higher-order corrections will include several Matsubara summations, as $\sum_{\omega_n} \sum_{\omega_m} f(\omega_n, \omega_m)$. One may naively think that, taking into account only the terms with $\omega_n = 0$, $\omega_m = 0$ will give the desired contribution for $T = 0$. However, this is not the case! Actually, there are two alternative ways of performing zero temperature calculations:

(1) You have to perform analytically all Matsubara summations up to the end, and then set there $f_B(x) = 0$, like $W_k = 1/2 + f_B \to 1/2$;
(2) Or, directly start with the following action in Minkowski space: [4]

$$S[\psi, \psi^+] = \int dt \int d\mathbf{r}\{\psi^+(t, \mathbf{r})[i\partial_t + \frac{\nabla^2}{2m} + \mu]\psi(t, \mathbf{r}) - \frac{g}{2}[\psi^+(t, \mathbf{r})\psi(t, \mathbf{r})]^2\}, \tag{17.3}$$

In this case, we have to deal with integration by ω, instead of summation.

Thus, following the second way, we present the zero temperature thermodynamic potential as follows

$$\Omega(T = 0) = \frac{i}{\mathcal{T}} \ln(Z), \tag{17.4}$$

with the partition function

$$Z = \int \mathcal{D}\psi^+ \mathcal{D}\psi e^{iS}. \tag{17.5}$$

Note that here $\mathcal{T}$ is not the temperature, but the total time interval: $\mathcal{T} = \int_0^{\mathcal{T}} dt$. Now in accordance with the general ideology of OPT, we make the Bogoluybov shift

$$\psi(\tau, \mathbf{r}) = \sqrt{\rho_0} + \tilde{\psi}(\tau, \mathbf{r}), \qquad \psi^+(\tau, \mathbf{r}) = \sqrt{\rho_0} + \tilde{\psi}^\dagger(\tau, \mathbf{r}), \tag{17.6}$$

introduce the auxiliary parameter δ through $g \to \delta g$, and add to the action the term

$$S_\delta = (\delta - 1) \int dt \int d\mathbf{r} \left[\Sigma_n(\tilde{\psi}^+\tilde{\psi}) + \frac{\Sigma_{an}}{2}(\tilde{\psi}^+\tilde{\psi}^+ + \tilde{\psi}\tilde{\psi}), \right] \tag{17.7}$$

where Σ_n and Σ_{an} are still variational parameters, which are supposed to be fixed in δ^2 order. Next, we pass to the real field representation, $\{\psi_1, \psi_2\}$ introduced by the Eq. (4.55), to separate the "free" and "interacting" parts of the action as follows:

$$S = S_0 + S_{\text{free}} + \delta S_{\text{int}},$$
$$S_0 = v_0 T V,$$
$$S_{\text{free}} = \frac{1}{2} \int dx \sum_{a,b\in(1,2)} \psi_a(x) \left[-\varepsilon_{ab}\partial_t + (\frac{\nabla^2}{2m} - X_a)\delta_{ab} \right] \psi_b(x),$$
$$S_{\text{int}} = S_1 + S_2 + S_3 + S_4,$$
$$S_1 = v_1 \int dx\,\psi_1(x), \tag{17.8}$$
$$S_2 = \int dx[v_2\psi_1^2(x) + u_2\psi_2^2(x)],$$
$$S_3 = v_3 \int dx\,\psi_1(x)[\psi_1^2(x) + \psi_2^2(x)],$$
$$S_4 = v_4 \int dx[\psi_1^2(x) + \psi_2^2(x)]^2,$$

where we introduced short notation $x = (t, \mathbf{r})$ and the vertices:

$$v_0 = -\frac{g\rho_0^2}{2} + \mu_0\rho_0, \qquad v_1 = -g\sqrt{2}\rho_0^{3/2}, \qquad v_2 = \frac{X_1 + \mu_1 - 3g\rho_0}{2},$$
$$u_2 = \frac{X_2 + \mu_1 - g\rho_0}{2}, \qquad v_3 = -\frac{g\sqrt{\rho_0}}{\sqrt{2}}, \qquad v_4 = -\frac{g}{8}. \tag{17.9}$$

The new variational parameters X_1 and X_2 are the linear combinations of Σ_n and Σ_{an}:

$$X_1 = \Sigma_n + \Sigma_{an} - \mu_1, \qquad X_2 = \Sigma_n - \Sigma_{an} - \mu_1. \tag{17.10}$$

The term S_{free} in (17.8) gives rise to following *free* partition function:

$$Z_0 = \int \mathcal{D}\psi_1 \mathcal{D}\psi_2 e^{i\,S_{\text{free}}}$$
$$= \int \mathcal{D}\psi_1 \mathcal{D}\psi_2 \exp \left\{ -\frac{1}{2V} \int \frac{d\omega}{(2\pi)} \frac{d\mathbf{k}}{(2\pi)^3} \sum_a \psi_a(\omega, \mathbf{k}) D_{ab}^{-1}(\omega, \mathbf{k})\psi_b(\omega, \mathbf{k}) \right\},$$
$$\tag{17.11}$$

where the zero temperature inverse propagator in momentum space with

$$\psi_a(x) = \frac{1}{V} \sum_k \int \frac{d\omega}{2\pi} e^{i\omega t - i\mathbf{kr}} \psi_a(\omega, \mathbf{k}), \tag{17.12}$$

is given by [4]:

$$D^{-1}(\omega, \mathbf{k}) = -i \begin{pmatrix} \varepsilon_k + X_1 & -i\omega \\ i\omega & \varepsilon_k + X_2 \end{pmatrix}, \tag{17.13}$$

The perturbation series are performed in orders of δ with the vertices defined in Eqs. (17.9) and the propagator

$$D(\omega, \mathbf{k}) = \frac{i}{(\omega^2 - E_k^2 + i\epsilon)} \begin{pmatrix} \varepsilon_k + X_2 & i\omega \\ -i\omega & \varepsilon_k + X_1 \end{pmatrix}, \tag{17.14}$$

where

$$E_k = \sqrt{(\varepsilon_k + X_1)(\varepsilon_k + X_2)}. \tag{17.15}$$

as usual. For a uniform system, the propagator in coordinate space depends only on the difference $x_1 - x_2$:

$$D_{ab}(x_1, x_2) = \frac{1}{V} \sum_k \int \frac{d\omega}{2\pi} e^{i\omega(t_1 - t_2) - i\mathbf{k}(\mathbf{r}_1 - \mathbf{r}_2)} D_{ab}(\omega, \mathbf{k}). \tag{17.16}$$

Therefore, the whole partition function has the form

$$Z = e^{i S_0} \int \mathcal{D}\psi_1 \mathcal{D}\psi_2 e^{i S_{\text{free}}} e^{i \delta S_{\text{int}}}, \tag{17.17}$$

where

$$i S_{\text{free}} = -\frac{1}{2V} \int \frac{d\omega}{2\pi} \sum_k \sum_{a,b} \psi_a(\omega, \mathbf{k}) D_{ab}^{-1}(\omega, \mathbf{k}) \psi_b(\omega, \mathbf{k}). \tag{17.18}$$

Now, making formal expansion in (17.17)

$$e^{i \delta S_{\text{int}}} = 1 + i \delta S_{\text{int}} - \frac{\delta^2 [S_{\text{int}}]^2}{2} + \mathcal{O}(\delta^3) \tag{17.19}$$

one may present Ω as follows

$$\Omega = \frac{i}{T} \ln Z \approx \frac{i}{T} \ln e^{i S_0} + \frac{i}{T} \ln \left\{ \int \mathcal{D}\psi_1 \mathcal{D}\psi_2 e^{i S_{\text{free}}} \left[1 + i \delta S_{\text{int}} - \frac{\delta^2 [S_{\text{int}}]^2}{2} \right] \right\}. \tag{17.20}$$

The first term is simple:

$$\Omega_0 \equiv \frac{i}{T} \ln e^{i S_0} = V \left[\frac{g \rho_0^2}{2} - \mu_0 \rho_0 \right]. \tag{17.21}$$

The path integral in the second term of Eq. (17.20) can be rewritten as

$$\int \mathcal{D}\psi_1 \mathcal{D}\psi_2 e^{i S_{\text{free}}} \left[1 + i\delta S_{\text{int}} - \frac{\delta^2 [S_{\text{int}}]^2}{2} \right]$$

$$= Z_0 \left\{ 1 + i\delta \left[\langle S_1 \rangle + \langle S_2 \rangle + \langle S_3 \rangle + \langle S_4 \rangle \right] - \frac{\delta^2}{2} \langle [S_1 + S_2 + S_3 + S_4]^2 \rangle \right\}, \tag{17.22}$$

where the mean value of an operator $\mathcal{O}$ is defined as usual

$$\langle \mathcal{O} \rangle = \frac{1}{Z_0} \int \mathcal{D}\psi_1 \mathcal{D}\psi_2 e^{i S_{\text{free}}} \mathcal{O}[\psi_1, \psi_2], \tag{17.23}$$

with

$$Z_0 = (Det\, D^{-1})^{-1/2}, \tag{17.24}$$

(see Eq. (8.16)).

It is clear that, in (17.22) $\langle S_1 \rangle = \langle S_3 \rangle = 0$. Therefore, the second term of Eq. (17.20) can be simplified as

$$\frac{i}{T} \ln Z_0 \left\{ 1 + i\delta \left[\langle S_2 \rangle + \langle S_4 \rangle \right] - \frac{\delta^2 [S_{\text{int}}]^2}{2} \right\}$$

$$\approx \frac{i}{T} \ln Z_0 - \frac{\delta}{T} \left[\langle S_2 \rangle + \langle S_4 \rangle \right] - \frac{i\delta^2 [S_{\text{int}}]^2}{2T}, \tag{17.25}$$

where we expanded the logarithm in a power series of δ. Further, as it has been explained in Chap. 6, $\ln Z_0$ in (17.25) can be evaluated easily, resulting in:

$$\Omega_{ln} = \frac{i}{T} \ln Z_0 = \frac{1}{2} \sum_k (E_k - \varepsilon_k), \tag{17.26}$$

where we performed contour integration by ω, using the formulas [5]

$$\int \frac{d\omega}{2\pi} \ln(\omega^2 - E_k^2) = i E_k, \qquad \int \frac{d\omega}{2\pi (\omega^2 - E_k^2 + i\epsilon)} = -\frac{i}{2E_k} \tag{17.27}$$

Therefore for Ω we have

$$\Omega = \Omega_0 + \Omega_{ln} - \frac{\delta}{T} \left[\langle S_2 \rangle + \langle S_4 \rangle \right] - \frac{i\delta^2 [S_{\text{int}}]^2}{2T} + \mathcal{O}(\delta^3). \tag{17.28}$$

Now we discuss δ and δ^2 orders separately.

17.3 $\Omega(T = 0)$ in the First Order of δ

To verify our zero temperature approach, below we make an attempt to derive again the expression for the free energy brought in Chap. 8. For this purpose, we first bring the following equations:

$$D_{11}(0) \equiv D_{11}(x_1 - x_2)|_{x_1 \to x_2} = \frac{1}{V} \sum_k \int \frac{d\omega}{2\pi} D_{11}(\omega, \mathbf{k})$$

$$= \frac{i}{V} \sum_k \int \frac{d\omega}{2\pi} \frac{\varepsilon_k + X_2}{\omega^2 - E_k^2 + i\epsilon} = \frac{1}{2V} \sum_k \frac{\varepsilon_k + X_2}{E_k} \equiv \frac{A_1}{V},$$

$$D_{22}(0) \equiv D_{22}(x_1 - x_2)|_{x_1 \to x_2} = \frac{1}{V} \sum_k \int \frac{d\omega}{2\pi} D_{22}(\omega, \mathbf{k})$$

$$= \frac{1}{2V} \sum_k \frac{\varepsilon_k + X_1}{E_k} \equiv \frac{A_2}{V},$$

$$D_{12}(0) \equiv D_{12}(x_1 - x_2)|_{x_1 \to x_2} = 0, \quad D_{21}(0) \equiv D_{21}(x_1 - x_2)|_{x_1 \to x_2} = 0,$$
$$\tag{17.29}$$

which can be obtained by using Eqs. (17.14), (17.16), (17.27) and

$$\int_{-\infty}^{\infty} \frac{\omega d\omega}{2\pi(\omega^2 - E_k^2 + i\epsilon)} = 0. \tag{17.30}$$

The mean values such as $\langle S_2 \rangle$ and $\langle S_4 \rangle$ in Eq. (17.28) can be directly evaluated by using Eq. (6.94):

$$\langle S_2 \rangle = \int dx [v_2 \langle \psi_1^2(x) \rangle + u_2 \langle \psi_2^2(x) \rangle],$$
$$\tag{17.31}$$
$$\langle S_4 \rangle = v_4 \int dx [\langle \psi_1^4(x) \rangle + \langle \psi_2^4(x) \rangle + 2\langle \psi_1^2(x)\psi_2^2(x) \rangle],$$

here, e.g. $\langle \psi_1^2(x) \rangle = D_{11}(0)$. Therefore, putting this into (17.28) and using Eq. (17.29) we finally obtain

$$\Omega^{(1)} = \Omega_0 + \Omega_{ln} + \Omega_2 + \Omega_4, \quad \Omega_2 = -\frac{1}{2}[v_2 A_1 + u_2 A_2],$$
$$\tag{17.32}$$
$$\Omega_4 = -\frac{v_4}{V}[3A_1^2 + 2A_1 A_2 + 3A_2^2],$$

which exactly coincides with (8.22) at zero temperature, bearing in mind the vertices defined in Eqs. (17.9). Minimization of $\Omega^{(1)}$ with respect to $X_{1,2}$ leads to the familiar equations, obtained in Chap. 8:

$$X_1^{(1)} = -\mu_1 + 3g\rho_0 + 2g\rho_1 + g\sigma, \quad X_2^{(2)} = -\mu_1 + g\rho_0 + 2g\rho_1 - g\sigma,$$
$$\tag{17.33}$$

where $\rho_1 = (A_1 + A_2)/2V$, $\sigma = (A_1 - A_2)/2V$, $\rho_0 = \rho - \rho_1$. The Feynman diagrams, contributing to the effective potential $\Omega^{(1)}$ are illustrated in Fig. 17.1.

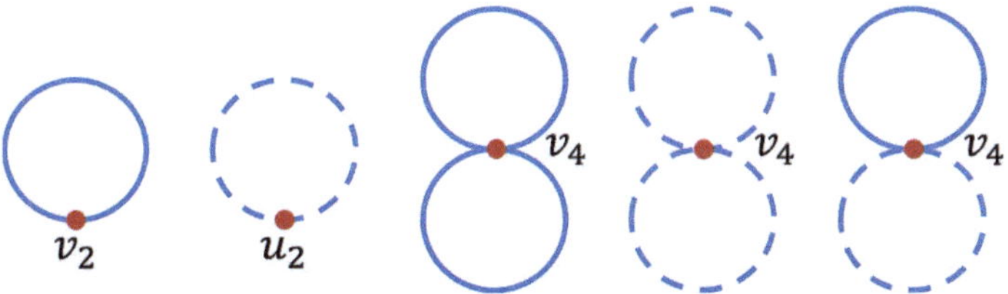

Fig. 17.1 Feynman diagrams contributing to Ω in the first order of δ. Here and below, solid and dashed lines correspond to the propagators D_{11} and D_{22}, respectively

17.4 The Second Order Calculations

Following the Eq. (17.28) the second order contribution to Ω has the form:

$$\Delta\Omega^{(2)} = -\frac{i}{2T}[\langle S_1 S_1\rangle + \langle S_2 S_2\rangle + \langle S_3 S_3\rangle + \langle S_4 S_4\rangle + 2\langle S_1 S_2\rangle \\ + 2\langle S_1 S_3\rangle + 2\langle S_1 S_4\rangle + 2\langle S_2 S_3\rangle + 2\langle S_2 S_4\rangle + 2\langle S_3 S_4\rangle], \tag{17.34}$$

such that

$$\Omega^{(2)} = \Omega^{(1)} + \Delta\Omega^{(2)}, \tag{17.35}$$

In (17.34), e.g.

$$\langle S_1 S_2\rangle = v_1 \int dx_1 dx_2[v_2\langle\psi_1(x_1)\psi_1^2(x_2)\rangle + u_2\langle\psi_1(x_1)\psi_2^2(x_2)\rangle]. \tag{17.36}$$

Starting to analyze each term of the long-expression in (17.34), one should keep in mind the following rules:

- A term including odd numbers of ψ vanishes exactly. Therefore, $\langle S_1 S_2\rangle = 0$, $\langle S_1 S_4\rangle = 0$, $\langle S_2 S_3\rangle = 0$ and $\langle S_3 S_4\rangle = 0$;
- Only one particle irreducible diagram (1PI) should be taken into account. Actually, by definition 1PI diagram is a diagram that can not be split in two by removing a single line. For example, it can be seen that $\langle S_1 S_1\rangle$ and $\langle S_1 S_3\rangle$ do not obey this rule.
- Disconnected diagrams should also be omitted (see Fig. 17.2). Actually, in δ^2 order we have two vertex points, x_1 and x_2. So, any term which does not include at least one $D_{ij}(x_1, x_2)$, that is, a line connecting these two points, is classified as a disconnected one;
- Any diagram, including $D_{1,2}$ or $D_{2,1}$ with equal arguments, e.g. $D_{2,1}(x_1, x_1)$ must be directly neglected due to the Eq. (17.29);

Proof of the first rule may be found in Appendix C, while the rest ones elsewhere (see e.g. the textbook by Fetter and Walechka [6]).

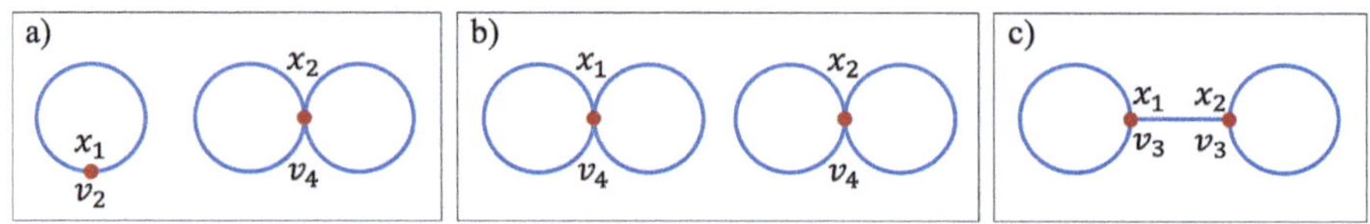

Fig. 17.2 Examples of disconnected (**a**, **b**) and one particle reducible diagrams (**c**) in δ^2-order

All and all, we are left with only four terms given below:

$$\Delta\Omega^{(2)} = -\frac{i}{2T}[\langle S_2 S_2\rangle + \langle S_3 S_3\rangle + \langle S_4 S_4\rangle + 2\langle S_2 S_4\rangle] \tag{17.37}$$

where each S_i ($i = 2, 3, 4$) includes i number of psis. For this reason, direct evaluation of $\langle S_i S_j\rangle$ using Wick's theorem is a rather tedious business.

It is not difficult to generate a computer code in MAPLE which performs such a procedure by functional differentiation, which provides correct combinatorial weights (see Appendix C). For demonstration, we consider, for example, the term $\langle S_2 S_4\rangle$, which is written explicitly as follows:

$$\langle S_2 S_4\rangle = v_4 \int dx_1 dx_2 \langle [\psi_1^2(x_1) + \psi_2^2(x_1)]^2 [v_2\psi_1^2(x_2) + u_2\psi_2^2(x_2)]\rangle. \tag{17.38}$$

Now, opening the brackets, consider the term

$$P_{111111} = \langle \psi_1^4(x_1)\psi_1^2(x_2)\rangle. \tag{17.39}$$

Using the rules described in the Appendix C one obtains

$$P_{111111} = 3D_{11}(x_2, x_2)D_{11}^2(x_1, x_1) + 12D_{11}^2(x_1, x_2)D_{11}(x_1, x_1) \tag{17.40}$$

It is clear that the first term is disconnected and hence should be omitted. So, we are left with

$$\langle \psi_1^4(x_1)\psi_1^2(x_2)\rangle = 12D_{11}^2(x_1, x_2)D_{11}(0) = 12D_{11}^2(x_1, x_2)A_1/V. \tag{17.41}$$

Task 17.1. Explain why the following terms arose from $\langle S_2 S_4\rangle$, do not contribute $\Omega^{(2)}$: $D_{12}^2(x_2, x_2)D_{11}^2(x_1, x_1)$ and $D_{22}(x_2, x_2)D_{11}(x_2, x_2)D_{11}^2(x_1, x_1)$.

Thus, after some algebraic manipulations, we obtain the following final expressions

$$\langle S_2 S_2\rangle = 2\int dx_1 dx_2 \left\{ v_2^2 D_{11}^2(x_1, x_2) + u_2^2 D_{22}^2(x_1, x_2) + 2v_2 u_2 D_{12}^2(x_1, x_2)\right\}, \tag{17.42}$$

$$\langle S_2 S_4 \rangle = 4v_4 V^{-1} \int dx_1 dx_2 \{ [v_2 D_{11}^2(x_1, x_2) + u_2 D_{12}^2(x_1, x_2)][3A_1 + A_2]$$

$$+ [u_2 D_{22}^2(x_1, x_2) + v_2 D_{12}^2(x_1, x_2)][3A_2 + A_1] \}, \tag{17.43}$$

$$\langle S_3 S_3 \rangle = v_3^2 \int dx_1 dx_2 \{ 6D_{11}(x_1, x_2)[D_{11}^2(x_1, x_2) + 2D_{12}^2(x_1, x_2)]$$

$$+ 2D_{22}(x_1, x_2)[D_{11}(x_1, x_2)D_{22}(x_1, x_2) - 2D_{12}^2(x_1, x_2)] \}, \tag{17.44}$$

$$\langle S_4 S_4 \rangle = 8v_4^2 \int dx_1 dx_2 \{ S_{44}^a(x_1, x_2) + \frac{S_{44}^b(x_1, x_2)}{V^2} \},$$

$$S_{44}^a = 3D_{11}^4(x_1, x_2) + [2D_{22}^2(x_1, x_2) + 12D_{12}^2(x_1, x_2)]D_{11}^2(x_1, x_2)$$

$$- 8D_{12}^2(x_1, x_2)D_{22}(x_1, x_2)D_{11}(x_1, x_2)$$

$$+ 12D_{22}^2(x_1, x_2)D_{12}^2(x_1, x_2) + 8D_{12}^4(x_1, x_2) + 3D_{22}^4(x_1, x_2), \tag{17.45}$$

$$S_{44}^b = [D_{22}^2(x_1, x_2) + 9D_{11}^2(x_1, x_2) + 6D_{12}^2(x_1, x_2)]A_1^2$$

$$+ [6D_{11}^2(x_1, x_2) + 20D_{12}^2(x_1, x_2) + 6D_{22}^2(x_1, x_2)]A_1 A_2$$

$$+ [6D_{12}^2(x_1, x_2) + 9D_{22}^2(x_1, x_2) + D_{11}^2(x_1, x_2)]A_2^2,$$

where we used the relation $D_{21}(x, y) = -D_{1,2}(x, y)$, which directly follows from
Eq. (17.14). Feynman diagrams, contributing to the thermodynamic potential in δ^2
order are presented in Fig. 17.3.

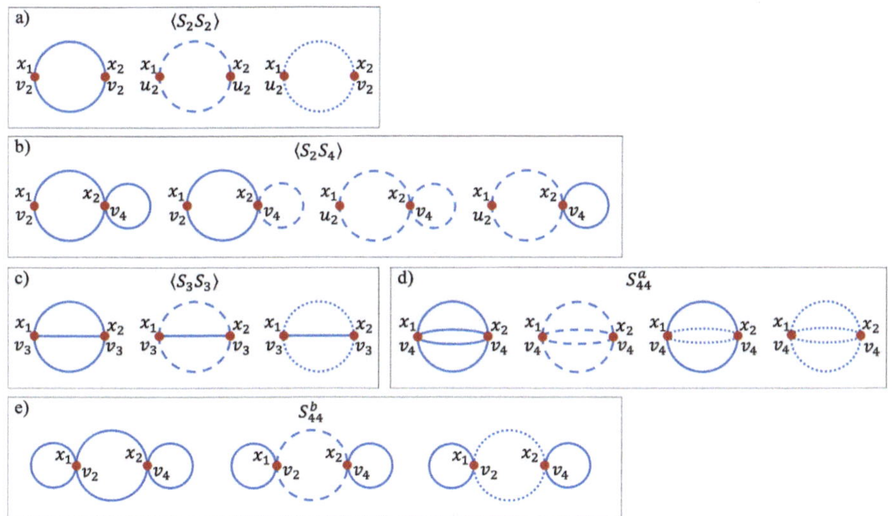

Fig. 17.3 Examples of Feynman diagrams contributing to Ω in δ^2-order. They are correspondingly given in Eqs. (17.42)–(17.45)

17.5　Momentum Space

These diagrams look more elegant in momentum space. Using Eq. (17.16) we may present e.g. the integral in the first term of (17.42) as follows:

$$
\int dx_1 dx_2 D_{11}^2(x_1, x_2) = \mathcal{T}V \int dt d\mathbf{r} D_{11}^2(t, \mathbf{r})
$$

$$
= \frac{\mathcal{T}V}{V^2(2\pi)^2} \sum_{k,p} \int d\omega_1 \omega_2 e^{i(\omega_1+\omega_2)t} e^{-i(\mathbf{k}+\mathbf{p})\mathbf{r}} D_{11}(\omega_1, \mathbf{k}) D_{11}(\omega_2, \mathbf{p})
$$

$$
= \mathcal{T} \sum_k \int \frac{d\omega}{2\pi} D_{11}(\omega, \mathbf{k}) D_{11}(-\omega, -\mathbf{k}) \tag{17.46}
$$

$$
= i^2 \mathcal{T} V \int \frac{d\omega}{2\pi} \int \frac{d\mathbf{k}}{(2\pi)^3} \frac{(\varepsilon_k + X_2)^2}{(\omega^2 - E_k^2 + i\epsilon)^2},
$$

where we used well-known textbook formulas:

$$
\int dt = \mathcal{T}, \qquad \int d\mathbf{r} = V, \qquad \int d\mathbf{r} e^{i(\mathbf{k}-\mathbf{p})\mathbf{r}} = (2\pi)^3 \delta(\mathbf{k}-\mathbf{p}),
$$

$$
\sum_k f(\mathbf{k}) \delta(\mathbf{k}-\mathbf{p}) = \frac{V f(\mathbf{p})}{(2\pi)^3}, \qquad \int d\mathbf{r} \delta(\mathbf{r}) = \int d\mathbf{k} \delta(\mathbf{k}) = 1,
$$

$$
\int dt e^{i(\omega-\omega')t} = (2\pi)\delta(\omega-\omega'), \tag{17.47}
$$

$$
\int d\omega' \delta(\omega'-\omega) f(\omega, \omega') = f(\omega, \omega).
$$

Note that, throughout this book, we consider only uniform systems, where the equality $D_{ij}(x_1, x_2) = D_{ij}(x_1 - x_2)$ holds. Similarly, one particularly obtains

$$
\int dx_1 dx_2 D_{11}(x_1, x_2) D_{12}^2(x_1, x_2) =
$$

$$
\frac{\mathcal{T}}{V} \sum_k \sum_p \int \frac{d\omega d\omega'}{(2\pi)^2} D_{11}(\omega, \mathbf{k}) D_{12}(\omega', \mathbf{p}) \tag{17.48}
$$

$$
\times D_{12}(-\omega - \omega', -\mathbf{k} - \mathbf{p}),
$$

$$
\int dx_1 dx_2 D_{11}(x_1, x_2) D_{22}(x_1, x_2) D_{12}^2(x_1, x_2) =
$$

$$
\frac{\mathcal{T}}{V^2} \sum_{k,p,q} \int \frac{d\omega d\omega' d\omega''}{(2\pi)^3} D_{11}(\omega, \mathbf{k}) D_{22}(\omega', \mathbf{p}) \tag{17.49}
$$

$$
\times D_{12}(\omega'', \mathbf{q}) D_{12}(-\omega - \omega' - \omega'', -\mathbf{k} - \mathbf{p} - \mathbf{q}).
$$

These are illustrated in Fig. 17.4.

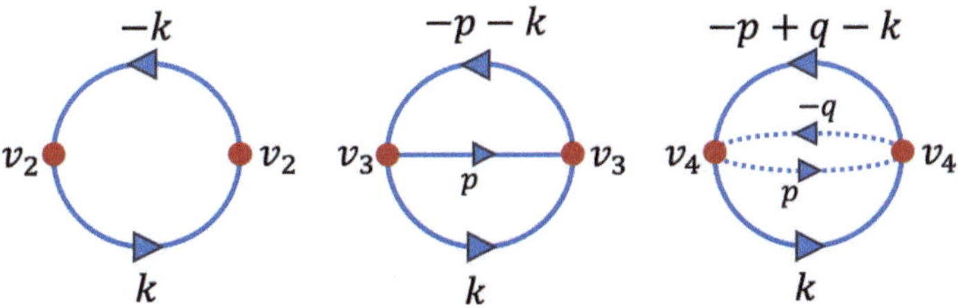

Fig. 17.4 Illustration of Feynman diagrams in momentum space

It is seen that any diagram has, in general, multiple integrals, both energy and momentum ones. The energy integrals can be evaluated, in principle, by using residue theory.

Task 17.2 Using residue theory, or directly using Mathematica evaluate the following integrals:

(1)

$$\int \frac{d\omega}{2\pi} \frac{1}{(\omega^2 - E_k^2 + i\epsilon)^2},\tag{17.50}$$

(2)

$$\int \frac{d\omega d\omega'}{(2\pi)^2} \frac{1}{(\omega^2 - E_k^2 + i\epsilon)} \frac{1}{((\omega + \omega')^2 - E_q^2 + i\epsilon)}.\tag{17.51}$$

However, calculating momentum integrals takes a lot of time and labour. In addition, most of them contain UV and IR divergences. For this reason, to save time and space, here we don't discuss this item, referring the reader to the original articles [7–10]. Note that, after the difficulties with integrals are overcome, one may fix the trial parameters $X_{1,2}$ by minimization of the whole $\Omega = \Omega^{(1)} + \Delta\Omega^{(2)}$, where the terms are given by Eqs. (17.32) and (17.37), respectively.

At the end of this section, we outline Feynman rules for constructing Ω in the nth order of δ:

- Any diagram of δ^n order contains n pieces of points, say x_1, x_2 ... x_n.
- Provide any point with a vertex v_i, $(i = 1 \ldots 4)$, which has i pieces of legs, e.g. v_4 has four legs.
- Connect each pair of the points with a line, which represents a propagator.
- Generate a code, using the rules given in Appendix C, to fix combinatorial coefficients.
- Exclude all disconnected and one particle reducible diagrams.
- Integrate over all points: $(dt_1 dt_2 \cdots dt_n d\mathbf{r}_1 d\mathbf{r}_2 \cdots d\mathbf{r}_n)$. Or introduce Matsubara summations for finite temperature.

17.6 Summary

HFB approximation, which corresponds to OPT of first order of δ fails to describe systems of bosons with large γ. To improve the theory higher orders of OPT should be developed. We have outlined the procedure of construction of the thermodynamic potential in δ^2 order and formulated corresponding Feynman rules. Although high-order corrections include many loop integrals, they can be evaluated, at least numerically, by using powerful computers.

References

1. C. Chin, R. Grimm, P. Julienne, E. Tiesinga, Feshbach resonances in ultracold gases. Rev. Mod. Phys. **82**, 1225–1286 (2010). https://link.aps.org/doi/10.1103/RevModPhys.82.1225
2. P.A. Whitlock, D.M. Ceperley, G.V. Chester, M.H. Kalos, Properties of liquid and solid ^{4}He. Phys. Rev. B **19**, 5598–5633 (1979). https://link.aps.org/doi/10.1103/PhysRevB.19.5598
3. M. Rossi, L. Salasnich, Path-integral ground state and superfluid hydrodynamics of a bosonic gas of hard spheres. Phys. Rev. A **88**, 053617 (2013). https://link.aps.org/doi/10.1103/PhysRevA.88.053617
4. J. Blaizot, G. Ripka, J.W. Negele, H. Orland, G.E. Brown, Quantum theory of finite systems and quantum many? particle systems. Phys. Today **41**(9), 106–107 (1988). https://doi.org/10.1063/1.2811565
5. J.O. Andersen, Theory of the weakly interacting Bose gas. Rev. Mod. Phys. **76**, 599–639 (2004). https://link.aps.org/doi/10.1103/RevModPhys.76.599
6. A. Fetter, J. Walecka, *Quantum Theory of Many-particle Systems*, Dover Books on Physics (Dover Publications, 2003). https://books.google.co.uz/books?id=0wekf1s83b0C
7. E. Braaten, A. Nieto, Quantum corrections to the ground state of a trapped Bose-Einstein condensate. Phys. Rev. B **56**, 14 745–14 765 (1997). https://link.aps.org/doi/10.1103/PhysRevB.56.14745
8. J.O. Andersen, E. Braaten, Semiclassical corrections to a static Bose-Einstein condensate at zero temperature. Phys. Rev. A **60**, 2330–2345 (1999). https://link.aps.org/doi/10.1103/PhysRevA.60.2330
9. H. Kleinert, S. Schmidt, A. Pelster, Quantum phase diagram for homogeneous Bose-Einstein condensate. Annalen der Physik **517**(4), 214–230 (2005). https://onlinelibrary.wiley.com/doi/abs/10.1002/andp.20055170402
10. T. Haugset, H. Haugerud, F. Ravndal, Thermodynamics of a weakly interacting Bose–Einstein gas. Ann. Phys. **266**(1), 27–62 (1998). https://www.sciencedirect.com/science/article/pii/S0003491698957955

Appendix A: Chemical Potential in Boltzmann Distribution

A

All's well that ends well

We start with the Avogadro law, in accordance with the molar volume of any gas under standard conditions: ($T = 273.15$ K, $P = 760$mm/Hg) $V_m = 22.4L/mol$ and consists of $N_{av} \approx 6. * 10^{23}$ atoms. This means that its number density $\rho = N/V = 2.7 \cdot 10^{25}/m^3 = 2.7 \cdot 10^{-2}/nm^3$. Now introducing the fugacity $z = \exp(\beta\mu)$, ($\beta = 1/T$) we rewrite the Boltzmann distribution as

$$z\frac{N}{V} = z\rho \equiv I = \frac{4\pi}{(2\pi\hbar)^3} \int_0^\infty dk k^2 \exp(-\varepsilon_k \beta) = \frac{4\pi\sqrt{2}(mT)^{3/2}I_0}{(2\pi\hbar)^3} = \frac{\sqrt{2}I_0 c_0}{2\pi^2},$$

$$(A.1)$$

where $I_0 = \int_0^\infty dx \sqrt{x} \exp(-x) = \sqrt{\pi}/2$. As to $c_0 = (mT)^{3/2}/\hbar^3$, it can cause trouble for a beginner, who has no experience with dimensional units. Below we show that c_0 has a dimension of $1/volume$ as it is expected. Multiplying the numerator and denominator of c_0 by c^3 (c is the speed of light) and using the following numerical values $m = Am_p, m_p c^2 \approx 940 MeV = 940 \cdot 10^6 ev, \hbar c = 197.33 ev \cdot nm, k_B = 8.617 \cdot 10^{-5} ev/K$ leads to

$$c_0 = \frac{(mk_B T)^{3/2}}{\hbar^3} = \frac{(AT_k m_p c^2 k_B)^{3/2}}{(\hbar c)^3} \approx 3(AT_k)^{3/2}\frac{1}{nm^3}, \qquad (A.2)$$

where m_p is the proton mass, A is the atomic weight, $m_{atom} \approx Am_p$ and T_k is the value of temperature in Kelvin. Now using (A.2) and (A.1) we obtain

$$\mu = T_k k_B [-1.66 + \ln \frac{(AT_k)^{3/2}}{\tilde{\rho}}], \qquad (A.3)$$

where $\rho = \tilde{\rho}/nm^3$. In particular, for ^4He gas $A = 4$, $T_k = 273.15$ with $\tilde{\rho} = 0.027$ one finds $\mu = 0.3ev$.

A. Rakhimov and S. Mardonov, *Theory of Quantum Bose Liquids Beyond Bogoliubov Approximation*, Lecture Notes in Physics 1045, https://doi.org/10.1007/978-3-032-05096-0

Appendix B: Laser Cooling

Nearly 50 years ago, physicists knew mainly two methods of obtaining ultracold matter: Cooling with liquid helium and then proceeding with magnetic cooling. For diluting atomic gases by using the first method, they could reach $T \sim 4K$, and further, they could not advance significantly. The point is that magnetic cooling is not effective for neutral atoms. This was one of the main reasons why BEC had been still existed only theoretically, being predicted long ago in 1925 [1]. At last, in 1975, two research groups [2,3] separately proposed a new method - laser cooling, which was experimentally realized later on [4]. The 1997 Nobel Prize in Physics was awarded to Claude Cohen-Tannoudji, Steven Chu, and William Daniel Phillips "For the development of methods to cool and trap atoms with laser light". Below we briefly explain the essence of this method.

Laser cooling is based on the Doppler effect. Let's consider two levels of an atom: the ground state $|0\rangle$ and the nearest excited one $|1\rangle$. We denote the "distance" between them as ϵ_{01}. Let the laser beam has the frequency ω, which is tuned slightly below the frequency of the transition $|0\rangle \leftrightarrow |1\rangle$ i.e. $\omega < \epsilon_{01}$. Such a situation is referred to as "red" tuning. Clearly, a resting atom ($v = 0$) cannot absorb this photon. In reality, atoms in a gas have non-zero velocity due to thermal effects. Among them, in a moment, there exists at least one atom that moves toward the light. Thanks to the Doppler shift, this atom will be the best candidate to absorb the photon going into the excited state $|1\rangle$. After a short time, it reemits another photon in a random direction, going back into the ground state. The atom momentum vectors would add to the original if they were in the same direction, but they are not. So, the atom loses its kinetic energy and slows down.[1]

There is a large probability that the energy of the emitted photon will be larger than that of the absorbed one: $\omega' > \omega$. Where does it take the additional energy?

---

[1] Note that $T \sim mv^2/2$.

A. Rakhimov and S. Mardonov, *Theory of Quantum Bose Liquids Beyond Bogoliubov Approximation*, Lecture Notes in Physics 1045, https://doi.org/10.1007/978-3-032-05096-0

Actually, it was taken from the kinetic energy of the atom. Due to the conservation of the total energy

$$E_{atom} + \omega = E'_{atom} + \omega', \quad E_{atom} = E_{kin} + E_0, \quad E'_{atom} = E'_{kin} + E_0, \quad (B.1)$$

where E_0 and E_{kin} are the internal and kinetic energies, respectively. So, it is not prohibited that if $\omega' > \omega$ then $E'_{kin} < E_{kin}$. In other words, translational kinetic energy can be transferred from the gas to the scattered light, until the atomic velocity is reduced by the ratio of the Doppler width to the natural line width.

This brilliant method makes it possible to cool atoms down to ~ 240 microkelvins. Since atoms in the gas move chaotically, any other "fast" atom may turn toward the light to loose its kinetic energy. This procedure goes on until every atom in the gas is slowed down.

When it is necessary, further cooling may be proceeded by the well-known evaporative cooling [5,6] which is similar to the cooling of water during evaporation.

Appendix C: Functionals and Functional Derivatives

C

What is the difference between function, functional and operator? Although a mathematician would give a rather complicated answer, the physical community may give a simplified answer to this question at a student level. So, in a nutshell, a function $f(x)$ maps a given c-number[2] into another c-number: $c' = f(x = c)$; a functional maps a given function $f(x)$ into a c-number: $c = F[f(x)]$; and an operator maps a given function $f(x)$ into another function, say, $\phi(x)$: $\phi(x) = \hat{F} f(x)$. This is illustrated in Fig. C.1. For example, the energy is a c-number, the classical Hamiltonian $H = \mathbf{k}^2/2m + U(x)$ is a function, while the quantum Hamiltonian $\hat{H} = -\nabla^2/2m + U(x)$ is an operator, and the action $S = S[\psi, \psi^\dagger]$ is a functional. Another widely used functional is $\delta_y[f(x)] = f(y)$, which returns the value of the function at point y and defines the Dirac delta-function $\delta(x - y)$:

$$\delta_y[f(x)] = \int_{-\infty}^{\infty} dx\, \delta(x - y) f(x) = f(y). \tag{C.1}$$

Particularly, any integral of a given function $f(x)$ in the range $[a, b]$ may be also considered as a functional: $c = \int_a^b f(x) dx$.

Similarly to the derivative of a function $df(x)/dx = \lim_{\epsilon \to 0}[f(x + \epsilon) - f(x)]/\epsilon$ one may define a functional derivative of the functional $F[f(x)]$ with respect to the function $f(x)$ as

$$\frac{\delta F[f(y)]}{\delta f(x)} = \lim_{\epsilon \to 0} \frac{F[f(y) + \epsilon\delta(y - x)] - F[f(y)]}{\epsilon}. \tag{C.2}$$

---

[2] We use the term c-number as an old nomenclature, introduced by Dirac, which refers to real and complex numbers.

A. Rakhimov and S. Mardonov, *Theory of Quantum Bose Liquids Beyond Bogoliubov Approximation*, Lecture Notes in Physics 1045, https://doi.org/10.1007/978-3-032-05096-0

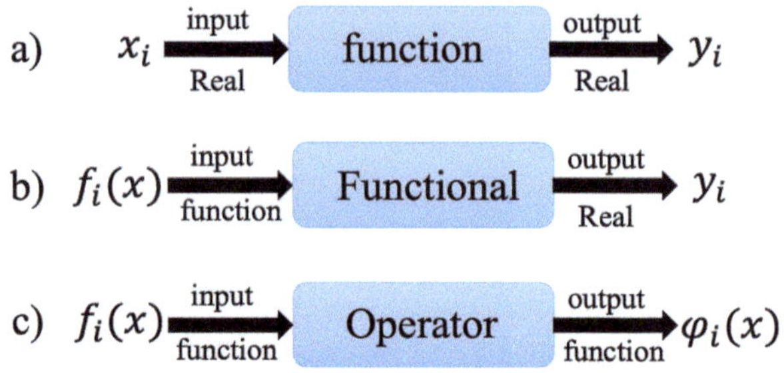

Fig. C.1 Schematic definition of a function (**a**), a functional (**b**) and an operator (**c**)

It has similar properties as a usual derivative, such as

$$\frac{\delta(F_1[f]) + F_2[f]}{\delta f(x)} = \frac{\delta F_1[f]}{\delta f(x)} + \frac{\delta F_2[f]}{\delta f(x)}. \tag{C.3}$$

$$\frac{\delta(F_1[f]F_2[f])}{\delta f(x)} = \frac{\delta F_1[f]}{\delta f(x)} F_2[f] + \frac{\delta F_2[f]}{\delta f(x)} F_1[f]. \tag{C.4}$$

$$\frac{\delta(F[\phi(f(x))]}{\delta f(x)} = \frac{d\phi}{df} \frac{\delta F[\phi]}{\delta \phi}. \tag{C.5}$$

The last equation is an analogy of the chain rule for the derivative of the composition of two differentiable functions f and ϕ. For example,

$$\frac{\delta(F[f^2(x)]}{\delta f(x)} = 2 \frac{\delta F[f]}{\delta f(x)}. \tag{C.6}$$

In practical calculations following identities are widely used

$$\frac{\delta f(x)}{\delta f(y)} = \delta(x - y), \qquad \frac{\delta}{\delta f(y)} \int_{-\infty}^{\infty} dx f(x)\phi(x) = \phi(y). \tag{C.7}$$

More complicated cases can be overcome by using the above Eqs. (C.3)–(C.7). We illustrate this with the following example:

$$\frac{\delta}{\delta f(y)} \exp\left(\alpha \int_{-\infty}^{\infty} dx f^2(x)\phi(x)\right) = I \frac{\delta}{\delta f(y)} \alpha \int_{-\infty}^{\infty} dx f^2(x)\phi(x)$$
$$= 2I\alpha \int_{-\infty}^{\infty} dx \frac{\delta f(x)}{\delta f(y)} \phi(x) = 2I\alpha \int_{-\infty}^{\infty} dx \delta(x - y)\phi(x) = 2I\alpha\phi(y), \tag{C.8}$$

where we introduced notation $I = \exp(\alpha \int_{-\infty}^{\infty} dx f^2(x)\phi(x))$ and α is a constant.

Note that, above, to facilitate the explanation, we adopted one-dimensional integrals $D = 1$. The case of $D > 1$ is straightforward.

Task A3.1. Let

$$W_0 = \exp\left(\frac{1}{2}\int_{-\infty}^{\infty} dx\,dy\,j(x)A(x,y)j(y)\right), \tag{C.9}$$

where $A(x,y) = A(y,x)$ is a symmetric function.

Prove (or derive) the following equations

$$1)\ \frac{\delta W_0}{\delta j(x_1)} = W_0 \int_{-\infty}^{\infty} dx\,A(x_1,x)j(x) \equiv W_0'(x_1). \tag{C.10}$$

$$2)\ \frac{\delta^2 W_0}{\delta j(x_1)\delta j(x_2)} = W_0\left(A(x_1,x_2) + \int_{-\infty}^{\infty} dx\,dy\,A(x_1,x)j(x)A(x_2,y)j(y)\right.$$
$$\left. \equiv W_0''(x_1,x_2)\right). \tag{C.11}$$

$$3)\ \frac{\delta^3 W_0}{\delta j(x_1)\delta j(x_2)\delta j(x_3)} = W_0''(x_1,x_2)\int_{-\infty}^{\infty} dx\,A(x_3,x)j(x)$$
$$+ \left(A(x_1,x_3)W_0'(x_2) + A(x_2,x_3)W_0'(x_1)\right). \tag{C.12}$$

$$4)\ \frac{\delta^4 W_0}{\delta j(x_1)\delta j(x_2)\delta j(x_3)\delta j(x_4)} = A(x_3,x_4)W_0''(x_1,x_2)$$
$$+ W_0'''(x_1,x_2,x_3)\int_{-\infty}^{\infty} dx\,A(x_3,x)j(x)$$
$$+ \left(A(x_1,x_3)W_0''(x_2,x_4) + A(x_2,x_3)W_0''(x_1,x_4)\right). \tag{C.13}$$

Here we note that functional derivatives may be realized by Maple also (the authors may share their code with a reader by demand).

The code may be generated by the algorithm, which uses the relation

$$\langle\psi_{n_1}(x_1)\psi_{n_2}(x_2)\cdots\psi_{n_N}(x_N)\rangle = \frac{\delta}{\delta J_1(x_1)}\frac{\delta}{\delta J_2(x_2)}\frac{\delta}{\delta J_N(x_N)}$$
$$\times \exp\left[\sum_{a,b=1,2}\int dx\,dy\,J_a(x)D_{a,b}(x,y)J_b(y)\right]_{J_1=0,\,J_2=0\cdots J_N=0,} \tag{C.14}$$

and the equations, similar to (C.10)–(C.13). It makes functional differentiation one by one, starting from the first derivative. If it works correctly, then the particularly following output must be produced:

$$\langle\psi_{n_1}(x_1)\rangle = \langle\psi_{n_1}(x_1)\psi_{n_2}(x_2)\psi_{n_3}(x_3)\rangle = 0,$$
$$\langle\psi_{n_1}(x_1)\psi_{n_2}(x_2)\rangle = D_{n_1,n_2}(x_1,x_2),$$
$$\langle\psi_{n_1}(x_1)\psi_{n_2}(x_2)\psi_{n_3}(x_3)\psi_{n_4}(x_4)\rangle = D_{n_1,n_2}(x_1,x_2)D_{n_3,n_4}(x_3,x_4) \tag{C.15}$$
$$+ D_{n_1,n_3}(x_1,x_3)D_{n_2,n_4}(x_2,x_4)$$
$$+ D_{n_1,n_4}(x_1,x_4)D_{n_2,n_3}(x_2,x_4)$$

which leads e.g. for the following equation:

$$\langle \psi_1(x_1)\psi_1(x_1)\psi_2(x_2)\psi_2(x_2)\rangle = D_{1,1}(x_1, x_1)D_{2,2}(x_2, x_2)$$
$$+ D_{1,2}(x_1, x_2)D_{1,2}(x_1, x_2) \tag{C.16}$$
$$+ D_{1,2}(x_1, x_2)D_{1,2}(x_1, x_2),$$

where the first term should be neglected as a disconnected one.

Appendix D: The Green Function and Densities in Two-Component Bose Mixtures

D

Here we present the explicit expressions for the Green function $D_{ij}(\omega_n, \mathbf{k})$ ($i, j = 1 \div 4$) and the related densities introduced in Chap. 11. By inversion of $\mathcal{D}_{ij}^{-1}(\omega_n, \mathbf{k})$ given in Eq. (11.15) one obtains

$$
\begin{aligned}
\mathcal{D}_{11}(\omega_n, \mathbf{k}) &= \frac{W_2 W_3 W_4 + W_2 \omega_n^2 - X_6^2 W_3}{\tilde{D}}, \\[2mm]
\mathcal{D}_{12}(\omega_n, \mathbf{k}) &= -\frac{(W_3 W_4 + \omega_n^2 + X_5 X_6)\omega_n}{\tilde{D}}, \\[2mm]
\mathcal{D}_{13}(\omega_n, \mathbf{k}) &= -\frac{W_2 X_5 W_4 - X_6^2 X_5 - \omega_n^2 X_6}{\tilde{D}}, \\[2mm]
\mathcal{D}_{14}(\omega_n, \mathbf{k}) &= -\frac{(X_5 W_2 + W_3 X_6)\omega_n}{\tilde{D}}, \\[2mm]
\mathcal{D}_{22}(\omega_n, \mathbf{k}) &= \frac{W_1 W_3 W_4 + W_1 \omega_n^2 - W_4 X_5^2}{\tilde{D}}, \\[2mm]
\mathcal{D}_{23}(\omega_n, \mathbf{k}) &= -\frac{(X_6 W_1 + X_5 W_4)\omega_n}{\tilde{D}}, \\[2mm]
\mathcal{D}_{24}(\omega_n, \mathbf{k}) &= -\frac{W_3 X_6 W_1 - X_6 X_5^2 - \omega_n^2 X_5}{\tilde{D}}, \\[2mm]
\mathcal{D}_{33}(\omega_n, \mathbf{k}) &= \frac{W_4 W_1 W_2 + W_4 \omega^2 - X_6^2 W_1}{\tilde{D}}, \\[2mm]
\mathcal{D}_{34}(\omega_n, \mathbf{k}) &= -\frac{(W_1 W_2 + \omega_n^2 + X_5 X_6)\omega_n}{\tilde{D}}, \\[2mm]
\mathcal{D}_{44}(\omega_n, \mathbf{k}) &= \frac{W_3 W_1 W_2 + W_3 \omega_n^2 - X_5^2 W_2}{\tilde{D}},
\end{aligned}
\tag{D.1}
$$

© The Editor(s) (if applicable) and The Author(s), under exclusive license to Springer Nature Switzerland AG 2026
A. Rakhimov and S. Mardonov, *Theory of Quantum Bose Liquids Beyond Bogoliubov Approximation*, Lecture Notes in Physics 1045,
https://doi.org/10.1007/978-3-032-05096-0

where $\tilde{D} = (\omega_n^2 + \omega_1^2)(\omega_n^2 + \omega_2^2)$, $\omega_n = 2\pi n T$; $\omega_{1,2}$ are given in the main text and $W_1 = \varepsilon_a(k) + X_1$, $W_2 = \varepsilon_a(k) + X_2$, $W_3 = \varepsilon_b(k) + X_3$, $W_4 = \varepsilon_b(k) + X_4$, $X_5 = X_{13}$, $X_6 = X_{24}$.

Below, we present explicit expressions for the densities defined in Eqs. (11.21)–(11.23):

$$\rho_{1a} = \sum_k R_k \tilde{\rho}_{1a}(k), \quad \rho_{1b} = \sum_k R_k \tilde{\rho}_{1b}(k), \quad \sigma_a = \sum_k R_k \tilde{\sigma}_a(k),$$

$$\sigma_b = \sum_k R_k \tilde{\sigma}_b(k), \quad \rho_{ab} = 2\sum_k R_k \tilde{\rho}_{ab}(k), \quad \sigma_{ab} = 2\sum_k R_k \tilde{\sigma}_{ab}(k), \tag{D.2}$$

where

$$
\begin{aligned}
\tilde{\rho}_{1a}(k) = {}& \left(X_6^2 W_3 + W_2\omega_1^2(k) - W_2 W_3 W_4 - W_1 W_3 W_4 + X_5^2 W_4 + W_1\omega_1^2(k) \right) \\
& \qquad\qquad\qquad\qquad\qquad\qquad\qquad\qquad\qquad\qquad \times \omega_2(k)\tilde{W}_1(k) \\
& + \left(-X_6^2 W_3 - W_2\omega_2^2(k) + W_2 W_3 W_4 + W_1 W_3 W_4 - X_5^2 W_4 - W_1\omega_2^2(k) \right) \\
& \qquad\qquad\qquad\qquad\qquad\qquad\qquad\qquad\qquad\qquad \times \omega_1(k)\tilde{W}_2(k),
\end{aligned}
\tag{D.3}
$$

$$
\begin{aligned}
\tilde{\rho}_{1b}(k) = {}& \left(X_6^2 W_1 + W_4\omega_1^2(k) - W_1 W_2 W_3 - W_1 W_2 W_4 + X_5^2 W_2 + W_3\omega_1^2(k) \right) \\
& \qquad\qquad\qquad\qquad\qquad\qquad\qquad\qquad\qquad\qquad \times \omega_2(k)\tilde{W}_1(k) \\
& + \left(-X_6^2 W_1 - W_4\omega_2^2(k) + W_1 W_2 W_3 + W_1 W_2 W_4 - X_5^2 W_2 - W_3\omega_2^2(k) \right) \\
& \qquad\qquad\qquad\qquad\qquad\qquad\qquad\qquad\qquad\qquad \times \omega_1(k)\tilde{W}_2(k),
\end{aligned}
\tag{D.4}
$$

$$
\begin{aligned}
\tilde{\sigma}_a(k) = {}& \left(X_6^2 W_3 + W_2\omega_1^2(k) - W_2 W_3 W_4 + W_1 W_3 W_4 - X_5^2 W_4 - W_1\omega_1^2(k) \right) \\
& \qquad\qquad\qquad\qquad\qquad\qquad\qquad\qquad\qquad\qquad \times \omega_2(k)\tilde{W}_1(k) \\
& + \left(-X_6^2 W_3 - W_2\omega_2^2(k) - W_1 W_3 W_4 - W_2 W_3 W_4 - W_2\omega_2^2(k) + X_5^2 W_4 + W_1\omega_2^2(k) \right) \\
& \qquad\qquad\qquad\qquad\qquad\qquad\qquad\qquad\qquad\qquad \times \omega_1(k)\tilde{W}_2(k),
\end{aligned}
\tag{D.5}
$$

$$
\begin{aligned}
\tilde{\sigma}_b(k) = {}& \left(X_6^2 W_1 + W_4\omega_1^2(k) + W_1 W_2 W_3 - W_1 W_2 W_4 - X_5^2 W_2 - W_3\omega_1^2(k) \right) \\
& \qquad\qquad\qquad\qquad\qquad\qquad\qquad\qquad\qquad\qquad \omega_2(k)\tilde{W}_1(k) \\
& + \left(-X_6^2 W_1 - W_4\omega_2^2(k) - W_1 W_2 W_3 + W_1 W_2 W_4 + X_5^2 W_2 + W_3\omega_2^2(k) \right) \\
& \qquad\qquad\qquad\qquad\qquad\qquad\qquad\qquad\qquad\qquad \times \omega_1(k)\tilde{W}_2(k),
\end{aligned}
\tag{D.6}
$$

$$\tilde{\rho}_{ab}(k) = \left(-X_6^2 X_5 + W_1 W_3 X_6 + W_2 W_4 X_5 + \omega_1^2(k)X_6 + X_5\omega_1^2(k) - X_5^2 X_6\right)$$
$$\omega_2(k)\tilde{W}_1(k)$$
$$+ \left(X_6^2 X_5 - W_1 W_3 X_6 - W_2 W_4 X_5 - \omega_2^2(k)X_6 - X_5\omega_2^2(k) + X_5^2 X_6\right)$$
$$\times \omega_1(k)\tilde{W}_2(k),$$

$$(D.7)$$

$$\tilde{\sigma}_{ab}(k) = \left(-X_6^2 X_5 - W_1 W_3 X_6 + W_2 W_4 X_5 + \omega_1^2(k)X_6 - X_5\omega_1^2(k) + X_5^2 X_6\right)$$
$$\times \omega_2(k)\tilde{W}_1(k)$$
$$+ \left(X_6^2 X_5 + W_1 W_3 X_6 - W_2 W_4 X_5 - \omega_2^2(k)X_6 - X_5\omega_2^2(k) + X_5^2 X_6\right)$$
$$\times \omega_1(k)\tilde{W}_2(k),$$

$$(D.8)$$

In above equations $R_k = 1/2\sqrt{D}\omega_1(k)\omega_2(k)V$, $\tilde{W}_{1,2}(k) = 1/2 + f(\omega_{1,2}(k))$, and D, $\omega_{1,2}(k)$ are given in Eqs. (11.16).

Note that, in the practical calculation of momentum integrals, adequate counterterms should be included similar to $c_\rho = -1/2$ (for ρ_1), and $c_\sigma = \Delta/2\varepsilon_k$ (for σ) used in the one-component case.

Appendix E: Effective Hamiltonian in Bond-Operator Representation

E

Below we outline the derivation of the Hamiltonian in Eq. (13.8) from the basic spin Hamiltonian (13.3). Taking into account the dimerized nature of the ground state, it can be rewritten as follows

$$\mathcal{H} = \sum_i J_0 \mathbf{S}_{i1} \cdot \mathbf{S}_{i2} + \sum_i \sum_{m=1}^{9} J_m \mathbf{S}_{i,1} \cdot \mathbf{S}_{i+\delta\mathbf{r_m},1}, \qquad (E.1)$$

where i denotes locations of dimers. J_0 indicates the intradimer spin coupling while interdimer spin couplings are described by J_m ($m = 1, ..., 9$). The two spins within a dimer are labelled by subscripts 1 and 2. Since a dimer is made of two neighboring spins, the interaction between two dimers contains four different spin couplings.

The pair of one-half spins can be treated as a dimer connected by bonds. Sachdev and Bhatt [7] suggested the bond-operator representation for the pair of spins. Thus, following this representation, we introduce transformations:

$$S_1^x = \frac{1}{2}\left[s^\dagger t_x + t^\dagger_x s - i(t^\dagger_y t_z - t^\dagger_z t_y)\right],$$

$$S_1^y = \frac{1}{2}\left[s^\dagger t_y + t^\dagger_y s - i(t^\dagger_z t_x - t^\dagger_x t_z)\right], \qquad (E.2)$$

$$S_1^z = \frac{1}{2}\left[s^\dagger t_z + t^\dagger_z s - i(t^\dagger_x t_y - t^\dagger_y t_x)\right],$$

for the first spin, and as

© The Editor(s) (if applicable) and The Author(s), under exclusive license to Springer Nature Switzerland AG 2026
A. Rakhimov and S. Mardonov, *Theory of Quantum Bose Liquids Beyond Bogoliubov Approximation*, Lecture Notes in Physics 1045,
https://doi.org/10.1007/978-3-032-05096-0

$$S_2^x = -\frac{1}{2}\left[s^\dagger t_x + t^\dagger{}_x s + i(t^\dagger{}_y t_z - t^\dagger{}_z t_y)\right],$$

$$S_2^y = -\frac{1}{2}\left[s^\dagger t_y + t^\dagger{}_y s + i(t^\dagger{}_z t_x - t^\dagger{}_x t_z)\right], \tag{E.3}$$

$$S_2^z = -\frac{1}{2}\left[s^\dagger t_z + t^\dagger{}_z s + i(t^\dagger{}_x t_y - t^\dagger{}_y t_x)\right],$$

for the second spin. The auxiliary quasiparticles corresponding to the operators s and t can, therefore, be called singletons and triplons, respectively. The restriction on the physical states to be either singlets or triplets leads to the constraint

$$s^\dagger s + \sum_\alpha t_\alpha^\dagger t_\alpha = 1. \tag{E.4}$$

Bearing in mind the existence of the transverse magnetization, one may introduce the following canonical transformation:

$$t_x = \frac{1}{2}(t_+ + t_-), \qquad t_y = -\frac{i}{2}(t_+ - t_-), \qquad t_z = t_0, \tag{E.5}$$

Now we make a standard assumption that the triplon, described by the operator t_+ dominates due to the Zeeman effect as illustrated in Fig.13.4a. Moreover, we can ignore the singlet fluctuations by replacing the spin operators s with a complex number $s = < s_j >$. Then the bond-operator representation reduces to

$$S_{1j}^x = \frac{s}{2\sqrt{2}}(t^\dagger{}_j + t_j) = -2S_{2j}^x,$$

$$S_{1j}^y = \frac{is}{2\sqrt{2}}(t^\dagger{}_j - t_j) = -2S_{2j}^y, \tag{E.6}$$

$$S_{1j}^z = -\frac{1}{2}t^\dagger{}_j t_j = S_{2j}^z.$$

where j denotes the number of the lattice site of the dimers. Thus, inserting (E.6) into the spin Hamiltonian (E.1) and using constraints (E.4), after some algebraic manipulations, one arrives at the effective Bose Hamiltonian (13.8) [8–10].

References

1. A. Stone, *Chapter 24, The Indian Comet, in the book Einstein and the Quantum* (Princeton University Press, Princeton, NJ, 2013)
2. T. Hnsch, A. Schawlow, Cooling of gases by laser radiation. Opt. Commun. **13**(1), 68–69 (1975). https://www.sciencedirect.com/science/article/pii/0030401875901595
3. D. Wineland, H. Dehmelt., Proposed $1014 \Delta v < v$ laser fluorescence spectroscopy on Tl + Mono-Ion oscillator iii. Bull. Am. Phys. Soc. **20**(4), 637 (1975)
4. W.D. Phillips, H. Metcalf, Laser deceleration of an atomic beam. Phys. Rev. Lett. **48**, 596–599 (1982). https://doi.org/10.1103/PhysRevLett.48.596

5. J. Weiner, V.S. Bagnato, S. Zilio, P.S. Julienne, Experiments and theory in cold and ultracold collisions. Rev. Mod. Phys. **71**, 1–85 (1999). https://doi.org/10.1103/RevModPhys.71.1
6. W.D. Phillips, Nobel lecture: laser cooling and trapping of neutral atoms. Rev. Mod. Phys. **70**, 721–741 (1998). https://doi.org/10.1103/RevModPhys.70.721
7. S. Sachdev, Quantum phase transitions. Phys. World **12**(4), 33 (1999). https://doi.org/10.1088/2058-7058/12/4/23
8. V.I. Yukalov, Difference in Bose-Einstein condensation of conserved and unconserved particles. Laser Phys. **22**(7), 1145–1168 (2012). https://doi.org/10.1134/S1054660X12070171
9. J. Sirker, A. Weie, O.P. Sushkov, Consequences of spin-orbit coupling for the Bose-Einstein condensation of magnons. Europhys. Lett. **68**(2), 275 (2004). https://doi.org/10.1209/epl/i2004-10179-4
10. T. Dodds, B.-J. Yang, Y.B. Kim, Theory of magnetic-field-induced Bose-Einstein condensation of triplons in $Ba_3Cr_2O_8$. Phys. Rev. B **81**, 054412 (2010). https://doi.org/10.1103/PhysRevB.81.054412